bnes

ACS SYMPOSIUM SERIES **805**

Service Life Prediction

Methodology and Metrologies

Jonathan W. Martin, Editor
National Institute of Standards and Technology

David R. Bauer, Editor
Ford Motor Company

American Chemical Society, Washington, DC

Library of Congress Cataloging-in-Publication Data

Service Life Prediction : methodology and metrologies / Jonathan W. Martin, editor, David R. Bauer, editor.

  p. cm.—(ACS symposium series ; 805)

  Includes bibliographical references and index.

  ISBN 0–8412–3693–3

  1. Plastic coating—Deterioration—Congresses. II. Failure time data—Analysis—Congresses.

  I. Martin, Jonathan W. II. Bauer, David R. (David Robert), 1949- III. International Symposium on Service Life Prediction in Organic Coatings (2nd : 1999 : Monterey, Calif.) IV. Series.

TP1175.S6 S46   2001
667'.9—dc21   2001046446

The paper used in this publication meets the minimum requirements of American National Standard for Information Sciences—Permanence of Paper for Printed Library Materials, ANSI Z39.48–1984.

Foreword

The ACS Symposium Series was first published in 1974 to provide a mechanism for publishing symposia quickly in book form. The purpose of the series is to publish timely, comprehensive books developed from ACS sponsored symposia based on current scientific research. Occasionally, books are developed from symposia sponsored by other organizations when the topic is of keen interest to the chemistry audience.

Before agreeing to publish a book, the proposed table of contents is reviewed for appropriate and comprehensive coverage and for interest to the audience. Some papers may be excluded to better focus the book; others may be added to provide comprehensiveness. When appropriate, overview or introductory chapters are added. Drafts of chapters are peer-reviewed prior to final acceptance or rejection, and manuscripts are prepared in camera-ready format.

As a rule, only original research papers and original review papers are included in the volumes. Verbatim reproductions of previously published papers are not accepted.

ACS Books Department

Contents

Advances in Analytical Measurements

Mathematical Model

Preface

This book contains papers presented at the 2nd International Symposium on Service Life Prediction in Organic Coatings held in Monterey, California in November 1999. The papers include contributions from industry, national laboratories, and universities. This conference took place two and one half years after the first conference in Breckenridge, Colorado. Oxford University Press published papers from this conference in *Service Life Prediction of Organic Coatings*. The Breckenridge Conference concluded with a discussion of outstanding issues that had to be resolved in order to implement improved SLP methodologies in the coatings arena. The papers presented at Monterey provide a picture of the progress that has been made toward that goal during the past several years. To paraphrase one of the featured speakers at the conference, J. Gerlock (Ford), this book reports on the transition of service life prediction in coatings from an art to a science and finally to an engineering tool. As should become apparent, the coating community has made significant progress on the transition from art to science. While progress has also been made in developing engineering tools that can be routinely used to develop and specify coating systems, the job is far from complete. To appreciate what remains to be done, it is necessary to understand in detail what progress has been achieved and how it can be used.

The first area is in understanding the issues associated with developing relationships between in-service performance and exposure tests. After providing an overview of the basic approaches to service life prediction, J. Martin (NIST) details the various issues associated with interpretation of outdoor results caused by outdoor exposure variability. W. Meeker (Iowa State University) discusses other difficulties in relating accelerated tests and in-service performance. Of particular note is the combined effect of a non-linear dependence of degradation on exposure condition coupled with random variations in exposure condition on the variability of acceleration with different exposure runs. H. K. Hardcastle (Atlas Weathering Services) describes the effects of angle on exposure harshness, and also presents results on utilizing Emmaqua accelerated exposure conditions. One key factor in predicting service life is understanding actual in-service conditions. L. Kaetzel (K-Systems) presents an update on the growing solar UV monitoring network and describes how data will be gathered, stored, and accessed. P. Norberg (KTH-Sweden) describes the development of a novel tool for measuring surface moisture and time-of-wetness.

Just as it is important to be able to measure outdoor conditions, it is also critical to be able to reproduce controlled exposure conditions in the laboratory. Several novel exposure chambers have been developed and used to follow coating degradation. T. Nguyen (NIST) presents results from NIST's solar exposure chamber. This chamber is capable of exposing small samples to light at different wavelengths at controlled temperature and humidity. This device has been used to determine the quantum yield and dependence on humidity of oxidation and hydrolysis of a typical automotive coating. B. Dickens describes a convenient data-handling scheme for processing the vast quantity of spectra that can be obtained from this exposure chamber. G. Jorgensen (NREL) updates results from his concentrated sunlight exposure chamber where light intensities of up to 100 suns can be achieved. Measurement of the dependence of oxidation on light intensity lead both to an understanding of mechanism as well as to determining the limits of reliable acceleration. Finally, J. Chin (NIST) describes the latest NIST exposure chamber based on integrating sphere technology. This technology provides unprecedented control of a high intensity light exposure.

One of the key advances during the past several years has been the application of a variety of spectroscopic and other techniques to characterize coatings and their degradation chemistry. D. Bauer (Ford) and K. Adamsons (DuPont) review many of the techniques that can be applied to coating characterization and provide several examples of how the techniques can be used both to drive coating development as well as to develop improved specifications for performance. Depth profiling of degradation processes has proven to be particularly valuable in understanding coating degradation and stabilization. K. Adamsons presents results on depth-profiling experiments conducted at DuPont. J. Gerlock (Ford) discusses a number of techniques developed at Ford and describes their application to characterizing coating degradation and to determining the fate of photostabilizers (UVAs and HALS) on exposure. K. Williams (Colorado School of Mines) illustrates the use of field flow chromatography to characterize raw materials used in coatings.

Although chemical degradation often drives ultimate coating failure, it is also important to be able to characterize physical and mechanical behavior of coatings during degradation. Many coating failures are a loss of appearance. F. Hunt (NIST) describes a new technique for computer rendering of realistic images of coated surfaces. G. Bierwagon (North Dakota State University) presents results on the application of electro impedance spectroscopy to characterize corrosion processes in coatings. L. Floyd (PRA Labs) discusses the importance of fracture mechanics to service life prediction, and M. Nichols (Ford) describes a novel method for measuring fracture energy and presents results on the change of fracture energy with weathering. J. Jean (University of Missouri at Kansas City) utilizes positron annihilation lifetime spectroscopy to characterize changes in the microstructure of coatings on weathering. He correlates these changes with chemical changes. R. Ryntz (Visteon) discusses methods for understanding the relationships between plastic substrates, paint adhesion, and paint performance.

Models are critical to interpreting experimental results. J. Pickett (GE) describes a model for interpreting UVA loss experiments for both coatings and plastics. The model can extrapolate short-term measurements to predict long-term UVA lifetimes. A. DeBellis (Ciba) uses atomistic modeling to understand the relationship between UVA structure and UVA loss rates. S. Saunders discusses models for hydrolysis and provides an interpretation of degradation behavior observed in experiments described by T. Nguyen.

As at the Breckenridge Conference, the Monterey Conference ended with a group discussion of further research needs and procedures to implement the growing knowledge into practical engineering tools. L. Floyd summarizes this discussion. Further updated information is also available on the Service Life Prediction website: http://slp.nist.gov

Finally, we thank the Monterey organizing committee and all the authors and participants for making the conference an enjoyable success. We also thank the National Institute of Standards and Technology, the National Renewable Energy Laboratory, Wright-Patterson Air Force Base, the Federal Highway Administration, and the Forest Products Laboratory at Madison for their financial support of the Symposium.

Jonathan W. Martin
National Institute of Standards and Technology
Quince Orchard Boulevard
Route 270, Building 226, Room B350
Gaithersburg, MD 20899

David R. Bauer
Research Laboratory, MD-3182
Ford Motor Company
P.O. Box 2053
Dearborn, MI 48121

Advances in Field Exposure Experiments

Chapter 1

Repeatability and Reproducibility of Field Exposure Results

Jonathan W. Martin

National Institute of Standards and Technology, 100 Bureau Drive, Stop 8621,
Gaithersburg, MD 20899-8621

Field exposure results are almost always viewed as the de facto standard of performance against which laboratory results must match in estimating the service life performance of polymeric materials. Implicit in this view is that field exposure results are both repeatable and reproducible. A review of the literature was made in to identify evidence corroborating or refuting this premise. It was concluded from this review that substantial evidence exists refuting this premise and, correspondingly, little or no evidence exists supporting it; that is, the assumption that field exposure results are repeatable or reproducible does not appear to have any scientific merit. This lack of support draws into question the use of field exposure results as a de facto standard of performance. Alternative strategies are noted.

Introduction

Field exposure experiments play a crucial role in assessing the in-service performance of coatings and other polymeric materials. Results from these experiments are viewed as the "primary test" [1], the "real time exposure" [2-4], the "decisive test" [5], and thus the de facto standard of performance against which laboratory aging and fundamental mechanistic results are expected to match. Since field exposure results are viewed as a performance standard, these results should be expected to be both repeatable and reproducible. No support, however, could be found in the literature either corroborating or refuting these premises. The objective of this paper, therefore, was to review the published literature for such evidence.

Three sources of information were examined. They included the following:

1. testimonials from weathering researchers,
2. results from well-designed and executed field exposure experiments, and
3. trends and cycles in weather element data.

© 2002 American Chemical Society

Representative citations from each source are presented; a more in-depth presentation and a more comprehensive list of citations are given in Martin [6].

Testimonials

Over the years, many weathering researchers have published testimonials on the reproducibility and repeatability of field exposure experiments. Unfortunately, few of these researchers provided any quantitative data to support their statements. The testimonials generally deal with the effect of different variables on weathering results. These variables include 1) the year of exposure, 2) the time of year that the exposure commenced, 3) the duration of the exposure, 4) the angle of the exposure, and 5) the location of the exposure site. The consensus of opinion is that field exposure results are <u>neither repeatable nor reproducible</u> when specimens are exposed

1. at the same site, at the same angle of exposure, at the same time of the year, and for the same duration, but exposures begin on **different years** [7-11];
2. at the same site, at the same angle of exposure, during the same year, for the same duration, but exposures begin at **different times of the same year** [1-2, 8-9,12, 15-18];
3. at the same site, at the same angle of exposure, at the same time of the year, in the same year, but exposures are for **different durations** [19-22];
4. at the same site, at the same time of the year, in the same year, for the same duration, but exposures are made at **different exposure angles** [7, 14, 23-25];
5. at the same angle of exposure, at the same time of the year, in the same year, for the same duration, but exposures are at **different exposure sites** [1, 7, 9, 11-12, 18-23, 26-28].

The only citation found that contradicted the claim that field results are neither repeatable nor reproducible was by Dawson and Nutting [29].

Results from Well-Designed and Executed Field Exposure Experiments

From 1900 through 1970, a number of planned field exposure experiments were conducted and the results published in the open literature. In a few cases, the experiments included several thousands of specimens exposed over a number of years. By the end of the 1970's, however, the philosophy of field exposure experimental designs changed rather dramatically. Instead of designed experiments yielding vast amounts of quantitative data, field experiments were designed to make a simple comparison between two or more exposures. The degree of agreement between or among these exposures was assessed through a correlation coefficient, usually the Spearman rank correlation coefficient. Although these experiments are much easier to design and execute, almost all quantitative data, like "how much is coating A better than coating B", have been lost [6]. Since the purpose of this section is to quantitatively assess the repeatability and reproducibility of field exposure results, only results from well-designed experiments published prior to 1980 are described. They include the following:

a.   Oakley [8] exposed nominally identical coated panels at the Carlton Exposure Station in County Durham, England over a three-year period starting in 1957. The experiments were terminated after 36 weeks at which time the gloss loss of each panel was assessed. He concluded that panels exposed in 1959 exhibited about 15% greater gloss loss than did panels exposed in 1958 which, in turn, exhibited about 15% greater gloss loss than panels exposed in 1957.

b.   Mitton and Church [10] exposed nominally identical coatings in Miami, FL at different times from 1948 to 1953. They reported that the results were highly variable and that the majority of this variability was attributable to variations in the weather.

c.   Ramsbottom [13] exposed fabrics, used in the construction of dirigibles, in Farnborough, England starting on July 1, 1922 and the beginning of each month thereafter until July 1, 1923. The fabrics were exposed for one month; at the end of which, the fabrics were removed from exposure and failed in tension. Tensile strength loss of the fabrics ranged from 0% for fabrics exposed during November 1922 to 45% for fabrics exposed during July 1923. Ramsbottom concluded that summer exposures were more severe than winter exposures.

d.   Came [30] exposed two nominally identical sets of 50 spar varnishes at the National Bureau of Standards in Washington D.C. on two different start dates: January 10, 1929 and April 1, 1929. Exposures continued until almost all of the varnishes had failed. Came recorded the time to failure for each varnish and observed that specimens placed on exposure in January failed about twice as fast as did specimens that were exposed starting in April, 1929.

e.   Wirshing [15] initiated exposures of nominally identical sets of nitrocellulose coated panels on successive months during the same year and removed the panels from exposure whenever a fixed amount of degradation was observed. He observed that panels exposed at the beginning of April, May and June failed in 13 weeks, whereas panels exposed at the beginning of November took 21 weeks to fail. Thus, panels having spring or summer exposure start dates degraded much faster than panels having a winter start date. This observation is in direct opposition to that made by Came [30].

f.   Clark [1] exposed a series of vinyl films in Miami, FL and Bound Brook, NJ at the beginning of the spring, summer, fall and winter seasons. Exposures were continued for two years. Clark concluded that films having exposure start dates from May to August (summer) degraded about three times faster than did films having start dates from November to March (winter and spring). That is, summer start dates were more severe than winter start dates.

g. Melchore [16] initiated exposures of polyethylene films in Arizona at the beginning of each month for a year. The experiments were terminated whenever the total solar irradiance absorbed by a specimen reached 15000 Langleys at which time the percent carbonyl formation in the film was determined. Melchore observed that, even though all of the films absorbed the same number of Langleys, films placed on exposure in June and July degraded 7 times faster than those exposed in December and January. He concluded, "the Langley is a poor unit to measure ultraviolet radiation" and that films having a summer start date degraded much faster than films having a winter start date.

h. The Joint Services Research and Development Committee on Paints and Varnishes [17] exposed a variety of coatings in six different locations throughout England for approximately two years starting in the fall of 1959 and the spring of 1960. They observed that the degradation rate for the spring exposure was much greater than the rate for fall exposure.

i. Marshall et al. [23] exposed several thousands of coated panels in Miami, FL, Wilmington, DE, and Canyon, TX at 45° S and 90° S (i.e., vertical South) for 36 months in the early 1930s. The dominant failure mode was flaking. For all three-exposure sites, the rate of flaking was more severe for panels exposed at 45° S than it was for panels exposed at 90° S.

j. Evans [25] exposed pinewood panels (Pinus radiata) at 0° (horizontal), 45°, 60°, 70°, and 90° (vertical) for 50 days in Canberra, Australia (latitude 35° South) from February 18, 1987 to April 9, 1987 and monitored mass loss and chemical changes. Panels positioned horizontal to the sun experienced the greatest mass loss and the greatest chemical change.

k. Qayyum and Davis [24] exposed polysulphone films at 0° (horizontal), 20°, 45°, 60°, and 90° (vertical) South in Jeddah, Saudi Arabia (latitude 20° North) for a 12-month period starting in September 1981. Degradation was monitored by mass loss. They observed that mass loss was negatively correlated with total solar UV-irradiation; that is, the greater the total solar UV-irradiance, the lower the mass loss. Maximum mass loss was observed on films exposed vertically, 90° S, while minimum mass loss was observed for films exposed horizontally, 0°. This conclusion is in opposition to those made by both Marshall et al. [23] and Evans [25].

l. Marshall et al. [23] exposed several thousand coated panels in Miami, FL, Wilmington, DE, and Canyon, TX in the early 1930s. The exposures were terminated after 36 months, at which time, the amount of flaking was ascertained. They observed that the degradation response varied greatly among the three sites and concluded that knowledge of the time-to-flaking at one or even two of the exposure sites would not provide any useful information regarding the time-to-flaking at the third site.

m. Neville [26] exposed acrylic and alkyd coated panels in Carlton, County Durham, England and Miami, FL. Changes in gloss were reported as a function of total solar irradiance. Neville concluded that total solar irradiance was not a good metric for predicting gloss loss, since for the same total solar irradiance, gloss loss in Florida was much greater than it was in Carlton.

n. The Joint Services Research and Development Committee on Paints and Varnishes [17] exposed a number of coatings at six different locations throughout England for approximately two years. They observed that both the rate of weathering and the dominant failure mode changed from site-to-site.

o. Cutrone and Moulton [28] exposed nominally identical sets of coated specimens at 11 different field sites throughout the world during the same year. The computed Spearman rank correlation coefficients [31] ranged from 0 to 0.8. The authors concluded, "the results indicate another less-than-perfect correlation."

The results from planned field experiments quantitatively affirm the testimonials; that is, field exposure experiments are neither repeatable nor reproducible. For example, for exposures begun on contiguous years, results can vary by as much as a factor of 10. No planned experiment was found in the literature supporting the premise that field exposure results are either reproducible or repeatable.

Trends and Cycles in Weather Element Data

The weather and, thus, the elements of the weather influence weathering results. In weathering research, the weather elements of primary research interest are solar ultraviolet (UV) radiation, air surface temperature, relative humidity, precipitation and aerosols. In this section, the temporal and spatial stabilities of these weather elements are reviewed for trends and cyclic behavior.

Over the last hundred years, meteorologists have performed numerous and extensive statistical analyses aimed at determining the temporal and spatial stability of a wide variety of weather elements. Temporal stability is ascertained by determining if trends and cycles exist in the time series for each weather element. Spatial stability is assessed by aggregating the time series output from various subset of the meteorological network at different geographical scales and determining if the trends and cycles translate from one spatial to another.

A trend is present in a weather element time series if the moving average for the time series (usually the five or ten year moving average) is increasing, decreasing, or is level over some period of time (Figure 1). Cyclic behavior is discovered through the application of spectral analysis techniques, like Fourier analysis, to a time series. The presence of a cycle implies that the intensity of a weather element repeats over some time period (Figure 2). In the meteorological literature, a cycle is considered to be statistically significant whenever the spectral peak for this cycle explains at least 5% of the total variation in a weather element's behavior. For practical reasons, weathering cycles between 2 and 5 years are of particular interest in materials research, since this is the length of time that specimens are commonly exposed in the

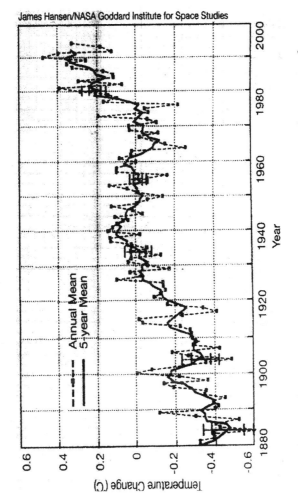

Figure 1. Global air-surface temperature, annual and 5-year-moving-average, from 1880 to the present.

field. Thus, the trend and cycle characteristics of a weather element that would tend to support repeatable and reproducible weathering results are those in which the trend line for a weather element displays a zero-slope and spectral analysis of the time series data indicates several statistically significant cycles between two and five years. Conversely, trend and cycle characteristics for a weather element, that would not tend to support repeatable and reproducible weathering results, include those in which the trend line for this weather element is either an increasing or decreasing and spectral analysis of the time series data indicates no statistically significant cycles between two and five years.

Spectral Ultraviolet Radiation

Radiation from the Sun provides essentially all of the energy driving the Earth's weather. Fortunately, solar radiation reaching the Earth's surface is greatly attentuated by the Earth's atmosphere. This is particularly true in the ultraviolet (UV) region, the radiation that is most photolytically effective in degrading materials. The Earth's stratospheric ozone layer, for example, effectively absorbs all solar ultraviolet radiation below about 290nm. Solar ultraviolet radiation transmitted through the Earth's atmosphere between 290 and 400 nm is known to be photolytically active in degrading polymeric materials. The photolytic effectiveness, however, varies greatly with wavelength. Wavelengths closer to 290 nm may be three to seven orders of magnitude more photolytically effective than are wavelengths near 400 nm. Photolytic effectiveness of radiation is not limited to the ultraviolet region. For some materials, like paper, wavelengths as high as 550 nm are known to cause photodegradation [32]. Thus, a full characterization of the solar spectral radiation between 290 nm and, let's say, 550 nm may be required to fully characterize the photodegradation effects of solar radiation.

Meterological stations for monitoring solar spectral ultraviolet radiation are a recent phenomena [33,34]. Weatherhead and Webb, for example, reported that the global solar spectral UV monitoring network consisted of five stations in 1992 and 250 stations in 1998. In the US, funding for most of these stations has been provided by the Departments of Energy, Agriculture, Commerce, Environmental Protection Agency, and the National Science Foundation under the auspices of the US Global Climate Change Research Program [35]. With two exceptions, the time series from this network are shorter than the 11 year solar cycle. The two exceptions are a worldwide network of Robertson-Berger (R-B) meters [36] and a filter wheel radiometer station located in Rockville, MD and operated by the Smithsonian Environmental Research Center (SERC) [37].

The Robertson-Berger meter is a broadband radiometer equipped with a filter that selectively transmits radiation approximating the erythemal absorption spectrum. The radiation transmitted through the filter is reported in terms of sunburn units. For 2 six-year periods starting in 1974 and again in 1980, the medical community instrumented 27 cities throughout the world with these instruments and correlated their output against the incidence of melanoma and non-melanoma cancers [36]. The usefulness of this meter for materials research, however, is limited since the exposure metric is tailored to the erythemal spectrum which is not useful absorption or quantum yield spectrum for commercial polymeric materials.

The SERC radiometer contains a filter wheel with 18 2nm-nominal band pass interference filters on its periphery and is equipped with a R-1657 solar blind photomultipler tube detector. The photomultipler tube is temperature regulated via a thermoelectric system. The filter wheel turns at 15 revolutions per minute and contains18 2nm full width half-maximum interference filters with nominal center wavelengths from 290nm through 324nm. Data from each interference filter is averaged over a 12 minute interval and this averaged value is stored in a computer as a observation in the time series for this filter. The unit operates 24 hours a day and is calibrated three or four times per year using NIST traceable calibration protocols. One unit has been in continuous operation in Rockville, MD or its vicinity since 1975; and, since 1998, similar units have been in operation in Miami, FL and Phoenix, AZ. Correll et al. [37] plotted erythema dosage (a unit of measure similar to sunburn units) for this unit from 1975 through 1990. For the Rockville, MD site (see Figure 3), erythemal dosage decreased 14% from 1975 through 1981; increased 40% from 1981 through 1987; and decreased 9% from 1987 through 1990. According to the authors these changes are consistent with TOMS satellite data and with changes in the total ozone column thickness recorded during these periods. The spectral UV trends observed by SERC are also consistent with short-term trends observed by other researchers; see, for example, McKenzie [38] and Kerr and McElroy [39]. To date, no spectral analysis research has been found for any solar spectral UV data.

Air-Surface Temperature

Time series for air-surface temperature are the longest and have the highest meteorological network station density for any weather element. The air-surface temperature time series for the British Isles, for example, dates back to 1659 [40]. Due to the length of the time series and the dense spatial distribution of monitoring stations, air-surface temperature time series are very attractive candidates for both trend and spectral analyses.

Representative air-surface temperature trends are presented in Table 2.1 over several geographical scales. For presentation purposes, these time series have been segmented into four independent time intervals: a) 1860 to 1900, b) 1900 to 1940, c) 1940 to 1975, and d) 1975 to the present. The trend in the air surface temperature for each time segment is indicated by one of the following arrows: 1) ↑ for increasing temperature trend, 2) ↓ for a decreasing temperature trend, and 3) → for a zero-sloped trend.

From Table 2.1, less than 25% of time segments are best described as having a zero-sloped trend. The trends for the remaining time segments are either increasing or decreasing. Most of the zero-sloped trends occurred during the time period from 1860 to 1900. From "1900 to 1940" and from "1975 to the present", the air surface temperature trendss for almost all geographical scales were increasing; while, from 1940 to 1975, the trends for almost almost all geographical scales were decreasing.

Weather cycles: real or imaginary?

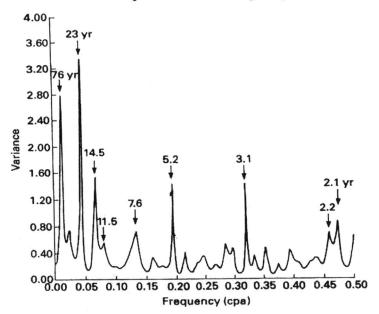

Figure 2. Typical spectral analysis output for a weather element. Spectral peaks 2.1 y, 5.1 y, and 20 y are statically significant at the 95% level (Fig. 2.taken from Spar and Mayer [67]).

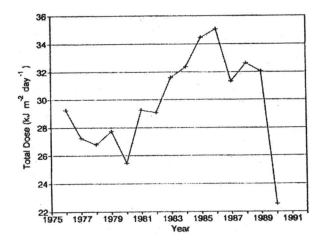

Figure 3. Smithsonian Environmental Research Center's Rockville, MD time series of annual mean daily total dosage of integrated 295-320 nm global irradiance [37]

TABLE I. Air-surface Temperature for Four Periods from 1860 to the Present at Decreasing Geographical Scales.

Spatial Scale	Author (year)	Trend 1860–1900	Trend 1900–1940	Trend 1940–1975	Trend 1975–present
Global	Jones [59]	↑	←		←
Southern Hemisphere	Jones [59]	↑	←	→	←
Northern Hemisphere	Jones [59]	←	←	←	←
Eurasia	Diaz and Bradley [60]	↑	←	↑	←
North America	Diaz and Bradley [60]	↑	←	→	←
Continental US	Diaz and Bradley [60]	→	←	→	←
Eastern US	Diaz and Bradley [60]	→	←	→	←
Western US	Diaz and Bradley [60]	↑	←	↑	←
Colorado and Platte River Basin	Diaz and Bradley [60]	↑		→	←

Spectral analysis results for air surface temperature are tabulated in Table 2.2. Only time series of length longer than 50 years and statistically significant cycles having a cycle time longer than one year are included. From Table 2.2, few statistically significant cycles have been identified and even fewer cycles translate from one geographical scale to another.

In conclusion, neither trend analysis nor spectral analysis for air surface temperature time series provide support for the repeatability and reproducibility of weathering results regardless of the time scale or the geographical location.

Precipitation

Moisture degrades polymeric materials through hydrolysis and acts together with other weather elements in degrading materials. In the hydrosphere, moisture is stored in five reservoirs: 1) the oceans, 2) the ice masses and snow deposits, 3) terrestrial waters, 4) the atmosphere, and 5) the biosphere. Although each reservoir could degrade polymeric materials, discussion will be limited to atmospheric moisture, specifically, precipitation and relative humidity.

Precipitation and air-surface temperature time series data are almost always collected concurrently; thus, it is not surprising that, like air-surface temperature, the time series for precipitation are of great length and the density of the monitoring sites are closely packed. The longest instrumented time series for precipitation dates back to 1727 and it covers most of England and Wales. A longer precipitation time series starting in 1470 AD is available for China, but this time series is largely descriptive and is considered by some researchers to be unsuitable for predictive purposes [76].

From extensive spectral and trend analyses on numerous precipitation time series, meteorologists have concluded that both temporal and spatial variabilities for the precipitation time series are significantly greater than they are air surface temperature. Bradley [45] performed trend analyses for precipitation data from 1850 to 1973 for the Northern Hemisphere using data collected from 1410 stations. He concluded that the precipitation trends were 1) decreasing from 1870 to 1920, 2) increasing from 1920 to 1950, and 3) decreasing from 1950 to 1973. Hense [46] reported precipitation data for the Earth's Equatorial Regions from 1965 to 1984 and concluded that precipitation increased an average of 0.005 mm/y in the Indo Pacific Equatorial Regions and increased an average of 0.05 mm/y in the Americas and African Equatorial Regions. Groisman [47] performed trend analysis on precipitation for the former USSR from 1890 to 1990 and observed annual increases of 6% per year for all regions except one.

Spectral analysis results for precipitation are tabulated in Table 2.3. Spectral peaks around 3.9 years and around 10 years appear in several time series, but these cycles do not transalate from one geographical scale to another.

Relative Humidity

Atmospheric moisture may be a greater contributor to weathering than precipitation. Most atmospheric relative humidity data are obtained from radiosondes or rawinsondes, that is, vertical ascending instrumented balloons used in monitoring atmospheric climatic variables. Currently, radiosondes are launched twice daily from

TABLE II. Spectral Analysis Results for a Variety of Temperature Time Series at Different Geographical Scales.

Spatial Scale	Author (year)	Time Series Start Year	Time Series End Year	Significant Spectral Peaks Greater than 1 y
Global	Dehsara and Cehak [61]	1901	1960	None
Global Sea Surface	Folland et al. [62]	1856	1981	16y, 83y
Northern Hemisphere	Borzenkova et al. [63]	1881	1975	None
Central England	Shaw [64]	1659	1960	None
Central England	Shapiro [65]	1659	1973	2.1 y
Central England	Mason [41]	1700	1950	2.1 y, 5.2y, 7.6y, 14.5y, 23y, 76y
Central England	Plaut [42]	1659	1994	5.2y, 7.7y, 14.2y, 25y
Netherlands	Shaw [64]	1735	1944	None
Eastern North America	Mock and Hibler III [43]	1800	1975	20y
Albany, NY	Thaler [66]	1821	1983	None
New York City	Thaler [66]	1822	1983	None
New York City	Spar and Mayer [67]	1822	1956	None
Woodstock, MD	Landsberg et al. [68]	1870	1959	50y

Table III. Spectral Analysis Results for a Variety of Precipitation Time Series.

Spatial Scale	Author (year)	Time Series		Significant Spectral Peaks Greater than 1 y
		Start Year	End Year	
England and Wales	Tabony [69]	1727	1975	2 y, 3.9 y
Kew, England	Tabony [69]	1697	1975	3.9 y, 6.1 y
Pode Hole, Lincolnshire, England	Tabony [69]	1726	1975	9.6 y
Manchester, England	Tabony [69]	1786	1975	3.9 y
Great Plains, US	Currie [70]	Unknown	Unknown	10 y, 18.6 y
US	Currie [70]	Unknown	Unknown	10 y, 18 y
Woodstock, MD, United States	Landsberg et al. [68]	1870	1959	None
South Africa	Burroughs [76]			3.5 y, 10-12 y, 18 y
South America	Burroughs [76]			3.8 y, 7 y, 20 y

over 1000 airports throughout the world [48,49]. The radiosonde network dates has a short duration, dating back to post World War II [50,51]. The shortest of these records make it almost impossible to assess the significance of any cycles in this data. The quality and homogeneity of relative humidity times series have also been questioned. Atmospheric moisture measurements have been plagued by a numerous measurement problems like instrument inhomogeneities and non-standard reporting and analysis. For example, radiosonde instrument packages still differ from country to country and from manufacturer to manufacturer [50]. These differences make it difficult to compare site data, that is, the data are not homogeneous. Atmospheric moisture is also a difficult gas to measure since it does not mix well in the troposphere and it has a short atmospheric residence time (approximately 10 days) [52]. Thus, the amount of moisture in the air changes dramatically over the diurnal cycle [53]. Finally, atmospheric moisture measurements are reported in non-standard formats. For example, measurements are commonly reported as relative humidity, specific humidity, saturation deficit (the difference between the specific humidity at saturation and the measured specific humidity), and vertically integrated amount of water vapor. Non-standard formats make it difficult for the lay person to compare data from different monitoring stations.

Trend analysis results for atmospheric moisture are tabulated in Table 2.4. All trends for all regions are monotonically increasing for the entire measurement history for this weather element. No spectral analyses of atmospheric moisture has been found; this is probably due to the shortness of the time series for atmospheric moisture.

In conclusion, trend analysis and spectral analysis results for precipitation and atmospheric moisture exhibit high temporal and spatial variability, non-zero trends and few statistically significant cycles between 2 and 5 years. Thus, this moisture data provide little support for the premise that field exposure results are repeatable and reproducible.

Aerosols

Aerosols are suspensions of liquids or solids in a gas. Aerosols include a wide-range of particles like dust, smoke, haze, SO_x and NO_x having diameters in the range of 1 nm to 100 μm. Anthropogenic aerosols, especially SO_x and NO_x, have been implicated in the acid etching of organic coatings [54-57] and in the crazing of poly (methyl methacrylate) [58]. The contribution of aerosols to polymeric degradation, however, is the least understood of all of the weather elements.

Table 2.5 contains a selection of trends for SO_x, NO_x, fogs, and smoke at different geographical scales. Spectral analyses for aerosols were not located in the literature. From Table 2.5, trends for SO_x generally increased throughout the world until the mid-1960, with the exception of London, England. In the mid-1960's, Clean Air regulations were enacted in the United States and other industrial nations leading to a decreasing trend. For non-industrial regions of the world, the trends for both SO_x and NO_x are increasing over the entire measurement history.

Table IV . Trends in Relative Humidity Reported for Different Geographical Spatial Scales.

Spatial Scale	Author (year)	Time Series		Moisture Content Variable	Trend Type (slope)
		Start Year	End Year		
Northern Hemisphere	Salstein, Rosen, and Peixoto [49]	1958	1970	Vertical integrated amount of water vapor	↑
Equatorial Regions	Flohn and Kapala [71]	1949	1979	Saturation deficit	←
Equatorial Regions --Americas and Africa	Hense et al [46]	1965	1984	Relative Humidity	←
Equatorial Regions—Indo Pacific	Hense et al. [46]	1965	1984	Relative Humidity	←
Equatorial Regions— Western Pacific	Gutzler [72]	1974	1988	Specific Humidity	←

Figure V. Trend Analysis for Aerosols and Tropospheric Ozone Reported as
Increasing ↑, Decreasing ↓, or No-change →

	Author (year)	Aerosol	Period 1	Period 2	Period 3
Asia	Digano and Hameed [73]	SO_x		1880-1965 ←	1965-1990 ←
Europe	Digano and Hameed [73]	SO_x		1880-1965 ←	1965-1990 ↑
North America	Digano and Hameed [73] and Karl et al. [74]	SO_x		1880-1965 ←	1965-1990 →
Former USSR	Digano and Hameed [73]	SO_x		1880-1965 ←	1965-1990 ←
London	Brimblecomb et al. [75]	SO_x	1700-1830 ↑	1830-1900 ←	1900-1990 →
London	Brimblecomb et al. [75]	Smoke	1700-1900 ←	1900-1950 →	
London	Brimblecomb et al. [75]	Fog	1700-1830 ↑	1830-1900 ←	1900-1950 →
London	Brimblecomb et al. [75]	NO_x			1976-1990 ↑

18

Summary

Field exposure results play a very important role in assessing the service life performance of polymeric materials. In particular, field exposure results are commonly viewed as the de facto standard of performance against which laboratory-aging experimental results must duplicate. As a standard of performance, field exposure results should be repeatable and reproducible The literature was reviewed to find support either corroborating or refuting this repeatability and reproducibility premise.

Three sources of information were reviewed including testimonials from weathering researchers, results from well-designed and executed field exposure experiments, and trends and cycles in weather element data. Data from all three sources provided stronge and consistent evidence refuting this premise and, correspondingly, provided little or no evidence corroborating the premise that field exposure results are either repeatable or reproducible.

All testimonials, except one, stated that field exposure results are neither repeatable nor reproducible for specimens exposed on different years, different times of the same year, for different exposure durations, for different exposure angles, or at different locations. Well-designed and executed field exposure experiments provided quantitative support for these claims indicating that exposure results could differ by as much as a factor of 10 for nominally identical specimens exposed on contiguous years at the same exposure site. The lack of reproducibility and repeatability of weathering results coincided with the lack of reproducibility and repeatability of all weather elements investigated. This later conclusion is consistent with conclusions from several recent studies [76-79]. Pittlock [80], for example, concluded that "most of the climatically important atmospheric and weather variables, be they temperatures, precipitation, or (say) ozone content, show day to day, seasonal, and year to year variations which are usually comparable with or larger than the variations in longer-term mean values".

Assuming that these conclusion are reaffirmed by other researchers, then the scientific validity of using field exposure results as the de facto standard of performance against which laboratory-aging results must duplicate must be questioned. It follows that service life prediction methodologies [81] that do not depend on field exposure results as a standard of performance should be more thoroughly investigated.

References

1.   Clark, F.G., Industrial and Engineering Chemistry 1952; Vol. 44, p 2697.

2.   Ellinger, M.L., Journal of Coatings Technology 1977; Vol. 49, p 44.

3.   Boxall, J., Journal of the Oil and Colour Chemists Association 1986; Vol. 15, p 7.

4.   Appleman, B.R., Journal of Coating Technology 1990; Vol. 62, p. 57.

5.   Association of Automobile Industries, Journal of Coatings Technology 1986; Vol. 58, p. 57.

6.   Martin, J.W., *Durability of Coatings and Plastics*, R.A. Ryntz [Ed.], Chapter 2, Hanser Publishers, New York, 2000.

7.   Troutman, R.E.; Vannoy, W.G., Journal of Industrial and Engineering Chemistry 1940; Vol. 32, p 232.

8.   Oakley, E., Journal of the Oil Colour Chemists Association 1960; Vol. 43, p 201.

9.   Morse, M.P., Official Digest 1964; Vol. 36, p 695.

10.   Mitton, P.B.; Church, R.L., Journal of Paint Technology 1964; Vol. 39, p 636.

11.   Epple, R., Journal of the Oil and Colour Chemist Association 1968; Vol. 51, p 213.

12.   Brighton, C.A., *Weathering and Degradation of Plastics*, Pinner, S.H. [Ed.], Columbine Press, London, 1966; p. 49

13.   Ramsbottom, J.E., Journal of the Royal Aeronautical Society 1924; Vol. 28, p 273.

14.   Ashman, G.W., Journal of Industrial and Engineering Chemistry 1936; Vol. 28, p 934.

15.   Wirshing, R.J., Journal of Industrial and Engineering Chemistry 1941; Vol. 33, p 234.

16.   Melchore, J.A., Journal of Industrial and Engineering Chemistry: Product Research and Development 1962; Vol. 1, p 232.

17.   Joint Services Research and Development Committee on Paints and Varnishes, Journal of the Oil and Colour Chemist Association 1964; Vol 47, p. 73.

18.   Singleton, R.W.; Kunkel, R.K.; Sprague, B.S., Textile Research Journal 1965; Vol. 35, p 228.

19.   Clark, J.E.; Green, N.E.; Giesecke, P., *National Bureau of Standards Report, NBS Report 9912*, 1969.

20.   Hoffmann, E.; Saracz, A., Journal of the Oil and Colour Chemist Association 1969, Vol 52, p 1130.

21.   Fischer, R.M., *SAE Technical Paper Series 841022*, 1984.

22.   Cleveland Society for Coatings Technology Technical Committee, Journal of Coatings Technology 1996; Vol. 68, p 47.

23.   Marshall, J.; Iliff, J.W.;Young, H.R., Journal of industrial and Engineering Chemistry 1935; Vol. 27, p 147.

24.   Qayyum, M.M.; Davis, A., Polymer Degradation and Stabilization 1984; Vol. 6, p 201.

25.   Evans, P.D., Polymer Degradation and Stabilization 1989; Vol. 24, p 81.

26. Neville, G.H.J., Journal of the Oil and Colour Chemists Association 1963; Vol. 46, p. 753.

27. Scott, J.L., Journal of the Oil Colour Chemists Association 1983; Vol. 66, p 129.

28. Cutrone, L.; Moulton, D.V., Journal of the Oil and Colour Chemist Association 1987; Vol. 70, p 225.

29. Dawson, D.H.; Nutting, R.D., Journal of Industrial and Engineering Chemistry 1940; Vol. 32, p 112.

30. Came, C.L., Journal of Research of the Bureau of Standards 1930; Vol. 4, p 247.

31. Lehmann, D.E., *Non Parametrics: Statistical Methods Based on Ranks*, Holden-Day, New York, 1978.

32. National Bureau of Standards, *National Bureau of Standards Circular 505*, 1951.

33. Ultraviolet Radiation Panel, *United States Global Change Research Program Publication USGCP-95-01*, 1995.

34. Weatherhead, E.C.; Webb, A.R., Radiation Protection Dosimetry 1997; Vol. 72, p 223.

35. Subcommittee on Global Change Research, *National Science and Technology Council Annual Report*, Washington, D.C, 1998.

36. Scotto, J.G.; Cotton, G.; Urbach, F.; Berger, D.; Fears, T., Science 1988; Vol. 239, p 762.

37. Correll, D.L.; Clark, C.O.; Goldberg, B.; Goodrich, V.R.; Hayes, D.R.; Klein, W.H.; Schecher, W.D., Journal of Geophysical Research 1992; Vol. 97(D7), p 7579.

38. McKenzie, R.; Connor, B.; Bodeker, G., Science 1999; Vol. 285, p 1709.

39. Kerr, J.B.; McElroy, C.T., Science 1993; Vol. 262, p. 1032.

40. Jones, P.D.; Hulme, M., *Climates of the British Isles: Present, Past, and Future*, Hulme, M. and Barrow, E. [Eds.], 1997; p. 173

41. Mason, B.J., Quarterly Journal of the Royal Meteorological Society, 1976; Vol. 102, p 473.

42. Plaut, G.; Ghil, M.; Vautard, R., Science 1995; Vol. 268, p 710.

43. Mock, S.J.; Hibler III, W.D., Nature 1976; Vol. 261, p 484.

44. Coleman, B.D., Journal of Applied Physics 1956; Vol. 27, p 862.

45. Bradley, R.S.; Diaz, H.F.; Eischeid, J.K.; Jones, P.D.; Kelly, P.M.; Goodess, C.M., Science 1987; Vol. 237, p 171.

46. Hense, A.; Krabe, P.; Flohn, H., Meteorology and Atmospheric Physics 1988; Vol. 38, p 215.

47. Groisman, P. Ya, *Greenhouse-Gas-Induced Climate Change: A Critical Appraisal of Simulations and Observations*, Schlesinger, M.E. [Ed.], Elsevier, New York, 1991; p. 297

48. Peixóto, J.P. ; Oort, A.H., *Variations in the Global Water Budget*, Street-Perrott, A, Beran, M., and Ratcliffe, R. [Eds.], D. Reidel Publishing Company, Boston, 1982; p 5.

49. Salstein, D.A.; Rosen, R.D.; Peixoto, J.P., Journal of the Atmospheric Sciences 1983; Vol 40, p 788.

50. Gaffen, D.J.; Barnett, T.P.; Elliott, W.P., Journal of Climate 1991; Vol. 4, p 989.

51. Peixóto, J.P.; Oort, A.H. *Physics of Climate*, American Institute of Physics, New York, 1992.

52. Elliott, W.P.; Smith, M.E.; Angell, J.K., *Greenhouse-Gas-Induced Climatic Change: A Critical Appraisal of Simulations and Observations*, Schlesinger, M.E. [Ed.], Elsevier, New York, 1991; p 311

53. Rasmusson, E.M., Water Resource Research 1966; Vol. 2, p 469.

54. Wolff, G.T.; Rodgers, W.R.; Collins, D.C.; Verma, M.H.; Wong, C.A., Journal of Air and Waste Management Association 1990; Vol 40, p 1638.

55. Schulz, U.; Trubiroha, P., *Durability Testing of Non-Metallic Materials*, ASTM STP 1294, R.J. Herling [Ed.], American Society for testing and Materials, Philadelphia, 1996.

56. Wernståhl, K.M.; Carlsson, B., Journal of Coatings Technology 1997; Vol. 69, p 69.

57. Rodgers, W.R.; Garner, D.P.; Cheever, G.D., Journal of Coatings Technology 1998; Vol. 70, p. 83.

58. Schulz, U.; Trubiroha, P.; Boettger, T.; Bolte, H., *Proceedings of the 8th International Conference on Durability of Building Materials and Components*, meeting held in Vancouver, B.C. from May 30 to June 3, 1999, NRC Research Press, Ottawa, Canada, 1999; p. 864.

59. Jones, P.D., *World Survey of Climatology: Future Climates of the World: A Modelling Perspective* , H.E. Landsberg [Ed.], Elsevier, New York, 1995; p. 151.

60. Diaz, H.F.; Bradley, R.S., *Natural Climatic Variability on Decade-to-Century Time Scales*, Climate Research Committee, National Academy Press, Washington, D.C., 1995; p 17.

61. Dehsara, M.; Cehak, K., Arch. Meteorolog. Geophysics Bioklimatol. Soc. Series B, 1970; Vol 19, p 269.

62. Folland, C.K.; Parker, D.E.; Kates, F.E., Nature 1984; Vol. 310, p 670.

63. Borzenkova, I.I.; Vinnikov, K. Ya.; Spirina, L.P.; Stechnovsky, D.I., Meteorol. Gidrol. No. 7, 1976; p 27.

64. Shaw, D., Journal of Geophysical Research 1965; Vol. 70, p 4997.

65. Shapiro, R., Quarterly Journal of the Royal Meteorological Society 1975; Vol. 101, p 679.

22

66. Thaler, J.S., *Climate: History, Periodicity, and Predictability*, Rampino, M.R., Sanders, J.E., Newman, W.S., and Königsson, L.K [Eds.], Van Nostrand Reinhold, 1987.

67. Spar, J.; Mayer, J.A., Weatherwise 1973; Vol. 26, p 128.

68. Landsberg, H.E.; Mitchell, J.; Crutcher, H., Monthly Weather Review 1959; Vol. 87, p 283.

69. Tabony, R.C., The Meteorological Magazine 1979; Vol. 108, p 97.

70. Currie, R.G., *Climate: History, Periodicity, and Predictability*, Rampino, M.R., Sanders, J.E., Newman, W.S., and Königsson, L.K [Eds.], Van Nostrand Reinhold, 1987.

71. Flohn, H.; Kapala, A., Nature, 1989; Vol. 338, p 244.

72. Gutzler, D.S., Geophysical Research Letters 1992; Vol. 19, p 1595.

73. Dignon, J.; Hameed, S., Journal of Air Pollution Control Association 1989; Vol. 39, p 180.

74. Karl, T.R.; Jones, P.D.; Knight, R.W.; Kukla, G.; Plummer, N.; Razuvayev, V.; Gallo, K.P.; Lindesay, J.A.; Charlson, R.J., *Natural Climate Variability on Decade-to-Century Time Scales*, Climate Research Committee of the National Research Council, National Academy Press, Washington, D.C., 1995; p. 80.

75. Brimblecombe, P.; Benthan, G., *Climates of the British Isles: Present, Past, and Future*, Hulme, M. and Barrow, E. [Eds.], Routledge, 1997; p. 243.

76. Burroughs, W.J., *Weather Cycles Real or Imagery?*, Cambridge University Press, New York, 1992. .

77. Climate Research Committee, *Natural Climate Variability on Decade-to-Century Time Scales*, National Academy Press, Washington, D.C., 1995.

78. National Research Council, *Decade-to-Century-Scale Climate Variability and Change,* National Academy Press, Washington, D.C., 1998.

79. Von Storch, H.; Navarra, A.[Eds.], *Analysis of Climate Variability,* Springer, New York, 1999.

80. Pittock, A.B., Review of Geophysical Space Physics 1978; Vol. 16, p 400.

81. Martin, J.W.; Saunders, S.C.; Floyd, F.L.; Wineburg, J.P., *Federation Series on Coatings Technology, Federation of Societies for Coatings Technology*, Blue Bell, PA, 1996.

Chapter 2

Surface Moisture and Time of Wetness Measurements

Peter Norberg[1]

Centre for Built Environment, Royal Institute of Technology, S. Sjo'fullsgatan 3,
SE-80176, Gävle, Sweden
[1]Current address: Centre for Built Environment, University of Gävle, Gävle, Sweden

Surface moisture plays an important role in the deterioration of building surfaces. The extent and duration of surface moisture is generally impossible to predict from meteorological data. The limitations of the ISO 9223 standard for estimating the time of wetness (TOW; RH>80%, T>0°C) is evident in climates with sub-zero temperatures, in environments with significant deposition of pollutants and salt, and in situations where the exchange of radiation between building surfaces and the surrounding environment creates large temperature differences. Consequently, direct measurement of TOW is essential, e.g. using the WETCORR method. This method is suitable for measurements of surface moisture and TOW on building materials in general. The actual sensor consists of an inert electrolytic cell with Au/Au-electrodes combined with a Pt-1000 surface temperature-sensing element

Background

The interest in surface moisture and time of wetness (TOW) has its origin in the field of atmospheric corrosion. Early on, Vernon (1) had shown that the corrosion rate of steel increased dramatically when a critical relative humidity (RH) of between 80 and 90% was exceeded. Understanding of the various mechanisms that, even under non-condensing conditions, can result in the build-up of significant amounts of moisture on metallic surfaces, and thereby cause corrosion, was essential for the continued research in this area. In addition, the electrochemical nature of the typical moist or wet atmospheric corrosion became more and more obvious (2). This also led to the adoption of electrochemical methods for studying the instantaneous rate of atmospheric corrosion with the prospects of replacing the traditional and time-consuming weight loss measurements done by long-term exposures of test coupons.

Mansfeld (3,4) has reviewed the early experience of electrochemical measurements of atmospheric corrosion and its relation to the concept of TOW. In this context TOW is commonly considered as the time for which the atmospheric conditions are such that electrochemical reactions of some magnitude can occur on the surface of the sensor. There is a general opinion among atmospheric corrosion scientists that surface moisture and TOW play a very important role in the corrosion of metals and alloys exposed to the atmosphere. Consequently, the idea of a critical RH determining TOW is very much reflected in the current standard ISO 9223 (5) defining TOW as the time for which RH is greater than 80% while the air temperature is above 0°C. As has been shown by many investigators this meteorological approach has its limitations, partly because the electrochemical reactions are in operation far below 0°C (6,7,8). It is also well known that the presence of hygroscopic salts on the surface (2,4) can considerably lower the humidity where wetting occurs. In addition, the difference between air and surface temperatures, as governed by the radiation conditions, is a very important factor to consider in relation to TOW (6,9,10).

In the very first attempt to study atmospheric corrosion by electrochemical methods, Tomashov and co-workers (2,11) used galvanic cells with alternate electrodes of different metals, e.g. Fe/Cu, Fe/Zn, Fe/Al and Cu/Al. When a film of moisture appeared on the surface of the electrode lamellae, a potential difference was produced between the terminals and the resulting external current was measured with a sensitive galvanometer. Sereda (12,13) used galvanic cells of the types Pt/Fe and Pt/Zn but measured the variation in voltage across an external resistor through which the galvanic current was flowing. TOW in these cases was defined as the time during which the galvanic current or voltage exceeded an arbitrary thres-hold value. Sereda et al (6) also made way for the ASTM standard (14) covering that particular method for the electrode combinations Au/Cu, Au/Zn and Pt/Ag. More recently, Hechler et al (7) have studied exposures of large sets of sensors following the ASTM procedure.

Kucera and Mattsson (15) and Mansfeld and co-workers (3,4,16,17) adopted the original concept of Tomashov using Cu/steel or Cu/Zn couples and studied the galvanic current, trying to relate this to the rate of atmospheric corrosion. Kucera and co-workers (15,18) and Mansfeld and co-workers (3,4,16,17,19) also used electrolytic cells of only one metal, e.g. Cu/Cu, steel/steel and Zn/Zn, to which an external constant voltage was applied. Kucera used voltages in the range 100-400 mV and the resulting current was only a vague measure of the corrosion rate, while Mansfeld limited the potential difference to ±30 mV in order to enable measurement of the corrosion current on the basis of the polarisation resistance technique.

The development of the electrolytic method originally proposed by Kucera and Mattsson (15) has continued in Scandinavia during the past 20-25 years, to a large extent within the frames of joint Nordic research programmes involving the Swedish Corrosion Institute (SCI) and the Norwegian Institute for Air Research (NILU), e.g.

Haagenrud et al (20). Further efforts made by NILU led to the so-called NILU WET-CORR (WETness and COrrosion Rate Recorder) method, involving an automatic six-channel current integrator and the use of miniature Cu/Cu cells, Haagenrud et al (21). A theoretical study of the electrochemical characteristics of the NILU/SCI sensor was done by Haagenrud et al (22) who showed, among other things, how the recorded current was depending on the thickness of the deposited moisture film.

More recent collaboration between NILU and the National Swedish Institute for Building Research (SIB) aimed at extending the NILU WETCORR concept to measurements of surface moisture and TOW on building materials and structures in general, Haagenrud et al (23) and Svennerstedt (24,25). The surface moisture studies by Lindberg (26) on paint and Yamasaki (27) on plastics should also be mentioned in this context as examples of TOW studies made on non-metallic materials. This generalisation of the view on TOW should have implications not only for the definition of the TOW concept as such but also for the measurement technique and the sensors used.

In the following a brief overview will be given of the experience gained in relation to the development and use of the WETCORR method, as seen from a Scandinavian perspective.

The WETCORR system

The WETCORR measurement technique, the way it is done today, was first introduced by Haagenrud et al (21) in 1984. Since that time several versions of instrument and sensor have been in use in Scandinavia. In the following sections the present version of the WETCORR system will be outlined with some reference to older devices.

Instrumentation

The most recent version of the WETCORR system, which has been in use since late 1994, is very similar to the one used previously, a description of which was given by Norberg (28). An overview of the present system is shown in Figure 1. The WET-CORR system can measure surface moisture and surface temperature from up to 64 sensors simultaneously. The actual measurements are conducted by up to 16 *sensor adapters*, each connecting a maximum of 4 sensors. The *system controller* communicates with the sensor adapters via an RS485-network and also provides the necessary power. The basic principle of the measuring technique involves excitation of the sensor with a DC voltage of normally 100 mV. To avoid net polarisation of the electrodes in the long perspective the polarity is reversed every 30 seconds. The absolute value of the resulting current is averaged over one voltage cycle, i.e. one minute. The main difference between the previous version of WETCORR and the present is that the temperature channels now require Pt-1000-elements instead of AD592AN-transducers. This improvement makes it possible to more accurately measure temperature in general and surface temperature in particular.

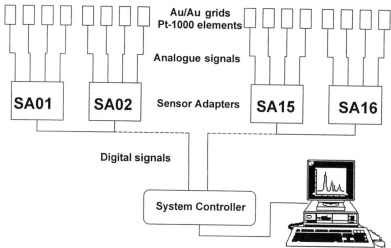

Figure 1. The WETCORR system

Sensor design

The present type of sensor is the third generation of the Au/Au-type of cell developed for the WETCORR system. The active grid measures 16 by 18 mm and the overall size is 22 by 30 by 0.7 mm, see Figure 2. The earlier generations of cell have been evaluated in (28,29). The main improvements include a better design of the electrode pattern, which should minimise interference between the measurements of moisture and temperature. In addition, for surface temperature measurements a very small Pt-1000-element was chosen, showing much better temperature adaptability and tolerance than the previous version. Preliminary experience in severe marine atmospheres has also indicated that this sensor is better suited to cope with salt films on the surface. This sensor is the latest in use and is produced by the Kongsberg Group in Kjeller, Norway.

Sensor characteristics

Haagenrud et al (21) studied the influence of alternating DC voltage on Cu/Cu-cells and found that a 5-6-fold increase in integrated current resulted compared with excitation using constant DC voltage. The influence of polarity reversal on the current response at different RH-levels was demonstrated in detail in (29), and the principal response of the current on reversal of the voltage is shown in Figure 3. The capacitance character of the electrochemical double layer governs the transient response of the current, especially at high moisture loads.

Figure 2. The latest WETCORR sensor.

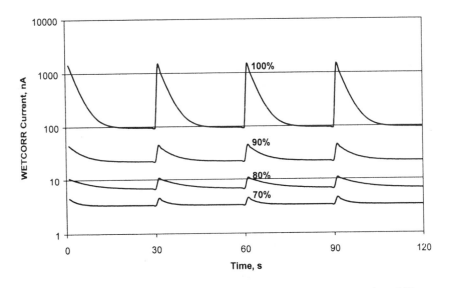

Figure 3. The response of the WETCORR current on polarity reversal at different relative humidities.

The principal shape of these curves and their relative appearance reveal that the TOW sensor, as a first approximation, may be described by an RC-circuit containing two resistors and one capacitor (30), see Figure 4.

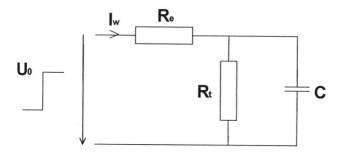

Figure 4. Equivalent electric circuit model for the WETCORR sensor.

R_e is the resistance of the electrolyte between the electrode fingers, R_t the resistance associated with the transfer of charges between the electrolyte and the electrode surface and C the capacitance of the electrochemical double layer in the vicinity of the electrode surface. The time dependence of the WETCORR current I_w after applying a step voltage U_0 may be written as follows, using the notation in Figure 4.

$$I_w = \frac{U_0}{R_e + R_t} \left(1 + \frac{R_t}{R_e} e^{-\frac{R_e + R_t}{R_e R_t C} t} \right)$$

The validity of this equation was tried out in (31) and it was confirmed that a satisfactory agreement with the measured curves was obtained and that the simple electronic circuit could explain the main features of the curves.

From this it may be concluded that the current measured with the WETCORR method not only reflects the resistance of the electrolyte, R_e, but also the impedance associated with the processes occurring at the electrode surfaces following the polarisation. However, a generalised view on the concept of TOW should be related to R_e rather than to the polarisation-induced impedance. This will be further discussed under the heading "Time of wetness".

Experience from measurements

There is quite a number of studies published that has involved WETCORR measurements. One reason for conducting such measurements is that the ISO 9223 standard (5) is insufficient in describing the actual TOW as experienced by an arbitrary building surface. The discrepancies between the two estimates depend on various factors of which the most important will be exemplified in the following.

Effects of deposition

The deposition of pollutants of various types greatly affects the response of surface moisture sensors, as have been noted in (32-36). It is also quite obvious that exposure positions sheltered from rain are more severely affected by degradation than open ones, particularly after longer exposures when sufficient amounts of aggressive species have accumulated. In order to illustrate the effects of salt deposition on TOW for sheltered exposures, some feature results extracted from Norberg et al (36) will be discussed below.

The results were recorded in January 1996 at the Water Board site at Flinders, Victoria, Australia. This site is situated a few kilometres from the sea (Bass Strait) and the average chloride deposition rate is of the order of 30 mg/m²day (37). Duplicate WETCORR sensors of the most recent design, as depicted in Figure 2, were employed. The hourly average of the current for the whole month was plotted against the surface RH (RH_{sur}). This variable was derived from the assumption that the vapour concentration close to the surface is the same as that found in the bulk air, i.e., $RH_{sur} \cdot v_s(T_{sur}) = RH_{air} \cdot v_s(T_{air})$, where v_s is the saturation vapour concentration depending mainly on the temperature T.

The data obtained under the shelter resulted in a swarm of points that were surprisingly well kept together, giving the impression that under these conditions the WETCORR sensor acted very much like a relative humidity sensor, see Figure 5.

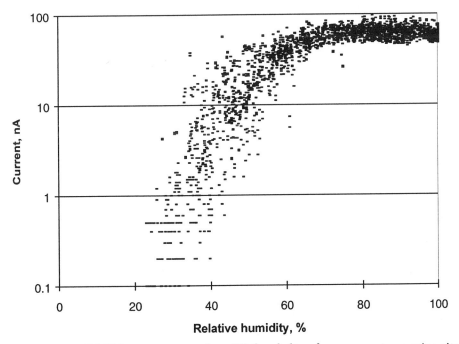

Figure 5. WETCORR current vs surface RH for sheltered exposure at a marine site during one month.

There seems to exist a limit around 30% RH below which the conductivity of the film becomes very low. This limit, and particularly the shape of the curve in Figure 5, can be understood from the hygroscopic properties of seawater relative to its composition and moisture content (31). Below approximately 33% the hydrate of one of the principal constituents of sea water besides NaCl, namely $MgCl_2$, will completely dry out and no longer conduct electric current. Since, in this case, rain and dew have little, if any, effect on the result, the current is directly related to the actual moisture content of the salt film, as determined by the surface RH.

The corresponding results obtained in the open (31) were quite scattered because the current not only derived from the conductivity of the salt film, as determined by its hygroscopicity, but also from episodes of dew and rain. The rain tended to wash off most of the salt at times and as a result the sensors temporarily resumed to the original, lower sensitivity.

Effects of temperature differences and radiation conditions

It is clear that surface temperature relative to ambient is a very important parameter in determining TOW. To illustrate this, further examples of the data from the Water Board site will be given in the following.

Figure 6 shows the current as a function of the temperature difference between the air and the surface for duplicate sensors kept under the specially designed glass shelter. As is obvious, the current was always at its maximum as soon as the surface temperature dropped to below that of the ambient air, even though the difference was not greater than 1°C. It should be noted that the WETCORR data appearing in Figures 5 and 6 are the same.

In the open, undercooling was more pronounced and frequent (31). However, since rain and dew sometimes washed the sensors in the open the sensitivity of the sensors to variations in RH was less. Consequently, the WETCORR current was comparatively low even when the surface temperature was below the ambient.

Time of wetness

When evaluating the current-time data obtained with the WETCORR equipment, a current criterion is normally chosen in order to estimate TOW. Typically, TOW is based on the time for which the average current across the sensor grid exceeds, say, 10 or 30 nA. The obtained TOW should ideally reflect the time when significant corrosion or degradation in general, takes place on a material surface being exposed to the same environment as the WETCORR sensor. From experience it may be stated that TOW is not as much a function of the amount of moisture deposited as it is of the conductivity of the moisture film. Recently, Elvedal et al (38) defined a critical current of 10 nA that should correspond to a substantial water film (~3μm) on the sensor surface. This evaluation was made in a fairly mild environment corresponding to corrosivity category C1 according to ISO 9223 (5). Should the same criterion be applied

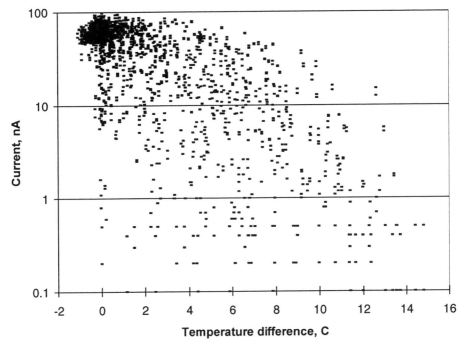

Figure 6. WETCORR current vs temperature difference between ambient air and surface for sheltered exposure at a marine site during one month.

to measurements in coastal or industrial areas, the deposition of salt and pollutants would considerably increase the conductivity and, as a consequence, the same current would be obtained for a much lower amount of surface moisture. In other words, the relation between the current and the thickness of the moisture film cannot be stated without taking the electrolytic conductivity into consideration. Presumably, there is also a better correlation between corrosivity and conductivity than between corrosivity and moisture-film thickness, particularly for natural environments.

So far, all estimates of TOW, whether made with galvanic or electrolytic cells, have been based on arbitrary criteria. This dilemma has since long been recognised by Mansfeld et al (17) and is also explicitly expressed in the ISO 9223 standard (5). In order to get around this problem a more generalised TOW concept can be introduced. This requires the use of inert electrolytic cells and a modified measuring technique.

New definition of TOW

The basic idea is to consider the measurement of surface moisture and TOW simply as a measurement of electrolytic conductivity. This is not a major deviation from the present situation but rather an adaptation to what is actually the case. For the sake of conformity this modification may be called TOC, *time of conduction*, or in the transferred sense for the case of atmospheric corrosion of metals, *time of corrosion*.

As shown in a preceding section on "Sensor characteristics", the response of the WETCORR sensor to the applied voltage can give information about the electrolytic conductivity and the nature of the electrode/electrolyte interface. The use of an inert sensor, such as the Au/Au sensor in the present study, implies that the electrodes involved will not significantly be influenced by corrosion. In addition, by eliminating the impedance effects of the polarised interface, the measurement of the cell resistance will be equivalent to a conventional measurement of the electrolytic conductivity. The impedance associated with the electrode/electrolyte interface may be eliminated by AC excitation of typically a few kHz or by DC polarisation during only a few milliseconds, including reversal of the polarity. In analogy with conventional measurements of conductivity, different cell geometries can be used and still be possible to compare via the cell constant. For a given geometry the cell constant constitutes the link between resistance and resistivity or their reciprocals, conductance and conductivity. In this way, a universal criterion for the limiting conductivity above which the sensor should be considered wet may be selected.

Such a criterion remains to be agreed upon but should, most likely, be related to typical conductivities found for precipitation and dew in relatively unpolluted environments. Thus, the time of conduction, TOC, may be defined as the length of time when the electrolytic conductivity is greater than x μS/cm, as measured on the surface of an inert electrolytic cell in thermal contact with the substrate material.

How to make use of TOW/TOC

TOW or TOC should *not* be considered as variables in the traditional sense. Since these entities are derived from variables via selected criteria they are rather delimiters or discriminators. Consequently, they should not be explicitly utilised in mathematical expressions such as dose-response functions of the type originally proposed by Guttman and Sereda (39). Instead TOW/TOC should be used to discriminate wet and dry periods from each other, i.e. to determine when the "wet" and "dry" model, respectively, should be used for calculating the extent of degradation (31). This procedure is most obvious for metals but should also be applicable to non-metallic materials.

Conclusions

The WETCORR technique has been developed into a versatile tool for making microclimate measurements in the built environment. The present version of the equipment can accommodate up to 64 surface moisture grids and just as many temperature sensors. The latest version of sensor is considered very suitable to estimating surface moisture loads and times and simultaneously to give a reasonably correct value of the surface temperature. The corrosion resistance of the combined moisture and temperature sensor has also improved considerably compared with previous versions.

The function of the Au/Au-type sensor under controlled conditions in the laboratory as well as in the field studies has shown very good reproducibility. A simple equivalent electric circuit that helps explaining the transient character of the measured WETCORR current on excitation can describe the working principle of the moisture sensor. This transient is attributed to the electrolytic conductance of the moisture film on the surface of the sensor grid. The asymptotic DC current, on the other hand, is the result also of electrode/electrolyte reactions. Measurement of moisture conditions and TOW should most likely be associated with the conductance of the electrolyte. To improve the sensitivity of the method with regard to moisture detection, AC excitation or short pulses of DC may be used to eliminate the capacitive properties of the cell.

There are numerous examples of the limitations of the ISO 9223 standard for estimating TOW. First, corrosion is not limited by temperatures below 0°C. Secondly, in most exposure environments, deposition of pollutants and salt will generally lower the RH above which wetting of the surface occurs due to the hygroscopicity of the deposits. For marine environments it is shown that the surface film remains conducting down to a surface RH of around 30%. This phenomenon is most pronounced for rain sheltered positions. Thirdly, radiation exchange between surfaces and the environment is not considered. Differences in temperature between a surface and the surrounding air may be considerable and cause both evaporation and condensation.

A generalised definition of TOW is proposed which takes into account the conductivity of the moisture film rather than its thickness. Under the assumption that significant degradation by corrosion or any other moisture related mechanism cannot occur below a certain conductivity, a universal criterion for TOW can be defined. This modified TOW is called time of conduction or time of corrosion, TOC, and is defined as the length of time when the electrolytic conductivity is greater than x μS/cm, as measured by an inert moisture sensor in thermal contact with a substrate material. The exact value of the limiting conductivity remains to be agreed upon. In compliance with conventional measurements of conductivity, any cell can be calibrated by determining the cell constant for the electrode configuration in a standard electrolyte.

The adoption of the TOC concept makes possible a more strict separation of time into "wet" and "dry" periods. In other words, TOC can be used to discriminate between periods when different dose-response relations should be applied to describing the degradation. This also implies that modelling of dose-response relations would not have to include TOW in the actual models.

Literature Cited

1. Vernon, W H J, Laboratory Study of the Atmospheric Corrosion of Metals, *Transactions of the Faraday Society*, Vol 31, 1935, pp 1678-1686.
2. Tomashov, N D, Atmospheric Corrosion of Metals, *Theory of Corrosion and Protection of Metals*, Chapter XIV, MacMillan, New York, 1966, pp 367-398.
3. Mansfeld, F, Evaluation of Electrochemical Techniques for Monitoring of Atmospheric Corrosion Phenomena, *Electrochemical Corrosion Testing, ASTM STP 727*, 1981, pp 215-237.

34

4. Mansfeld, F, Electrochemical Methods for Atmospheric Corrosion Studies, *Atmospheric Corrosion, The Electrochemical Society*, 1982, pp 139-160.

5. ISO 9223, Corrosion of Metals and Alloys, *Classification of Corrosivity Categories of Atmospheres*, 1992.

6. Sereda, P J, Croll, S G and Slade, H F, Measurement of the Time-of-Wetness by Moisture Sensors and Their Calibration, *Atmospheric Corrosion of Metals, ASTM STP 767*, 1982, pp 267-285.

7. Hechler, J J, Boulanger, J, Noël, D, Dufresne, R and Pinon, C, A Study of Large Sets of ASTM G 84 Time-of-Wetness Sensor, *Corrosion Testing and Evaluation:Silver Anniversary Volume, ASTM STP 1000*, 1990, pp 260-278.

8. King, G A and Duncan, J R, Some Apparent Limitations in Using the ISO Atmospheric Corrosivity Categories, *Corrosion & Materials*, Vol 23, No 1, 1998, pp 8-14 & 22-24.

9. Grossman, P R, Investigation of Atmospheric Exposure Factors that Determine Time-of-Wetness of Outdoor Structures, *Atmospheric Factors Affecting the Corrosion of Engineering Materials, ASTM STP 646*, 1978, pp 5-16.

10. Dean, S W and Reiser, D B, Time of Wetness and Dew Formation: A Model of Atmospheric Heat Transfer, Atmospheric Corrosion, ASTM STP 1239, 1994, pp 3-10.

11. Tomashov, N D, Berukshtis, G K and Lokotilov, A A, *Zavodskaja Laboratorija*, 22, (3), 1956, pp 345-349.

12. Sereda, P J, Measurement of Surface Moisture, A Progress Report, *ASTM Bulletin No 228*, Feb 1958, pp 53-55.

13. Sereda, P J, Measurement of Surface Moisture, Second Progress Report, *ASTM Bulletin No 258*, May 1959, pp 61-63.

14. ASTM G84-84, Standard Practice for Measurement of Time-of-Wetness on Surfaces Exposed to Wetting Conditions as in Atmospheric Corrosion Testing, *ASTM*, 1984.

15. Kucera, V and Mattsson, E, Electrochemical Technique for Determination of the Instantaneous Rate of Atmospheric Corrosion, *Corrosion in Natural Environments, ASTM STP 558*, 1974, pp 239-260.

16. Mansfeld, F and Tsai, S, Laboratory Studies of Atmospheric Corrosion — I. Weight Loss and Electrochemical Measurements, Corrosion Science, Vol 20, 1980, pp 853-872.

17. Mansfeld, F, Tsai, S, Jeanjaquet, S, Meyer, E, Fertig, K and Ogden, C, Reproducibility of Electrochemical Measurements of Atmospheric Corrosion Phenomena, *Atmospheric Corrosion of Metals, ASTM STP 767*, 1982, pp 309-338.

18. Kucera, V and Gullman, J, Practical Experience with an Electrochemical Technique for Atmospheric Corrosion Monitoring, *Electrochemical Corrosion Testing, ASTM STP 727*, 1981, pp 238-255.

19. Mansfeld, F, Monitoring of Atmospheric Corrosion Phenomena with Electrochemical Sensors, *J Electrochem Soc*, Vol 135, No 6, 1987, pp 1354-1358.

20. Haagenrud, S, Kucera, V and Gullman, J, Atmospheric Corrosion Testing with Electrolytic Cells in Norway and Sweden, *Atmospheric Corrosion, The Electrochemical Society*, 1982, pp 669-693.

21. Haagenrud, S E, Henriksen, J F, Danielsen, T and Rode, A, An Electrochemical Technique for Measurement of Time of Wetness, *Proc 3rd Int Conf on Durability of Building Materials and Components*, Espoo, 12-15 Aug 1984, pp 384-401.

22. Haagenrud, S E, Henriksen, J F and Wyzisk, Electrochemical Characteristics of the NILU/SCI Atmospheric Corrosion Monitor, *Corrosion 85, NACE*, Boston, 25-29 March, 1985, Paper 83.

23. Haagenrud, S, Henriksen, J and Svennerstedt, B, Våttidsmålinger på treplater — Prøvestudie med NILU WETCORR-metoden (Time of wetness measurements on wood — Pilot study of the NILU WETCORR method), *NILU OR 17/85*, 1985, (In Norwegian).

24. Svennerstedt, B, Time of Wetness Measurements in the Nordic Countries, *Proc 4th Int Conf on Durability of Building Materials and Components*, Singapore, 4-8 Nov 1987, pp 864-869.

25. Svennerstedt, B, Ytfukt på fasadmaterial (Surface moisture on facade materials), The National Swedish Institute for Building Research, *Research Report TN:16*, 1989, (In Swedish).

26. Lindberg, B, Mätning av våttid på målade byggnadsmaterial (Measurement of TOW on painted surfaces of building materials), Scandinavian Paint and Printing Ink Research Institute, *NIF-Report T 2-88 M*, 1988, (In Swedish).

27. Yamasaki, R S, Characterization of Wet and Dry Periods of Plastic Surfaces during Outdoor Exposure, *Durability of Building Materials*, Vol 2, No 2, 1984, pp 155-169.

28. Norberg, P, Evaluation of a New Surface Moisture Monitoring System, *Proc 6th Int Conf on Durability of Building Materials and Components*, Omiya, Japan, 26-29 Oct 1993, Paper 3.7, pp 637-646.

29. Norberg, P, Monitoring of Surface Moisture by Miniature Moisture Sensors, *Proc 5th Int Conf on Durability of Building Materials and Components*, Brighton, UK, 7-9 Nov 1990, pp 539-550.

30. Metals Handbook, Ninth Edition, *Volume 13, Corrosion*, ASM International, 1987, pp 212-228.

31. Norberg, P, Ph.D. thesis, Centre for Built Environment, Royal Institute of Technology, Gävle, Sweden, 1998.

32. Norberg, P, Sjöström, C, Kucera, V and Rendahl, B, Microenvironment Measurements and Materials Degradation at the Royal Palace in Stockholm, *Proc 6th Int Conf on Durability of Building Materials and Components*, Omiya, Japan, 26-29 Oct 1993, Paper 3.2, pp 589-597.

33. Norberg, P, King, G and O'Brien, D, Corrosivity and Microclimate Measurements in Open and Sheltered Marine Environments, *Proc 7th Int Conf on Durability of Building Materials and Components*, Stockholm, Sweden, 19-23 May 1996, Vol 1, Paper 18, pp 180-190.

34. Cole, I S, Norberg, P and Ganther, W D, Environmental Factors Promoting Corrosion in Building Microclimates, *13th International Corrosion Congress*, Melbourne, Australia, 25-29 Nov 1996, Paper 410.

35. Rendahl, B, Kucera V, Norberg, P and Sjöström, C, Våt- och torrdeponering av försurande luftföroreningar. Inverkan på byggnadsmaterial. (Wet and dry deposition of acidifying air pollutants. Effect on building materials), *KI-Report 1995:3*, Swedish Corrosion Institute, (In Swedish).

36. Norberg, P, King, G A, O'Brien, D J and Sherman, N, MIME Extension: A Project to Measure the Short Term Corrosion Rates of Steel, Zinc and Coated Steels and Correlate with Climatic Parameters and Time-of-Wetness, *CSIRO DBCE Doc. 96/23 (M)*, Mar 1996.

37. Norberg, P, King, G and O'Brien, D, Corrosivity Studies and Microclimate Measurements in Marine Environments. *ACA Annual Conference, Corrosion & Prevention 95*, Perth, Australia, 12-16 Nov 1995, Paper 36.

38. Elvedal, U, Henriksen, J F and Haagenrud, S, Project Deliverable No. 7.1 — Final Report WP2: Calibration of WETCORR Sensors in Field and Laboratory, EU-project ENV4-CT95-0110 Wood-Assess, *Wood-Assess Document No:WP2/Del7.1/NILU/1998.08.15*, 1998, (Confidential).

39. Guttman, H and Sereda, P J, Measurement of Atmospheric Factors Affecting the Corrosion of Metals, *"Metal Corrosion in the Atmosphere", ASTM STP 435*, 1968, pp 326-359.

Chapter 3

Overview of Exposure Angle Considerations for Service Life Prediction

Henry K. Hardcastle III

Atlas Weathering Services Group, Atlas Electric Devices Company,
17301 Okeechobee Road, Miami, FL 33018

This paper reviews selected considerations of exposure angle effect on materials weathering degradation. Calculation of solar angle of incidence is reviewed for surfaces with different slope angles. Historical UV irradiance measurement data is reviewed as a function of exposure angle. Empirical degradation as a function of exposure angle is reviewed for three materials.

It is important for treatments of Service Life Prediction (SLP) to account for the effect of exposure angle on materials degradation. This paper presents several considerations including solar angle of incidence, empirical solar radiation measurements at different angles, and degradation rates of several materials at different exposure angles. The presentation is grouped into three parts:

- Calculation of Solar Angle of Incidence
- Radiation Measurements as a Function of Exposure Angle
- Materials Degradation Measurements as a Function of Exposure Angle (Experimental)

Calculation of Solar Angle of Incidence

Irradiance represents one of the most important variables effecting weathering of materials in outdoor service. When materials are exposed at professional weathering laboratories, measurements of incident radiation are usually available. When exposures are conducted at remote locations, however, solar radiation impinging on surfaces must be estimated.

Models and calculations of solar irradiance are straightforward for simple situations but become more complex as exposure situations approach in-service conditions. The solar source itself has been well characterized and is considered stable for most purposes regarding materials performance. Most researchers accept the 1353 - 1373 W/m^2 solar constant (*1*) estimates for extraterrestrial environments. There is little evidence to suggest this estimate will change by an important amount in the near future.

However, the solar flux impinging on terrestrial surfaces varies according to daily and yearly cyclic causes. These effect solar radiation in two ways. First, as the angle of incidence of flux on the surface changes, the light density changes. Second, the effect of air mass filters the solar radiation. Thus, a materials orientation or angle in service environments is a critical consideration in service life prediction. The angle of incidence of sunlight for a surface at any angle, at any location on the earth, can be determined by the following calculation from Duffie and Beckman (1980) (*2*):

$$\cos \theta = \sin \delta \sin \phi \cos \beta - \sin \delta \cos \phi \sin \beta \cos \gamma + \cos \delta \cos \phi \cos \beta \cos \omega$$
$$+ \cos \delta \sin \phi \sin \beta \cos \gamma \cos \omega$$
$$+ \cos \delta \sin \beta \sin \gamma \sin \omega$$

Where ϕ represents latitude
        δ represents solar declination at solar noon
        β represents the slope of the surface
        γ represents the surface azimuth angle
        ω represents the angle of the sun east or west of local meridian
        θ represents angle of incidence

The declination, δ, can be found from the equation of Cooper (1969):

$$\delta = 23.45 \sin \left(\; 360 \; \left(\frac{284 + n}{365} \right) \right)$$

This calculation can be used to solve for the volume of angles of incidence "a" surface is exposed to throughout the year at a location. For instance, a surface oriented horizontally at Miami, FL experiences solar angles of incidence depicted in Figure 1. This figure shows the angle of incidence as an output due to time of day and day of year.

As a surface changes angle from horizontal, the family of angle of incidences changes significantly. For instance, Figure 2 shows angles of incidence of a 26 degree south exposure angle in Miami, FL, Figure 3 shows angles of incidence for a 45 degree south exposure and Figure 4 shows a vertical, south-facing surface's solar angle of incidence for a year. Figures 1-4 dramatically illustrate the different solar environments surfaces experience simply by altering the single variable of exposure angle.

39

Angle of Incidence - 0 degree South Slope Surface - 26 degree Latitude

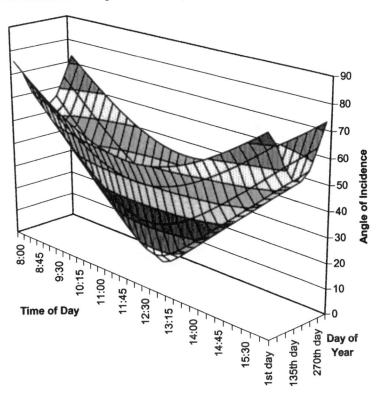

Figure 1. Miami Exposure, Horizontal

40

Angle of Incidence - 26 degree South Slope Surface - 26 degree Latitude

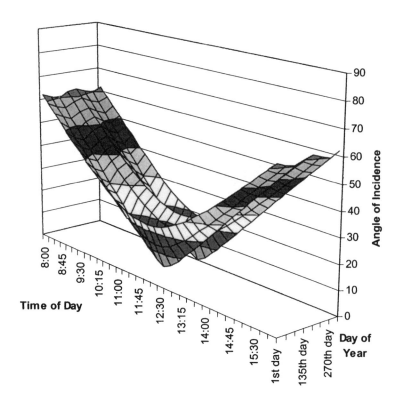

Figure 2. Miami Exposure, 26 degrees south

Angle of Incidence - 45 degree South Slope Surface - 26 degree Latitude

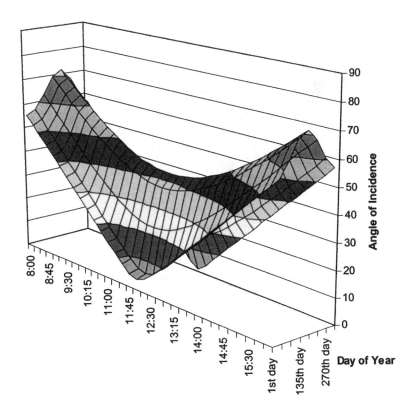

Figure 3. Miami Exposure, 45 degrees south

Angle of Incidence - 90 degree South Surface - 26 degree Latitude

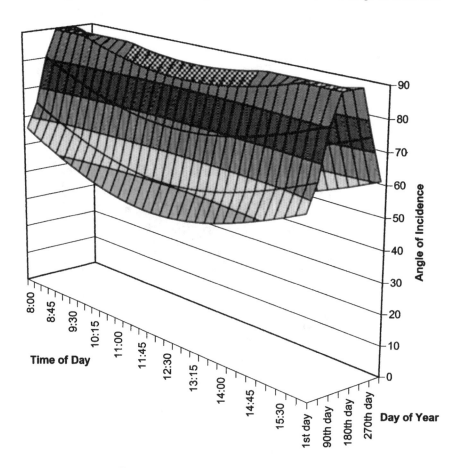

Figure 4. Miami Exposure, Vertical south

Changing a surface's exposure azimuth also dramatically changes a surface's solar environment. Figures 5 and 6 show a surface at 45 degrees slope angle facing southeast (135 degrees) and west (270 degrees), respectively. Not only are these two orientations subject to different solar irradiance variables, but an interaction between orientation angle and local climatic conditions may also exist. In south Florida over several months of the summer season, afternoons are significantly more cloudy than mornings. Thus, even if a good correlation exists with morning solar dosages, SLP models may require a weighting factor for afternoon irradiance estimates derived for western facing orientations.

Changing a surface's latitude also dramatically shifts the distribution of solar angles of irradiance. Figures 7 and 8 show angles of incidence for a 90 degree south-facing surface in Phoenix, Arizona and Moscow, Russia. A vertical surface in Moscow's summer encounters a very different set of incident angles than a similarly oriented surface in Phoenix or Miami.

The angle of incidence variable alone cannot be used as a robust co-variable for irradiance dosage. Coupled with each angle of incidence is a particular air mass through which the sunlight must travel. Air mass significantly effects the spectral power distribution of the solar source. The accepted standard extraterrestrial solar spectral power distribution published in ASTM E 490 shows irradiance values of $1074 W/m^2$ µm at 340nm. As the extraterrestrial radiation undergoes absorption, scattering, and other optical effects, the spectral power distribution is modified as represented in ASTM E 892 (3) which shows an irradiance value of $435.3 W/m^2$ µm at 340nm. Air mass is defined as the ratio of optical thickness of the atmosphere through which beam radiation passes to the optical thickness if the sun were at zenith (Air Mass = 1 when sun is directly overhead, Air Mass = 2 when sun is at 60° from overhead). This is why solar spectral distribution can be expected to change for different elevations if other variables are held constant.

Another important consideration for angle's effect on SLP models is that for each angle of incidence, there is a particular air mass associated with that angle of incidence and possibly a unique solar spectral power distribution. Thus it will be important for SLP models to account for the wavelength dependency of each materials' photodegradation rates in light of angle of incidence and air mass for a particular in-service orientation and location. It will also be necessary to account for the local environment's effect on air quality including the ratio of diffuse to direct beam component and, specifically, for clouds in the local environment. Partially hazy or cloudy days represent additional sources of variability that must be treated in theoretical models.

Radiation Measurements as a Function of Exposure Angle

Given the complex interactions of variables in converting angle of incidence to solar dose, it is often easier and more accurate to simply measure solar radiation in reference environments. Atlas Weathering Services Group measures solar radiation and maintains historical radiation data on a professional basis for a variety of surface orientations and locations.

44

Angle of Incidence - 45 degree Surface Facing 135 degree Azimuth - 26 degree Latitude

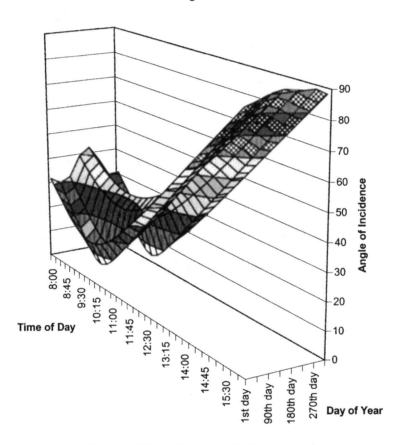

Figure 5. Miami Exposure, 45 degrees southeast

Angle of Incidence - 45 degree Surface Facing 270 degree Azimuth - 26 degree Latitude

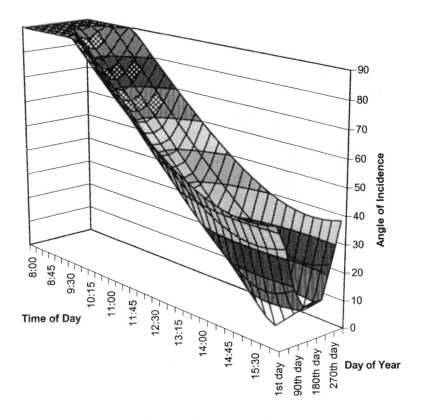

Figure 6. Miami Exposure, 45 degrees west

Angle of Incidence - Vertical South Surface - 34 degree Latitude

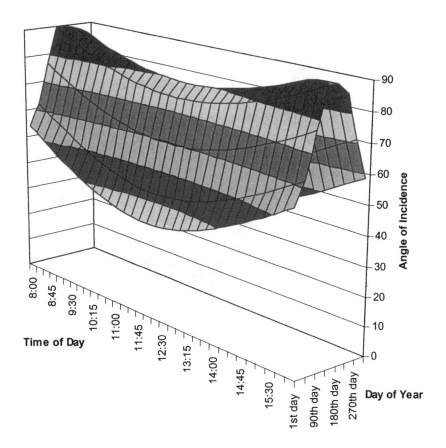

Figure 7. Phoenix Exposure, Vertical south

Angle of Incidence - Vertical South Surface - 56 degree Latitude

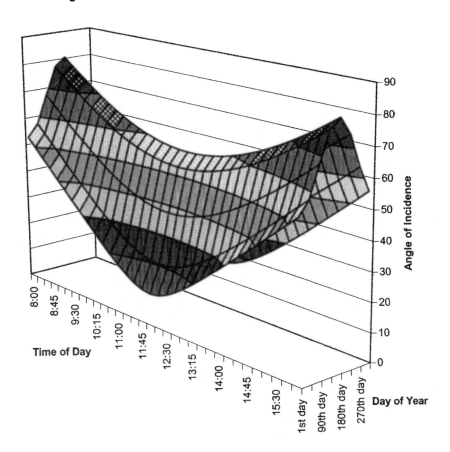

Figure 8. Moscow Exposure, Vertical south

Figure 9 shows historical daily radiation measurements for 26 degree oriented south surface from 1990 to 1998. These measurements were made with calibrated Epply Laboratories TUVR instruments and showed accumulated energy deposited from 295-385 nm. The figure plots median actual values, mean statistical values, and ± 1.96 standard deviations from the mean (95% confidence estimate) for the 1990 to 1998 time period. Interpreting this figure indicates that two different randomly selected days may differ anywhere from 0 to 1.6 MJ/m^2 UV dose. This level of variation may be important for in-service exposures and service life times of one or several days. Testing of skin response and specialty materials designed for rapid decomposition may be required to account for this type of daily variation. Many materials designed for the exterior service environment, however, undergo service lives involving much thicker time slices and require larger UV dosages to show significant photodegradation.

Figure 10 shows historical monthly total UV radiation measurements for a 26 degree south oriented surface from 1990 to 1998. Interpreting this figure indicates that exposures for two different Julys may differ from approximately only 22 to 32 MJ/m^2 UV dose within ± 1.96 standard deviations. Again, this variation may be important for those materials with service life times on the scale of one month. The focus of most materials SLP, however, is on materials with a multi-year in-service life expectancy.

Figure 11 shows historical yearly total UV radiation measurements for a 26 degree oriented surface from 1990 to 1997. Interpreting this figure indicates that exposures for any two different years may differ from approximately only 250 to 350 MJ/m^2 UV with in ± 1.96 standard deviations. The actual yearly totals from 1990 to 1997 show a moderately good agreement with the statistical mean for the 1990 to 1997 period. Actual totals for this data show no discernable trends or cyclic patterns. If the variation depicted in Figure 11 were considered analogous to a plot of output of a manufacturing process (i.e., control chart), most manufacturing process engineers would interpret this as evidence of a relatively stable process in a state of statistical control (4).

It is the multi-year time slice and predictability of solar irradiance dosage at the multi-year level of context that is important for the vast majority of materials designed for exterior in-service use conditions. Figures 12, 13, 14 and 15 show distributions of historical yearly total UV radiation dosages for a selection of angles facing due south measured at Atlas Laboratories in Phoenix, Arizona and Miami, Florida from calibrated and maintained Epply TUVR radiometers.

EvTL, FL - Daily 26 degree TUVR - '90-'98

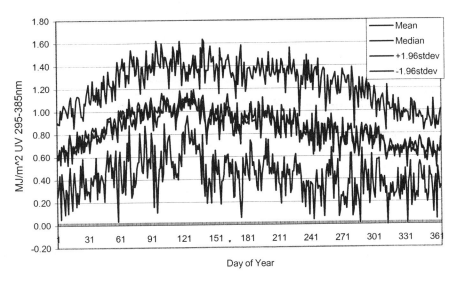

Figure 9. Daily Solar Radiation, Miami

EvTL, FL - 30 Day Totals 26 degree TUVR - '90-'98

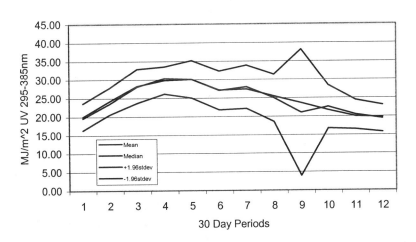

Figure 10. Monthly Solar Radiation, Miami (Note: The ±1.96 Standard Deviation lines widen in the ninth month due to an assignable cause in 1992 when hurricane Andrew passed over the measurement site. Radiation measurements were discontinued for many days reducing n and increasing the standard deviations.)

50

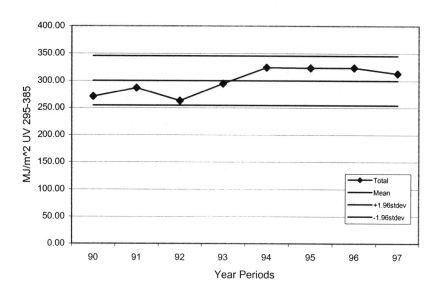

Figure 11. Yearly Solar Radiation, Miami, At Latitude

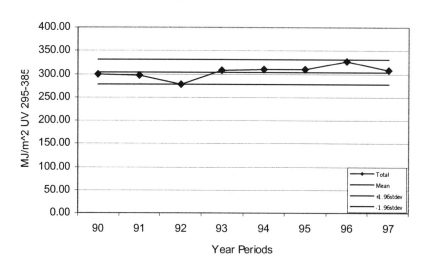

Figure 12. Yearly Solar Radiation, Miami, 45degrees

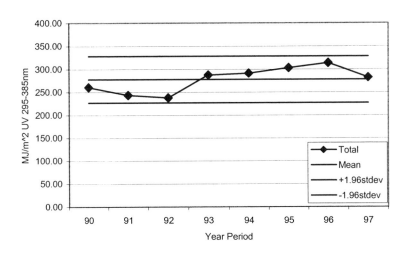

Figure 13. *Yearly Solar Radiation, Miami, 45 degrees*

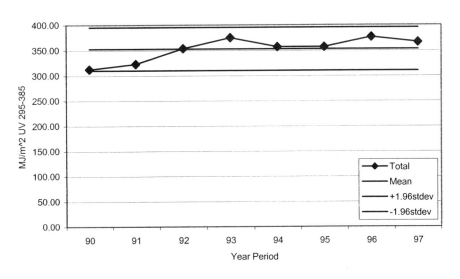

Figure 14. *Yearly Solar Radiation, Phoenix, At Latitude*

52

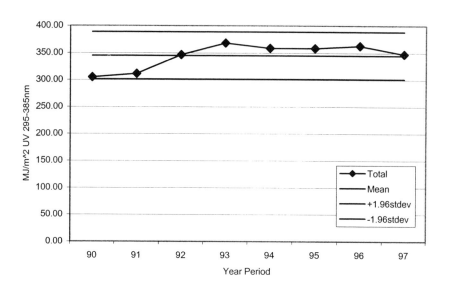

Figure 15. Yearly Solar Radiation, Phoenix, 45 degrees

Materials Degradation Measurements as a Function of Exposure Angle (Experimental)

If materials degradation were solely a function of UV dose, one could expect degradation rates to co-vary with the functions depicted in figures of the previous section. Experience and empirical data indicate, however, that a much more complex system of variables and interactions operates to degrade many materials. In order to gain some simple insights and appreciation for these complex systems, Atlas Weathering Services Group designed and conducted two very simple experiments involving the effect of exposure angle on materials degradation and one small empirical study involving measurements of finished construction components after exposure in the end-use environment.

Experiment A: Effect of Exposure Angle on Development of Yellowness in Clear Polystyrene Reference Materials.

Research Question:
    If a polystyrene reference material is exposed at different angles, will differences in rates of change of Yellowness Index correspond with different dosages of UV measured at these exposure angles?

Materials:
    Polystyrene plastic lightfastness standard material specimens were randomly selected from a single production lot obtained from Test Fabrics, Inc. This material is typically used as a standard reference material for monitoring controlled irradiance, water-cooled xenon arc apparatus. Two randomly selected specimen sets were placed on outdoor exposure at Atlas's SFTS Laboratory in Miami, Florida and Atlas's DSET Laboratory in Phoenix, Arizona during the summer of 1999. Replicate pairs were placed on different angles of exposure simultaneously. The degradation in yellowness index (ASTM E 313) was measured at periodic intervals of exposure from May to August 1999.
    During this exposure period, accumulated solar UV radiation dose was measured using Epply Laboratories TUVR radiometers at these sites oriented to the same exposure angles as the specimens.

Results:
    The data obtained for this experiment is shown in Figures 16 and 17.

Discussion:
    The results indicate a strong positive correlation between accumulated UV dose due to angle of exposure and Yellowness Index. Besides UV dose there also appears to be at least one variable that effects development of Yellowness Index that is linked to the different locations. The character of Yellowness Index development in the Florida exposure differs from that of the Arizona exposure. This difference is significant and important. It is these significant and important differences between

54

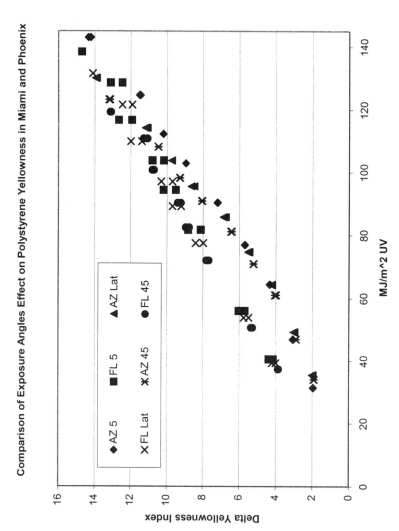

Figure 16. Yellowing of Polystyrene Based on UV Dose

55

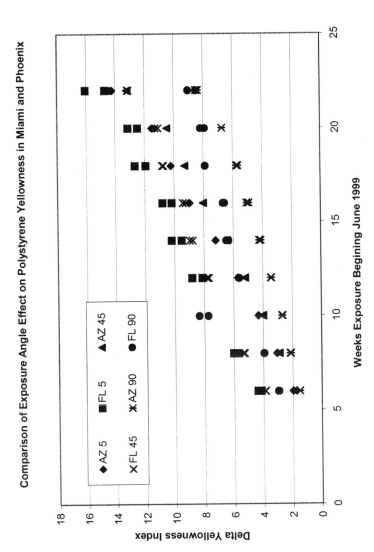

Figure 17. Yellowing of Polystyrene Based on Exposure Time

56

predicted values and empirical results that represent a critical limitation to theoretical SLP models now and possibly in the future.

Experiment B: Effect of Exposure Angle on Color and Gloss Change of Brown Coil Coated Cylinders.

Research Question:

If a short length of continuous coil coated metal is formed into a cylinder and placed on exposure, will color and gloss changes at different locations on the cylinder indicate the effect of exposure angle on materials' weathering characteristics?

Materials and Apparatus:

Lengths of commercially available brown coil coated construction material were obtained and fitted as cylinders around circular pipe substrates. Cylinders were placed on exposure with the long axis oriented east and west at Atlas' SFTS and DSET Laboratories in Miami, Florida and Phoenix, Arizona, respectively.

Each cylinder was carefully indexed so that repeated color and gloss measurements could be taken at specific target areas. The index marks corresponded to different angles of exposure. After one year, the coil coated material was removed from cylinders and measured for color (ASTM D2244, D65, 10 degree, CIE L*,a*,b* RSIN) and gloss (ASTM D523 60degree gloss).

Results:

The results of the 1-year exposure and measurements are shown in Figures 18-21.

Brown Coil Coated Cylinder - 1 Yr Phoenix Exposure - Color

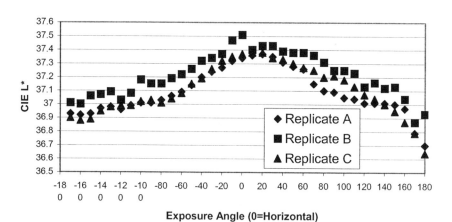

Figure 18. Phoenix Exposure Effect on Brown Coil Color

Brown Coil Coated Horizontal Cylinder - 1 Yr Phoenix Exposure - Gloss

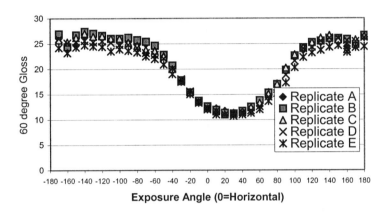

Figure 19. Phoenix Exposure Effect on Brown Coil Gloss

Brown Coil Coated Horizontal Cylinder - 1Yr Miami Exposure - Color

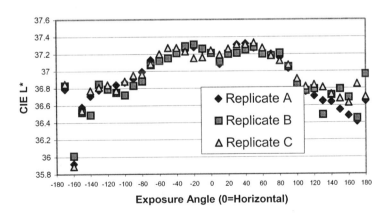

Figure 20. Miami Exposure Effect on Brown Coil Color

Brown Coil Coated Horizontal Cylinder - 1 Yr Miami Exposure - Gloss

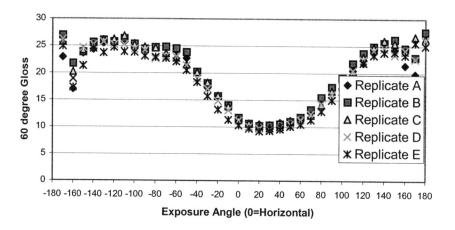

Figure 21. Miami Exposure Effect on Brown Coil Gloss

Discussion:

The results of the exposure and measurements show good characterization of the effect of weathering on a single specimen as a function of exposure angle. The Arizona exposed cylinder showed a peak in L* values approximately at the horizontal orientation. There did not appear to be important differences between areas within 30 degrees south of the horizontal orientation. The slope of the curve on the south-facing surface appeared more gradual than the slope of the north-facing orientations.

The Florida exposed cylinder showed a considerably different weathering behavior. At angles of incidence greater than 40 degrees from vertical, the change in L* value as a function of exposure angle behaved in a manner co-varying with solar irradiance. As exposure angles decreased to less than 40 degrees from vertical, the materials weathering behavior dramatically shifted. The angle considerations associated with service life prediction at angles less than 40 degrees from vertical do not appear to pertain to angles greater than 40 degrees from vertical for this material in specific environments. Failure of SLP models to account for this type of significant and important consideration of exposure angle (as well as other variables) may result in catastrophic unexpected failures for users of such insufficient models.

The Florida cylinder, however, did not show a shift in weathering behavior for gloss as it did for color. SLP models must consider different weathering behaviors. Not only must material dependencies be taken into account, but SLP models must also consider different material characteristics and attributes being measured. Degradation phenomena for color L* values is very different than degradation phenomena for 60 degrees gloss values for this material in Florida exposure.

Empirical Study: Finished Construction Components After Exposure in Service Environment

Research Question:
What are the degradation characteristics of a commercial material in end-use environments as a function of azimuth exposure angle?

Materials and Apparatus:
Due to the empirical nature of this study, material composition, length of exposure and initial values are unknown. Only the location and orientation of the construction unit is known. This may also be the case for many real world studies of materials in service use. A commercial parking lot in Fort Lauderdale, Florida which had sets of lampposts in relatively unobstructed areas was selected. These posts were octagonal vertical columns approximately 12 inches in cross section. The steel was coated with a dark green industrial coating of moderate gloss. These posts were oriented near vertically with one of the faces due south as shown in Figure 22.

Figure 22. Octagonal Lampposts in Ft. Lauderdale

60

Four of the posts were thoroughly measured using a hand-held Hunter Miniscan spherical colorimeter (LAV, RSIN, CIE L*,a*,b* 10 deg observer) for color and using a BYK Tri-Gloss hand-held glossmeter for 60 degree specular gloss. Each of the surfaces of the octagonal structure were measured at four haphazardly selected locations between 5 and 8 feet above ground with the mean reported. Measurements were conducted in June 1999 on a cloudy day to preclude large temperature differences on the different faces of the structures.

Results:

Figures 23, 24 and 25 show the data obtained for L*, b* and 60 degree gloss respectively.

Discussion:

The results of the measurements show a good characterization of the effect of weathering on specimens' color and gloss as a function of azimuth angle. It may be possible to use existing in-use examples such as this to check SLP model results in an empirical way. If a relationship between irradiance dosage as a function of angle is characterized, it may be possible to correlate the relationship to distributions such as those shown in figures 18 - 21 showing degradation as a function of angle. Understanding this correlation may allow a simple estimate of degradation due to solar irradiance dose vs. degradation due to other factors. In this way, researchers may gain insights as to the relative importance ranking of the solar irradiance variable with respect to other variables.

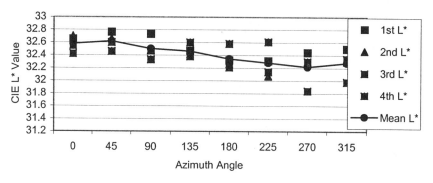

Figure 23. Distribution of L* Values on Ft. Lauderdale Lampposts

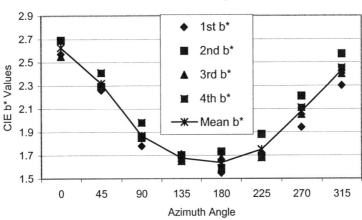

Figure 24. Distribution of b* Values on Ft. Lauderdale Lampposts

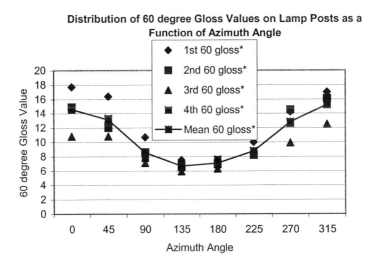

Figure 25. Distribution of Gloss Values on Ft. Lauderdale Lampposts

Summary

This paper has presented several contextual levels of considerations regarding exposure angle effects on service life prediction. Some of the effects of exposure angle on service life prediction can be considered very simple, straightforward effects predicted by Newtonian physics and geometry, such as solar angle of incidence. Other effects of exposure angles on service life predictions can be quite complex and involve special cases of material dependencies and local climates as seen with empirical data presented here within. A truly robust and useful SLP model should predict service life of specific materials in actual "in-service" conditions and not only in simple theoretical artificial or hypothetical simulations. Understanding real world considerations including effects of exposure angles along with these considerations in SLP models will improve the usefulness of SLP models.

References

1. ASTM E 490-73a (Reapproved 1992) Standard Solar Constant and Air Mass Zero Solar Spectral Irradiance Tables. 1992 Annual Book of ASTM Standards. Philadelphia, PA. American Society for Testing and Materials, 1992.
2. Duffie, J.A. and W.A. Beckman, *Solar Engineering of Thermal Processes*. John Wiley and Sons, New York, NY, 1977
3. ASTM E 892-87 (Reapproved 1992) Standard tables for Terrestrial Solar Spectral Irradiance at Air Mass 1.5 for a 37° Tilted Surface. 1992 Annual Book of ASTM Standards. Philadelphia, PA. American Society for Testing of Materials, 1992.
4. Juran's Quality Control Handbook, J.M. Juran and Frank M. Gryna, 4th ed., McGraw-Hill, New York, NY, 1988

Chapter 4

Fractional Factorial Approaches to Emmaqua Experiments

Henry K. Hardcastle III

Atlas Weathering Services Group, Atlas Electric Devices Company,
17301 Okeechobee Road, Miami, FL 33018

Sophistication of experimental designs for weathering research continues to evolve. The majority of current weathering experiments utilize simple designs which change few variables at a time. Conducting "Fractional Factorial" experiments before using traditional approaches focuses weathering research on the significant and important variables effecting material performance. This paper presents a methodology for applying "Screening Fractional Factorial" approaches to material performance research. It includes a case study and examples of weathering data.

Introduction

One may consider the recent history of the weathering discipline (the past 100 years or so) as an evolution of weathering experiments. This evolution trends from simpler, single variable experiments towards more stochastic and broader experimental approaches. All the different major types of weathering experiments represent important tools engineers can use for answering different types of questions regarding product development. Reviewing some of the levels of sophistication in this evolution provides an understanding of experimental tools available. This presentation reviews tools in the context of weathering experimentation. These tools also apply well to most aspects of experimental science and one observes their use in agriculture, health sciences, industrial engineering, service industries, etc. Some of the major levels of experimental sophistication can be organized as shown in Figure 1.

64

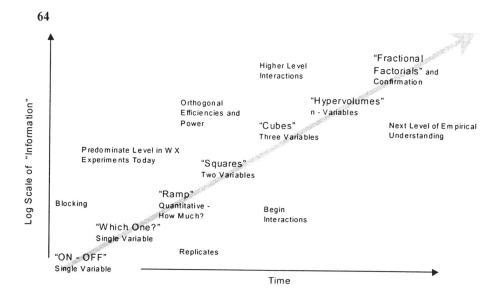

Figure 1. Evolution of Experimental Approaches

Theory

Single Variable – Interested in the Presence or Absence

A single variable or "on-off" trial probably represents the earliest levels of experiment. This type represents "Experiment" in it's simplest application, forms the foundation for all other levels of experimentation, enjoys wide use in the weathering industry today, and represents a good starting point for developing a lexicon describing weathering experimentation evolution. In this level, the experimenter wants to understand the effect of a single variable on a system. The system is tested with and without the variable applied. Other conditions are kept constant or "blocked"(1). Figure 2 shows a graphic representation of this simple design.

Application of this experimental design is simple. The experimenter prepares two sets of samples for trial: sample set "A" without the variable applied, and sample set "B" with the variable applied (often referred to as the "Input Variable" or the

"Independent Variable") and then exposes the two sets side by side to the weather. After a period of exposure, the experimenter measures the two sets and analyzes the measured values (often referred to as the "Output Variable" or "Dependent Variable"). The analysis looks for differences between the two sets.

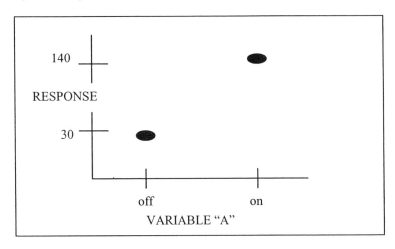

Figure 2. A Single Variable Design – "On – Off Trial"

Spin-off of the Single Variable Experiment: Which is better?

Formulators often use an important spin-off of the single variable experiment in weathering experiments, especially for alternative vendor evaluations or "Drop In" formulation components. In this design, one includes multiple pairs instead of a single pair of experimental samples to determine "which is better." Figure 3 shows a graphic representation of this simple design.

Application of this design is similar to the single variable. The experimenter prepares a separate sample set for each alternative candidate (usually including a standard set to compare performance against as a control). The experimenter measures the output variable(s) of all sets after exposure and analyzes for significance and importance.

Interested in "How Much" (Ramp)

The previous two approaches were interested in "does it" or "does it not"- presence or absence of a variable. The next level of sophistication involves a different research question. Process engineers often use a "ramp" experimental design to determine "how much" of an input variable effects performance. This design involves amounts of an input variable and represents a higher level of context than simple presence or absence. Currently, this design is the most widely used tool for optimizing formulation component levels today. Figure 4 shows a graphical representation of a single variable design with multiple levels of the variable.

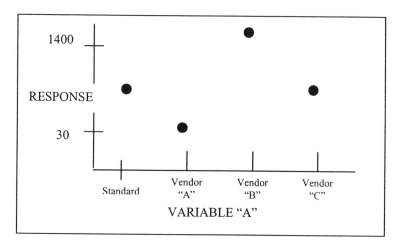

Figure 3. Single Variable "Which is Better"

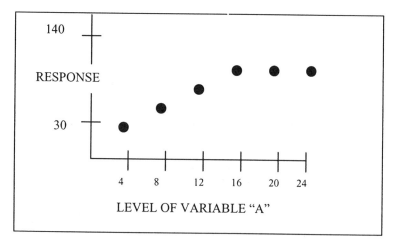

Figure 4. Single Variable – "Ramp"

The on-off trial and single variable ramp experimental designs with variations and combinations represent the overwhelming majority of weathering experiments performed today. Their application is easy and analysis is clearly communicated to technical and non-technical audiences. The application of these designs to weathering characterization is so widespread that dependence on these designs may represent a limiting paradigm constraining growth in understanding of weathering phenomena today. Exclusive use of these narrow focus designs represents important limitations for product designers and process engineers struggling with weathering research, as well as, manufacturing capability and control issues. Experimentation at these levels of context alone does not provide an efficient approach for comprehensive improvement efforts faced with the plethora of variables effecting product weathering performance. Supplementing these approaches with experimentation at higher levels of context represents a more stochastic approach. Performing multi-variable screening experiments prior to application of these simple, fundamental designs offers efficiencies for understanding weathering through experimentation.

Two Variables:

A jump in the sophistication of weathering experiments occurs when designs utilize more than one input variable. Rarely, if ever, does a product's manufacturing or in-service environment involve only a single variable. Two variable experiments involve different research questions than those of simpler designs including investigation of interactions between input variables (2). Often these experiments are represented as "Square" designs.

One of the most popular applications of these "Square" designs involves identifying a Low and High setting for each independent variable. Four trials are then conducted according to the following array:

Trial 1	Both Variables At Low Setting	
Trial 2	Variable "A" At Low Setting	Variable "B" at High Setting
Trial 3	Variable "A" At High Setting	Variable "B" at Low Setting
Trial 4	Both Variables At High Setting	

Multiple "identical" samples or replicates are often included in each trial to characterize the background variance within each trial. Output from each trial is often graphed on the same coordinate system for easy comparison with two low settings and two high settings compared for each variable. Figure 5 shows a graphical representation of this design. These graphs are read differently than traditional Cartesian coordinate systems. Often, intermediate input variable settings between the high and low settings are added to this design as additional trials and reveal intermediate topography. Sometimes this analysis is referred to as a "response surface."

These designs are also referred to as "Full Factorial" designs if each factor or input variable is tested at each setting. These designs are also said to be orthogonal or balanced; two trials with "A" set low, two trials with "A" set high, two trials with "B" set low, two trials with "B" set high. The othogonality is sometimes best understood by visual examination of the geometric layout of the "Square" design. The orthogonal characteristics of these designs result in experimental efficiency and power and they advance these designs to a higher level of sophistication than the traditional weathering experiments widespread throughout industry today.

68

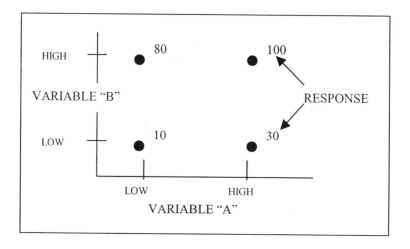

Figure 5. Two Variable Design

Three Variables

A simple enhancement to the two variable "Square" concept is to add a third variable resulting in a "Cube" experimental design. This simple advancement drives experimental sophistication to a new level of context. These designs are more appropriate for weathering experiments since very rarely are only two weathering variables acting simultaneously on a product in end-use. In this three-dimensional experimental volume resides the classic weathering interaction of temperature, irradiance, and moisture variables. A two-dimensional experimental design does not have the inherent level of context necessary to characterize temperature-irradiance-moisture interaction phenomenon. It is befuddling to this author why the vast majority of weathering research is conducted at less sophisticated levels of experimentation to gain understanding of systems that operate at higher levels of context.

Application of this "Cube" design follows similar procedures as for square designs. Eight Trials (2^3) are performed with individual variables set to high and low settings independently of other variable settings. The design is full factorial (each variable at each setting of high and low) and othogonally balanced. Intermediate input variable settings are often added to reveal intermediate topography. Figure 6 shows a graphical representation of a three variable design. Figure 7 shows empirical weathering data from a "Cube" design experiment involving exposure environment, wet time and exposure angle. The interactions between variables is clearly evident in this design.

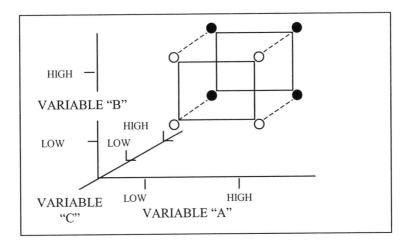

Figure 6. Three Variable Full Factorial Design

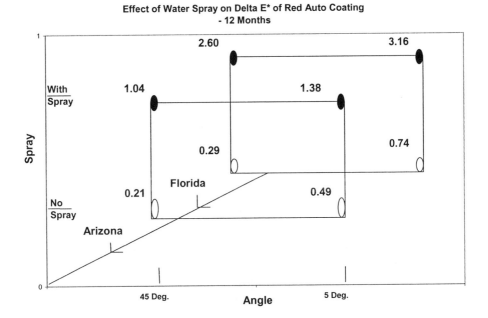

Figure 7. Summary of a Three Variable Full Factorial Weathering Experiment

Fractional Factorials

The orthogonality of previously discussed designs allows a jump to the next level of sophistication in weathering experimental designs. As we increase the number of variables in a full factorial design, the number of trials required dramatically increases: for a two variable orthogonal, four trials are required – two at high, two at low for each variable. In a three variable orthogonal; eight trials are required – four at high, four at low for each variable. In a four variable orthogonal; 16 trials – eight high, eight low. For any number of variables in a full factorial performed at two setting levels (high and low), the number of trials required equals the number of setting levels raised to the power of the number of variables investigated (for example, four variables at high and low settings = 2^4 = 32 trials = 16 at high setting, 16 at low setting). The experimenter may only need four trials at the high setting and four trials at the low setting for the required level of confidence needed to determine if a specific variable has a major effect on the system. Fractional Factorial designs allow the experimenter to delete some of the trials required by the full factorial design as long as the orthogonality of the design is maintained. The power of the orthogonality can be virtually exchanged for efficiency in the number of trials. The resulting design is termed a Fractional Factorial Screening Experiment.

Manufacturing Process Engineers widely use fractional factorial screening experiments to "screen out" the trivial many variables from the "important few" variables (3). A small confirmation experiment (typically two to four additional trials) follows a screening experiment to confirm the results. Once the manufacturing process engineer identifies the important variables, the engineer can focus process improvement efforts in an efficient manner on those variables that the process indicates are important. This analogy described for characterizing and improving manufacturing processes applies perfectly to improving weathering performance. Weathering processes are multi-variable complex processes that are highly material dependent. Weathering investigators can use fractional factorial screening experiments to "screen out" the trivial many variables from the "important few" variables. Once a screening experiment indicates the most important variables for a particular material, a small confirmation experiment always confirms the results. Once the weathering investigator identifies the important variables, he can effect the material formulation, processing variables, and in-service environments to improve weatherability. Equally important, the investigator can optimize variables identified as trivial by the process to reduce manufacturing costs! This advantage may be especially useful in raw material cost-cutting projects.The materials weathering research efforts can then be focused in an efficient manner on the variables that the process indicates are important.

The fractional factorial experimental design answers questions such as:

- Of the nine components in this vinyl formulation, which have the biggest effect on yellowing after five years Florida exposure, what is each component's order and magnitude of importance on yellowing, and which components can be optimized for cost without sacrificing weathering performance?
- Of the ten major production line variables the line operator can control, on which should I establish control charting to improve quality of weatherability, and approximately what mean and tolerances should I begin with?
- Of the major weathering agents this product will be exposed to (temperature, moisture, irradiance, pollution, abrasion, solvents, biologicals, cycling, etc.), which require research efforts to improve customer satisfaction for weathering performance?
- For the major weathering failure modes I have identified in the FMEA, what are the risks associated with each?
- For my material, which of these many weathering variables can be increased in order to accelerate weathering for test development?
- For this vendor's candidate material, which environmental variables have the biggest effect on the system's weatherability?

Clearly, the types of research questions addressed with fractional factorial screening approaches represent a different level of context than single variable and ramp level experiments which make up the majority of weathering experiments performed today.

Application

Application of fractional factorial screening and confirmation designs is relatively simple to implement but relatively complicated to describe thoroughly. This presentation describes a straightforward ten-step procedure leaving the "why" to more esoteric and involved statistical publications. The reader should become familiar with theoretical underpinnings of these designs. The reader should also practice application of these designs – beginning with inexpensive, simple, non-critical investigations – to gain experience with these techniques (2)(4). The reader should not become dissuaded by complex, restrictive, theoretical considerations that often become barriers to beginning these types of empirical applications in experimentation.

"The purpose of Mathematics is insight, not numbers." Sam Saunders, Ph.D., *02/08/99*

The author has used the following 10-step procedure effectively in many screening experiments, both for investigating manufacturing processes and weathering processes. This ten-step outline may need modification for specific applications and is not exhaustive in detail but will serve to identify the major components and sequence for most uses. This 10 step procedure was used to perform the case study described in the experimental section of this presentation.

This tool – like all others – if used improperly, can result in erroneous decisions with catastrophic results; or likewise, if used with skill, can enhance product performance and customer satisfaction. This treatment only reviews a tool available to the engineer. This treatment does not present application methodology details. Proper use of this tool includes reviewing procedures, cautions, and warnings as specified in appropriate standards.

Ten Step Procedure:
Step 1 - Write a simple, concise research question.

Step 2 – List the variables to be investigated. Check that variables are truly independent.

Step 3 – Select high and low settings (if a two-level experiment) or high, middle, low (if three-level experiment), etc. Check to make sure settings are not so far apart as to cause catastrophic failure if all variables are set high or low simultaneously. Check to make sure variable settings are far enough apart that they can be described as "significantly different."

Step 4 – Select an appropriate fractional factorial orthogonal array for the number of variables under investigation, L^{16}, L^8, L^{64}, etc. Use published arrays. Assign specific variables to specific columns in the design with regards to known interactions and alias concerns of the fractional factorial. Figure 8 shows an L16 Fractional Factorial Array (4)(5).

Step 5 – Perform trials according to the fractional factorial array schedule. Include multiple replicates within each trial as necessary, given within trial variability and gauge variability. Control the input variables to described levels for each trial in the array. Block other variables not designed into the experiment.

Step 6 – Measure the desired output variables from each trial.

Step 7 – Quality Control check all work. Recheck that all trials were performed in accordance with the array. Recheck all measurement values. Recheck all data entry. A simple error in variable settings, data entry, measurements, etc., may void the orthogonal basis resulting in erroneous decisions.

Step 8 – Analyze the output using samples collected, effects graphs, and ANOVA techniques.

Step 9 – Determine the significance and importance of each variable's effect on the output.

Step 10 – Confirm the conclusions. Run a minimum of two confirmation trials: one trial with the variables indicated as significant and important set to high levels, one trial with significant and important variables set to low levels. Set insignificant variables to cost effective settings for both of these trials. Check that the results of these two confirmation trials confirm predictions of the screening experiment in both direction and magnitude of differences. Additional confirmation trials can be added for additional understanding of main effects, interactions, and aliases.

	Temperature	Irradiance		DayTime Spray			Pretreat Soak-Freeze-	NightTime Soak			Pretreat Abrasion		Pretreat Oven	Pretreat Chemical	Pretreat High UV
Vari-ables	1	2	3	4	5	6	7	8	9	10	11	12	13	14	15
Trial No.															
1	1	1	1	1	1	1	1	1	1	1	1	1	1	1	1
2	1	1	1	1	1	1	1	2	2	2	2	2	2	2	2
3	1	1	1	2	2	2	2	1	1	1	1	2	2	2	2
4	1	1	1	2	2	2	2	2	2	2	2	1	1	1	1
5	1	2	2	1	1	2	2	1	1	2	2	1	1	2	2
6	1	2	2	1	1	2	2	2	2	1	1	2	2	1	1
7	1	2	2	2	2	1	1	1	1	2	2	2	2	1	1
8	1	2	2	2	2	1	1	2	2	1	1	1	1	2	2
9	2	1	2	1	2	1	2	1	2	1	2	1	2	1	2
10	2	1	2	1	2	1	2	2	1	2	1	2	1	2	1
11	2	1	2	2	1	2	1	1	2	1	2	2	1	2	1
12	2	1	2	2	1	2	1	2	1	2	1	1	2	1	2
13	2	2	1	1	2	2	1	1	2	2	1	1	2	2	1
14	2	2	1	1	2	2	1	2	1	1	2	2	1	1	2
15	2	2	1	2	1	1	2	1	2	2	1	2	1	1	2
16	2	2	1	2	1	1	2	2	1	1	2	1	2	2	1
	A	B	A B	C	A C	B C	A B C	D	A D	B D	A B D	C D	A C D	B C D	A B C D

Figure 8. An L^{16} Fractional Factorial Array: 1 = Variable at Low Setting, 2 = Variable at High Setting.

Experimental

During the summer of 1999, Atlas Weathering Services Group performed a fractional factorial weathering experiment on commercially available materials.

Step 1:
The general purpose of this investigation was to better understand the effect of specific pretreatment and weathering variables on appearance properties using the EMMA natural accelerated weathering device. Understanding how selected variables effect EMMA weathering may indicate: 1) which variables are most important for weathering different material types, 2) which variables are important for focused AWSG R&D efforts, and ultimately 3) which variables to control with advancements in the EMMA weathering process. Additionally, it was hoped this experiment would provide a check of "conventional wisdom" regarding weathering phenomena of these materials as described in the literature, by AWSG customers, and in previous parametric studies performed at DSET Laboratories.

Materials:
The materials for this experiment included current automotive paint systems, current coil coated building materials, colored in polyethylene sheet extrusion, clear polycarbonate sheet with a UV protected surface and injection molded clear polystyrene reference material used as indicators in Weather-O-Meters. All materials are commercially available at the time of this writing and were purchased from retail sources by AWSG. Although all the listed materials were subject to this experiment, this presentation includes results only for the Polystyrene reference material and the Brown Coil Coated construction material.

Clear polystyrene injection molded plaques were obtained from Test Fabrics, Inc., West Pittson, Pennsylvania. All specimens were from the same production lot No. 4. The specimens are properly denoted as "Polystyrene plastic lightfastness standard – Plastic chips for use in monitoring controlled irradiance water cooled xenon arc apparatus – specified in SAE test methods J1960 and J1885, 6/89".

Brown coil coated aluminum was obtained from a commercial retail source. All specimens were cut from within a single twelve foot length of coil. Individual specimens were randomized before assignment of codes and labels.

Step 2:
Nine independently controllable variables were selected for this experiment from two types: pretreatment variables and exposure variables. 2^9 or 512 trials would have been required to perform this analysis using full factorial approaches. Both the pretreatment and exposure variables were included as follows:

1. The temperature of the EMMA exposure under irradiance as indicated by metal black panel temperature instruments mounted on the EMMA target next to the

specimens (TEM). This variable was controlled by changing the amount of air flow available for cooling the specimens.

2. The strength of the radiant flux striking the surface and total radiant exposure from the solar spectral distribution. This variable will be referred to as "irradiance" for the rest of this presentation (IRR). This variable was controlled by changing the number of mirrors focused on the target area.

3. The application of a daytime spray cycle which wet the specimens surfaces during periods of irradiation (SPR). This variable was controlled by turning on or turning off the water source available for the daytime spray cycle.

4. The application of warm liquid water during periods when specimens were not being irradiated. This variable will be referred to henceforth as nighttime soak (NTS). Nighttime soak was controlled by removing the specimens from the target boards and placing them in a 40°C water bath each night and placing the specimens back on the target boards the next morning for daytime exposure.

5. Abrasion of the specimen's surface prior to exposure (hencefourth referred to as abrasion pretreatment). An AATCC Crockmeter device was modified to move the abrasive cloth back and fourth on the specimen's surface under highly repeatable conditions. The Polystyrene was subjected to five reciprocal strokes using an 8000 grit abrasive. The brown coil was subjected to seven reciprocal strokes using a 3600 grit abrasive. The abrasion pretreatment resulted in a light haze or scratching on the specimen's surface.

6. Thermo-mechanical cyclic stressing of the specimens using a soak - freeze - thaw pretreatment (SFT). Specimens were soaked for 16 hours in a 49°C de-ionized, re-circulating water bath followed by immediate placement into a –10°C freezer for two hours followed by a six hour thaw/dry in ambient office conditions. The cycle was repeated for five times. The soak – freeze – thaw pretreatment resulted in a slightly hazy or milky translucent appearance and the formation of micro cracks within the bulk of the Polystyrene injection molded plaques.

7. Chemical pretreatment of the specimens was accomplished by immersing the specimens in a 40°C bath of dilute hydrogen peroxide (CHM). Polystyrene chips were soaked for 24 hours. Brown coil specimens were soaked for six hours and developed very minor blisters during the soak. After immersion, the specimens were immediately flushed with de-ionized water for 15 minutes.

8. A high UV irradiance pretreatment by exposing specimens to a proprietary process reported to improve weathering resistance (ARC).

9. Oven pretreatment was accomplished by placing the specimens in an air circulating oven set to 70°C for 84 hours (OVN).

Pretreatments were performed in a specific sequence: abrasion preceded soak – freeze – thaw which preceded chemical pretreatment which preceded UV arc which preceded oven, all of which preceded EMMA exposure. Prior to initial measurements, all specimens were thoroughly washed using a 5% mild detergent solution, gentle hand-wipe and thorough rinsing with de-ionized water after being pretreated.

Step 3

High and low settings for each input variable were selected according to the schedule shown in Figure 9. This experiment utilized only two levels.

Variable	Low setting	High Setting
Temperature (TEM)	Nominal -7°C	Nominal + 7°C
Irradiance (IRR)	8 Mirrors (-10%)	10 Mirrors (+10%)
Daytime Spray (SPR)	No Daytime Spray	With Daytime Spray
Nighttime Soak (NTS)	No Nighttime Soak	With Nighttime Soak
Abrasion Pretreatment (ABR)	Not Abraded	Abraded
Soak–Freeze–Thaw Pretreat (SFT)	No Soak-Freeze-Thaw Pretreat	Soak-Freeze-Thaw Pretreatment
Chemical Pretreatment (CHM)	No Chemical Pretreatment	Chemical Pretreatment
High UV Pretreatment (ARC)	No High UV Pretreatment	High UV Pretreatment
Oven Pretreatment (OVN)	No Oven Pretreatment	Oven Pretreatment

Figure 9. High and Low Variable Settings

Step 4

An L^{16} fractional factorial array was selected for this experiment. Although the L^{16} can theoretically handle up to 15 independent variables, it is not appropriate to fully saturate the array with variables. The nine variables identified for this investigation fit into the L16 while leaving six columns blank. The blank columns were used in the analysis for estimating the background variance, to check for significance of results, and to check for some interactions.

Assigning each of the input variables to specific columns in the array requires some judgment and understanding of aliases in the fractional design. For instance, we believe there is a reasonable likelihood of interaction between temperature and irradiance in this experiment. We want to understand the independent effects of these two variables as well as the synergy between them. The temperature variable is assigned to the first column since the first column will reveal temperature effects alone. The irradiance variable is assigned to the second column since the second column will reveal irradiance effects alone. The third column is left blank since the third column's settings will reveal any effects due to interaction between the temperature and irradiance variables. If we did assign a variable to the third column in this design, we would create a confound. We would not be able to tell if the effects tested by the third column were due to a synergy between temperature and irradiance or were due to the variable we assigned to the third column. This confound is often referred to as an "alias." The remaining variables are assigned to columns with similar justification; daytime spray is assigned to Column 4. Column 5 will reveal a significant interaction between daytime spray and temperature settings. Interaction between daytime spray and irradiance will be revealed by Column 6 and Column 7 will reveal a three-way interaction between temperature, irradiance and daytime spray. We consider it highly unlikely that a three-way interaction would show up from treatments in Column 7 without showing significant differences from Columns 3, 5 and 6 (the two-way interaction indicators). We chose to assign a variable and alias Column 7 with the soak – freeze – thaw pretreatment variable. The remaining variables for this experiment were assigned using similar justification. The final array with variable assignments is shown in Figure 8. This schedule details 16 trials with unique settings of nine different variables.

Step 5

The 16 trials prescribed by the experimental array were performed simultaneously on 16 different EMMA (Equatorial Mount with Mirrors for Acceleration) machines at DSET Laboratories from May 21, 1999 to July 29, 1999 (6). The machines utilized were quality checked throughout the exposure. Proper variable settings were maintained for each trial. Great care was utilized to insure all variables outside the scope of the experimental design were blocked across all 16 trials. At several intervals throughout the exposure, black panel temperatures were measured in the target area. A graph representative of the temperature differences is shown in Figure 10. Note that the target black panel temperature variable was controlled independently of the irradiance variable for the 16 trials.

Step 6

After the exposure period, specimens were removed from exposure, measured for appearance properties and compared to their initial values before exposure. Polystyrene chips were measured for transmittance yellowness index according to ASTM E 313-96 (7). Brown coil material was measured for color (CIE L*, D65, Spherical, Specular included) using ASTM D2244-93 (8) and ASTM E 308-95 (9). Brown coil was also measured for 60 degree gloss using ASTM D523-89 (10). Polystyrene specimens were measured three times across the exposed surface and the mean reported for each specimen. Brown coil specimens were measured five times across the exposed surface and the mean reported for each specimen. Two specimen replicates were included in each trial. The values reported are calculated deltas between the initial measurements before exposure and final measurements after

78

MQ L16 - Black Panel temperatures on 05/19/99

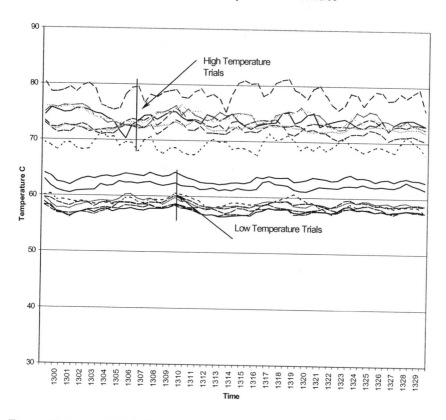

Figure 10. Low and High Black Panel Temperature Distributions Achieved on 16 EMMAs

exposure. The delta values for the two specimens included in each trial are shown in Figure 11. The trial number for the output values corresponds to the trial numbers used in the fractional factorial array schedule in Figure 8.

Step 7

A thorough check of exposure conditions was performed weekly during the exposure. The scheduled pretreatments for specific trials were confirmed. All work was recorded as performed in engineering logs and written checklists were utilized throughout the work effort. Measurements were confirmed based on comparison of replicate values.

Trial Number	Delta Yellowness Index of Polystyrene Replicate "A"	Delta Yellowness Index of Polystyrene Replicate "B"	Delta L* of Brown Coil Replicate "A"	Delta L* of Brown Coil Replicate "B"	Delta 60 degree Gloss of Brown Coil Replicate "A"	Delta 60 degree Gloss of Brown Coil Replicate "B"
T01	12.92	13.26	0.90	0.89	11	11
T02	23.19	22.79	2.44	2.29	14	11
T03	11.95	12.34	0.61	0.60	22	19
T04	24.80	26.00	1.60	1.72	18	15
T05	18.62	17.99	0.59	0.64	23	23
T06	25.54	26.38	3.20'	3.19	16	17
T07	20.05	19.18	1.01	1.00	13	15
T08	22.45	20.56	2.03	2.19	18	18
T09	15.53	14.78	0.48	0.47	18	22
T10	24.58	24.63	2.43	2.23	19	20
T11	16.48	17.73	1.12	1.09	17	19
T12	16.43	17.84	1.31	1.28	17	17
T13	17.93	17.63	1.30	1.23	22	18
T14	26.6	26.57	1.99	2.02	20	22
T15	13.13	12.86	0.68	0.74	28	27
T16	29.04	29.71	2.63	2.33	23	20

Figure 11. Delta Yellowness Index, Color and Gloss for Exposed Replicates

Step 8

Analysis of the output data was performed at three levels: 1) Visual review of exposed specimen collection, 2) review of experimental grouping using graphical techniques, and 3) ANOVA.

1) Visual Review

A visual review was performed by laying out exposed specimens in groups. Groups were formed using the orthogonal array for each variable. For the first layout, all specimens that were exposed at higher temperatures were grouped together. All specimens that were exposed to lower temperatures were grouped together. The two groups were compared for overall appearance. Next, the eight sets of specimens that were exposed to high irradiance were grouped together and compared to a group composed of the eight sets of specimens exposed to low irradiance. This grouping procedure was continued for each variable in the experimental design. The groupings that revealed the most apparent differences between high and low settings were identified.

2) Graphical Technique

A similar analysis to the visual grouping was performed using the Delta Yellowness Index values. A mean was calculated for the eight sets of specimens exposed to high temperature. A mean was calculated for the eight sets of specimens exposed to the low temperature condition. These two mean values were plotted on a graph and connected with a line. Next, a mean was calculated for all specimens exposed to high irradiance.

80

A mean was calculated for all specimens exposed to low irradiance and these two means were plotted next to the temperature variable means. This procedure was continued for all the variables included in this design. This graphing technique allowed the effects of each variable to be compared with the effects of all other variables with a single graph. Using this analysis, it was quite simple to determine which variables had the largest effect on yellowing of the polystyrene and the magnitude of the effect compared to that of the other variables. Figures 12, 13, and 14 show the graphs obtained.

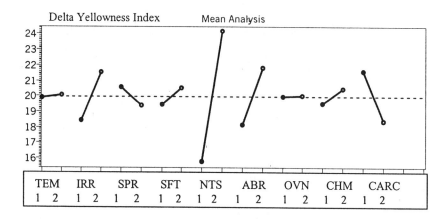

Figure 12. Main Effects Graph for Polystyrene Yellowness Index Values

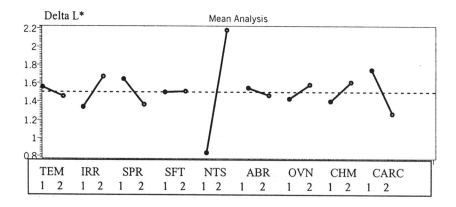

Figure 13. Main Effects Graph for Brown Coil Coated Material L* Values

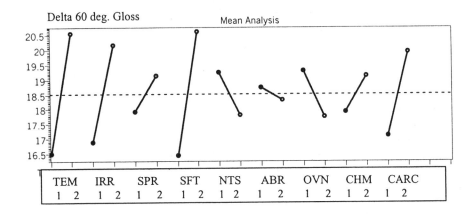

Figure 14. Main Effects Graph for Brown Coil Coated Material 60° Gloss Values

3) ANOVA

Because the two techniques so far do not sufficiently account for within and between trial variance, an ANOVA analysis was performed using a software package from ASD, Inc. ANOVA Analysis of Fractional Factorial Screening experiments allows experimenters to complete the following table:

Variables Tested That Are Insignificant	Variables Tested That Are Significant But Unimportant	Variables Tested That Are Significant And Important
1.	1.	1.
2.	2.	2.
3.	3.	3.
4.	4.	4.

The ANOVA performs an analysis using the F test. In an F test, an F ratio is calculated comparing variance due to treatment variables to variance due to experimental noise. The F ratio is often described as having the between column variance in the numerator and the within column variance in the denominator (11).

For this analysis, we simply compare the effects caused by input variables to the background variation of the experiment. As this ratio approaches one, we say that the effects due to the input variables are not significantly different than the background variation of the experiment or that the effect of input variables is not significant.

However, as the F ratio becomes larger and larger, the effects due to the input variables become more different than background experimental noise; a large F ratio indicates the effect of the input variables is significant.

The variation due to input variables is easily traced back through the orthogonal array. The ANOVA uses outputs from the columns to which the input variables were assigned. An estimate of the variation due to experimental noise (experimental error, background variation) comes from a combination of three sources for this analysis. First, if replicate samples are used in the experiment (n>1), the sample to sample variation in the results can be a good estimate of within treatment variation. Second, the columns that were left blank in the orthogonal design represent a rich source for estimating experimental error once interaction effects are ruled out. By incorporating these blank columns and using unsaturated designs during the design steps of the experiment, we are now rewarded with a very robust and statistically valid estimate of experimental error. Third, any of the input variables that are not significant can also be pooled into the estimate of experimental variance. By using all three sources to develop a pooled estimate of experimental error, a very robust appropriate experimental error term can be developed for the denominator of the calculated F ratio. It is with this experimental error term that the effects of treatments are compared to determine significance.

Once the F ratio is calculated from the experimental data, it may be compared to values found in the standard F distributions given the degrees of freedom for the numerator and denominator and the desired confidence level. If the data generated F ratio value exceeds the critical F ratio value (from the table of standard F distribution values) the effect of the input variable can be said to be significant (the output due to that variable exceeds what would normally be expected due to random experimental noise).

For this analysis, the ANOVA table, treatment variables, pooled sources of experimental variance, and calculated F ratios are shown in Figure 15, Figure 16, and Figure 17. Using the F column and r column in the ANOVA table, we can begin to assign a rank order to the significant input variables and understand the magnitude of effects compared to each other for polystyrene yellowness index as follows:

Variables Tested That Are **Insignificant**	Variables Tested That Are **Significant But Unimportant**	Variables Tested That Are **Significant And Important**
1. Temperature	1. Daytime Spray	1.
2. Oven Pretreatment	2. Soak – Freeze – Thaw Pretreat	2.
3.	3. Chemical Pretreat	3.
4.	4.	4.

It is important to compare the ANOVA table with the goals from Step 1. ANOVA only indicates significance, not direction. For instance, the F ratio for ARC pretreatment shows the effect of high UV pretreatment on Delta Yellowness Index after 68 days EMMA exposure is significant. Only the graphs, however, indicate that no pretreatment causes greater degradation while pretreating with high UV retards degradation of yellowness.

The ANOVA table should also be inspected for effects of interactions between input variables. Recall that these interactions should appear in the columns that were left blank. The interaction between temperature and irradiance for this experiment should show up in Column 3, the interaction between temperature and daytime spray should show up in Column 5, and so forth. In this experiment, the F ratio values for these interactions are small indicating no significant interactions in this experiment. After inspecting the F ratios of these interaction columns for significant effects on the output, if no significance is found, we can contribute these blank columns to our estimate of experimental variation.

Source	Pool	DF	S	V	F	S'	r
TEM		1	0.3692	0.3692	0.5883	-0.2583	-0.03
IRR		1	74.9608	74.9608	119.4663	74.3334	8.49
	Y	1	0.1172	0.1172			
SPR		1	10.5536	10.5536	16.8194	9.9261	1.13
	Y	1	4.2464	4.2464			
	Y	1	0.1454	0.1454			
SFT		1	8.2855	8.2855	13.2047	7.6580	0.87
NTS		1	567.1459	567.1459	903.8700	566.5185	64.70
	Y	1	0.4795	0.04795			
	Y	1	0.5452	0.5452			
ABR		1	107.3735	107.3735	171.1230	106.7461	12.19
	Y	1	2.2562	2.2562			
OVN		1	0.0408	0.0408	0.0651	-0.5866	-0.07
CHM		1	7.7648	7.7648	12.3748	7.1373	0.82
ARC		1	85.2916	85.2916	135.9307	84.6641	9.67
e1		1					
e2	Y	16	6.0142	0.3759			
(e)		22	13.8042	0.6275		19.4514	2.22
Total		31	875.5899	28.2448			

Figure 15. ANOVA Table for Polystyrene Delta Yellowness Index

Source	Pool	DF	S	V	F	S'	r
TEM		1	0.0770	0.0770	0.9385	-0.0051	-0.03
IRR		1	0.8811	0.8811	10.7352	0.7991	3.96
	Y	1	0.0026	0.0026			
SPR		1	0.5913	0.5913	7.2045	0.5093	2.52
	Y	1	0.1815	0.1815			
	Y	1	0.0488	0.0488			
SFT		1	0.0001	0.0001	0.0010	-0.0820	-0.41
NTS		1	14.4857	14.4857	176.4862	14.4036	71.32
	Y	1	0.3424	0.3424			
	Y	1	0.3301	0.3301			
ABR		1	0.0604	0.0604	0.7356	-0.0217	-0.11
	Y	1	0.7970	0.7970			
OVN		1	0.1969	0.1969	2.3987	0.1148	0.57
CHM		1	0.3342	0.3342	4.0712	0.2521	1.25
ARC		1	1.7625	1.7625	21.4735	1.6804	8.32
e1		1					
e2	Y	16	0.1033	0.0065			
(e)		22	1.8057	0.0821		2.5444	12.60
Total		31	20.1948	0.6514			

Figure 16. ANOVA Table for Brown Coil Delta L

Source	Pool	DF	S	V	F	S'	r
TEM		1	132.0313	132.0313	51.4673	129.4659	23.98
IRR		1	87.7813	87.7813	34.2182	85.2159	15.78
	Y	1	2.5313	2.5313			
SPR		1	11.2813	11.2813	4.3976	8.7159	1.61
	Y	1	0.7813	0.7813			
	Y	1	9.0313	9.0313			
SFT		1	140.2813	140.2813	54.6833	137.7159	25.50
NTS		1	16.5313	16.5313	6.4441	13.9659	2.59
	Y	1	0.2813	0.2813			
	Y	1	1.5313	1.5313			
ABR		1	1.5313	1.5313	0.5969	-1.0341	-0.19
	Y	1	0.7813	0.7813			
OVN		1	19.5313	19.5313	7.6135	16.9659	3.14
CHM		1	11.2813	11.2813	4.3976	8.7159	1.61
ARC		1	63.2813	63.2813	24.6678	60.7159	11.24
e1		1					
e2	Y	16	41.500	2.5938			
(e)		22	56.4375	2.5653		79.5256	14.73
Total		31	539.9688	17.4183			

Figure 17. ANOVA Table for Brown Coil 60° Gloss

Step 9

Once the significance of the individual treatments is understood, it is appropriate to decide which input variables are important. Significance involves a statistical exercise. Importance is a human judgment exercise and often depends on other sources of data beyond the scope the experiment. For instance, in this experiment, a 20% difference in irradiance resulted in a difference in mean Yellowness Index of about three units (21.5 at −10% to 24.5 at +10%). In some type of end-uses, this magnitude of difference may cause great user consequences, including formulation and/or process changes. In other end-uses, this magnitude of difference might not be important to the end-use at all. Prior to beginning this study, the end-user of these polystyrene chips was interviewed regarding criteria for importance. The customer identified differences exceeding approximately two Delta Yellowness Index units at this experimental level to be important. Similar justification was used for determining the importance for Brown Coil color and gloss. Based on this information, the significant and important information can be completed as shown in Figures 18, 19 and 20.

Variables Tested That Are **Insignificant**	Variables Tested That Are **Significant But Unimportant**	Variables Tested That Are **Significant And Important**
1. Temperature	1. Daytime Spray	1. Nighttime Soak
2. Oven Pretreatment	2. Soak – Freeze – Thaw Pretreat	2. Abrasion
	3. Chemical Pretreatment	3. High UV Pretreatment
		4. Irradiance

Figure 18. Summary Results for Polystyrene Delta Yellowness Index

Variables Tested That Are **Insignificant**	Variables Tested That Are **Significant But Unimportant**	Variables Tested That Are **Significant And Important**
1. Temperature	1. Irradiance	1. Nighttime Soak
2. Oven Pretreatment		2. High UV Pretreatment
3. Soak – Freeze – Thaw Pretreat		
4. Chemical Pretreatment		
5. Daytime Spray		
6. Abrasion Pretreatment		

Figure 19. Summary Results for Brown Coil Delta L*

Variables Tested That Are Insignificant	Variables Tested That Are Significant But Unimportant	Variables Tested That Are Significant And Important
1. Daytime Spray	1. High UV Pretreatment	1. Soak – Freeze – Thaw Pretreat
2. Oven Pretreatment		2. Temperature
3. Abrasion		3. Irradiance
4. Chemical Pretreatment		
5. Soak – Freeze – Thaw Pretreat		

Figure 20. Summary Results for Brown Coil Delta 60° Gloss

Step 10
Confirmation trials should always be conducted in conjunction with fractional factorial screening experiments. Confirmation trials should also be considered as a critical part of the screening experiment. This is especially important if high levels of input variable saturation are designed into the orthogonal array and where significant interactions are identified between several input variables. Only confirmation trials can decode alias characteristic of the fractional array. Two confirmation trials were conducted with input variables set as shown in Figure 21.

Confirmation Trials:	
#1 – Least Degradation Predicted	**#2 – Most Degradation Predicted**
• No Nighttime Soak	• With Nighttime Soak
• No Abrasion Pretreatment	• With Abrasion Pretreatment
• No Soak Freeze Thaw Pretreatment	• With Soak Freeze Thaw Pretreatment
• Low Irradiance	• High Irradiance
• With High UV Pretreatment	• No High UV Pretreatment

Figure 21. Confirmation Trials

The remaining variables investigated in the screening experiment that were identified as not important or not significant were set to optimal levels for cost for both trials as shown in Figure 22.

Confirmation Trials: Variables optimized for cost
• Low Temperature Exposure
• No Oven Pretreatment
• No Chemical Pretreatment
• No Daytime Spray

Figure 22. Confirmation Trials Optimized for Cost

These confirmation trials were then conducted for polystyrene. The data obtained was compared to the screening experiment results. The confirmation trials at 220 MJ/m^2 UV dose compared directly with the predictions made in the original screening experiment for polystyrene. From this confirmation trial it was concluded that no hidden interactions, confounds or aliases were operating in the original screening experiment. Follow-on efforts can now be focused on those variables the weathering process has indicated are significant and important. Confirmation trials for the brown coil will be analyzed after publication of this paper.

Summary

There has been an evolution in the sophistication of experimental designs for weathering tests. The vast majority of current weathering exposures utilize more fundamental designs effecting few variables. This type of experimental design requires far more trials and thus more cost, less information, and poorer quality than more sophisticated approaches using screening fractional factorial experiments. Preceding fundamental level, few variable weathering trials with fractional factorial screening and confirmation experiments represent an efficient, stochastic, powerful approach for improving knowledge regarding weathering's n-dimensional hyper-volume of environmental effects on materials degradation.

References

1. Duncan, R.C. *Introductory Biostatistics for The Health Sciences*; John Wiley and Sons: NY, 1977.
2. Montgomery, D.C. *Design and Analysis of Experiments*, 3rd ed.; John Wiley and Sons: NY, 1991.
3. Luftig, J.T. *Experimental Design and Industrial Statistics*, Level I-IV; Luftig and Associates, Inc.: Farmington Hills, NJ, 1987.

4. Phadke, M. *Quality Engineering Using Robust Design*; Prentice-Hall: Englewood Cliffs, NJ, 1989.

5. Taguchi, G.; Konishi, S. *Ortogonal Arrays and Linear Graphs*; ASI Press: Dearborne, MI, 1987.

6. ASTM G90-94 Practice for Performing Accelerated Outdoor Weathering of Nonmetallic Materials Using Concentrated Natural Sunlight. 1994 Annual Book of ASTM Standards, vol. 14.02. Philadelphia, PA: American Society for Testing and Materials, 1994.

7. ASTM E 313-96 Practice for Calculating Yellowness and Whiteness Indicies from Instrumentally Measured Color Coordinates. 1996 Annual Book of ASTM Standards, vol. 6.01. Philadelphia, PA: American Society for Testing and Materials, 1994.

8. ASTM D 2244-93 Standard Test Method for Calculation of Color Differences from Instrumentally Measured Color Coordinates. 1998 Annual Book of ASTM Standards, vol. 6.01. West Conshohocken, PA: American Society for Testing and Materials, 1998.

9. ASTM E 308-96 Standard Practice for Computing the Color of Objects by Using The CIE System. 1998 Annual Book of ASTM Standards, vol. 6.01. West Conshohocken, PA: American Society for Testing and Materials, 1998.

10. ASTM D 523-89 Standard Test Method for Specular Gloss. 1998 Annual Book of ASTM Standards, vol. 6.01. West Conshohocken, PA: American Society for Testing and Materials, 1998.

11. Levin, R.I. *Statistics for Management*, 4th ed.; Prentice-Hall: Englewood Cliffs, NJ, 1987.

Chapter 5

Data Management and a Spectral Solar UV Network

Lawrence J. Kaetzel

K-Systems, 1805 Rohreraville Road, Brownsville, MD 21715

Quantitative data of known precision and accuracy are essential for characterizing the weather factors causing polymeric construction materials exposed in the field to degrade. One such factor is solar spectral ultraviolet radiation. A small solar spectral ultraviolet radiation network has been established by the National Institute of Standards and Technology and the Smithsonian Environmental Research Center to provide this data. This network serves another, equally important function, in that it provides an opportunity to perfect the network's data management system. Hugh volumes of data are collected and an efficient data management system is needed to handle this large volume of data. This paper describes the data management system including data acquisition, cataloguing, processing, storage, retrieval and analysis.

Introduction

Solar ultraviolet (UV) radiation, along with other weather factors, has long been identified by the scientific community as causing the weathering of polymeric construction materials. The solar ultraviolet radiation reaching the Earth's surface falls between 290 nm and 400 nm and, by convention radiation, ultraviolet radiation between 290 nm and 315 nm is referred to as UV-B ultraviolet radiation while ultraviolet radiation falling between 315 nm and 400 nm is referred to as UV-A ultraviolet radiation. UV-B radiation is much more photolytically active than is UV-C radiation. At present, only a few sites throughout the US are equipped with solar ultraviolet radiometers and even fewer are equipped with solar spectral ultraviolet radiometers. One such network is the National Institute of Standards (NIST) and Technology and the Smithsonian Environmental Research Center (SERC) solar

89

spectral UV-B network (hereinafter called the NIST/SERC UV-B network). This network is small, consisting of 4 sites, and monitors the photolytically active spectral UV-B component of solar radiation and these monitoring sites are the only ones located at sites that polymeric construction materials are also being weathered.

The current NIST/SERC UV-B network is an ideal size for perfecting the networks data management system, the heart and soul of the network. The data management system manages, catalogues, processes, stores, retrieves, and analyzes the volumes upon volumes data collected at each site and integrates the collected knowledge between the sites. For logistical reasons, it is much easier to master the data management tasks on a small-scale network than it is to tackle the volumes of data generated by a larger network.

Data Requirements for the NIST Solar UV Data Network

First and foremost, the NIST Solar UV Data Network has been established to provide relevant and value added scientific data for predicting service life of polymer coatings. The network conforms to guidelines established by the NIST, Standard Reference Data Program (SRDP) [7]. Support for establishing the data network is provided by the SRDP. As with other solar UV data networks, the NIST Solar Spectral UV Data Network relies on a specific instrument for measuring the solar UV radiation flux. An 18 channel scanning spectral radiometer (SR18) developed by the Smithsonian Environmental Research Center (SERC) is used. The instrument measures solar UV-B radiation in the 280-324 nm range. Eighteen interference filters are used at 2nm nominal bandwidth. Intercomparison tests conducted by the National Institute of Standards and Technology, Optical Technology Division on "The 1995 North American Interagency Intercomparison of Ultraviolet Monitoring Spectroradiometers" and reported in the NIST Journal of Research [8] revealed the high level of accuracy of the SR18 instrument. This aspect was important and ensured data traceable to NIST. In addition to SR18 measurements, additional weather parameters were identified and used for service life prediction computer models. They include ambient temperature, ambient relative humidity, wind speed, precipitation, and sky temperature. Table 1 summarizes the environmental data elements that are represented in the NIST Solar UV-B Data Network.

Additional information is stored in the database that helps in characterizing the exposure site. This information includes site management, and site location (latitude, longitude, and elevation).

Data Network Operation and Quality Control

Electronic communication of data is emphasized throughout the data network. A strong line of communication is maintained through the use of Internet facilities. These include message communication between network management at

Table 1: NIST Solar UV-B data network data elements

Data element	Unit
Solar radiation	W/m^2
Ambient temperature	°C
Wind speed	m/s
Ambient relative humidity	%
Sky temperature	°C
Precipitation	Cm

NIST and SERC and site managers. Currently, there are 4 data acquisition sites operational on the network. Figure 1 shows their locations.

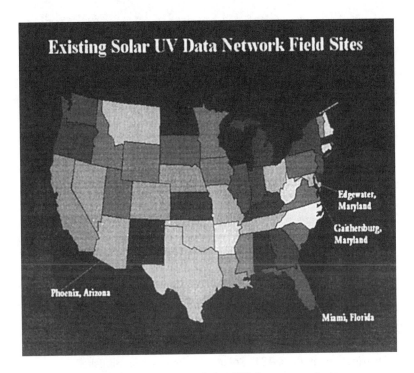

Figure 1: Existing NIST Solar UV data network sites.

Each remote data site is equipped with a data acquisition computer, a SR18 instrument, and computer software and communications capability. Data is acquired from the SR18 at 1 minute intervals; stored locally on the data acquisition computer; then periodically transmitted to the SERC for processing. SR18 instruments are rotated on a quarterly basis for calibration and data analysis. NIST host computers are used for network messaging and data storage. Figure 2 shows a configuration of the data network and its communications capabilities.

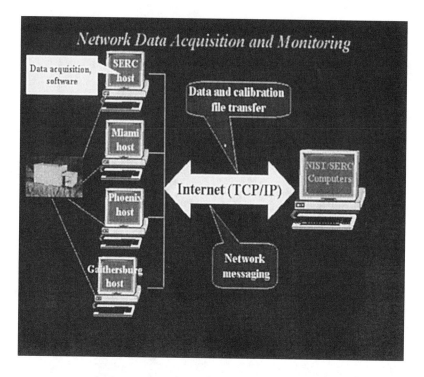

Figure 2: Data network configuration and communication.

SERC personnel are responsible for instrument service, calibration and data processing. Data Processing involves transmitting the acquired SR18 data, applying the calibration constants and producing a final archive data set. The finalized data sets are represented by 12 minutes averages in W/m^2. The last step in creating the archival database involves reformatting and storage in a data management system maintained at NIST. Figure 3 shows the processing steps.

Data Distribution

Distribution of the NIST Solar UV data is accomplished by writing CD-ROM disks and through the use of the Internet World Wide Web. Information technology now provides an efficient means for providing Internet access to scientific data and this method will be used exclusively in the future. Data from the network is provided initially to members of the NIST/Industry/University then, after 2 years, this data will be released to the public. An Internet World Wide Web site has been created as a prototype for data distribution [9]. Figure 4 shows a display of the home page for the site. The site's contents include:

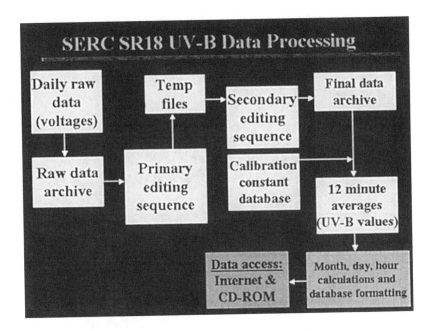

Figure 3: SR18 data processing steps.

- Network information
- Data sets
- Data analysis software
- Site information
- Events
- Published literature
- Links to other relevant Internet sites

A component of the "Data" content is a data dictionary. The data dictionary is intended to provide the data consumer with information on the data tables, data elements, units of measure that are contained in the database. In addition to data elements, summary data are included in the database. These take the form of yearly averages for temperature, and total solar UV-B for specific sites. Figures 5, shows an example of the database tables and data elements. Figures 6, and 7 show examples of the data database summaries.

The Importance of Data Sharing

The fundamental design of the NIST Solar UV-B data network promotes future data sharing. By providing information on database design and quality control measures in data collection and data analysis, increased application of the data sets should result. It will be possible to better characterize exposure sites and improve the integration of the data sets. Also, as the data repository grows in depth and breadth, it

94

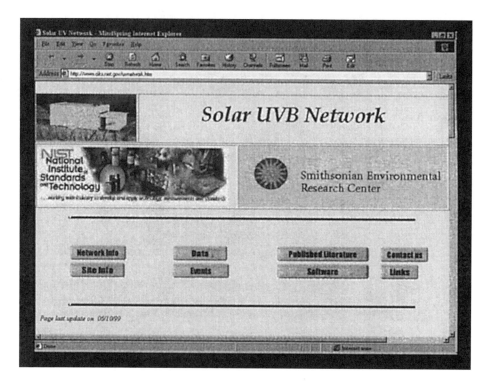

Figure 4: Display of the NIST Solar UV-B data network home page.

will be possible to mine (retrieve relevant data based on known or unknown criteria). Examples of this type of data query are already being used in the semiconductor and health sectors through a process defined as "data mining." However, unless a consensus is reached on data formats, data collection procedures, data applications and data analysis by industry and other interested parties, these data sets will find little use beyond that of the originator of the data." A suggested model that summarizes the data sharing-data integration process is shown in figure 8. It is realized that there may be different views of data (e.g., public or private). It is also anticipated that an evolution will occur, beginning with data (e.g., experimental data sets—x/y pairs). From this data, information will develop (e.g., published information that forms a context—organized by site location or time period). Lastly, knowledge will be developed. Knowledge represents the highest level, where rules can be expressed. Rules are derived from known facts, such as those that link coating material properties and performance with conditions and effects (e.g., environmental factors). Knowledge then becomes an improved element for high-level decision support systems used by product manufactures and users.

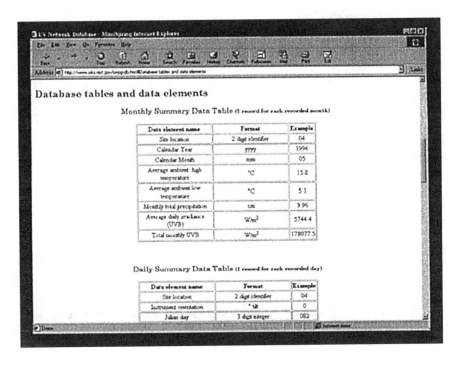

Figure 5: Example of database tables and data elements.

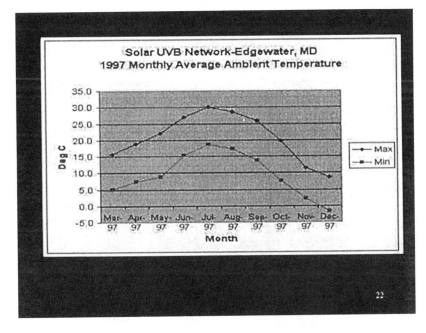

Figure 6: Summary data of yearly average ambient temperature.

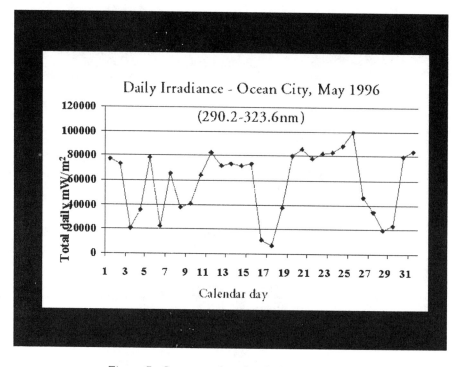

Figure 7: Summary data for daily irradiance.

Future Network Expansion and improvements

Expanded capabilities involving the NIST Solar UV-B data network will focus on an expansion of field sites, improved and more automated data collection and processing, and improvements to the data access capabilities of the Internet World Wide Web site. These capabilities will be implemented during the 2000-2001 timeframe. Additional SR18 instruments currently in operation at U.S. Department of Agriculture field sites will be included as well as a site at the Forest Products Laboratory in Madison, Wisconsin. This expands the coverage and provides a more comprehensive data repository. Improvements to the World Wide Web site will include improved query capabilities for remote site data and interactive graphics capabilities to allow summary and processed data to be displayed in graphical form, at the user's desktop. Improved computing resources will include a more powerful database management system and intuitive user interface, use of the NIST SRDP data server, and improved security features.

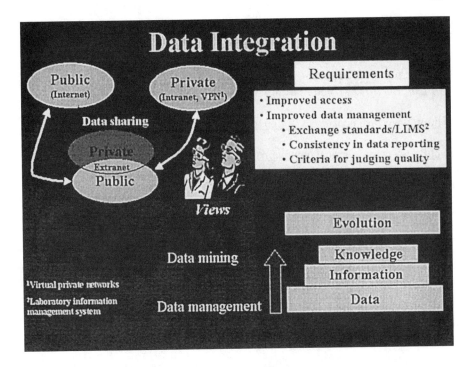

Figure 8: Model for data sharing—data integration.

Summary

This paper has presented an overview of the design, data management, and data resources for the NIST Solar UV-B data network. It is has described critical components used in establishing a data repository to aid researchers, product manufacturers, and product users in determining the service life for polymer coatings. An attempt has been made to describe the importance of publishing database design, providing easy access, and features of quality control for scientific data. Adoption of these methods by the service life community should improve the access and understanding of relevant data and improve the performance of coating materials.

References

[1] *Spectral Ultraviolet-B Radiation Fluxes at the Earth's Surface: Long-Term Variations at 39°N, 77°W,* Correll, D.L., et.al., Journal of Geophysical Research, Vol. 97, No. D7, pg 7579-7591 (May 1992).

[2] *A Systems Approach to the Service Life Prediction Problem for Coating Systems,* Martin, J.W., ACS Symposium Series 722, pg. 1-20.

[3] *Predicting the Temperature and Relative Humidity of Polymer Coatings in the Field,* Burch, D.M. and Martin, J.W., ACS Symposium Series 722, pg. 85-107.

[4] http://www.srrb.noaa.gov/UV/country/usa.html

[5] http://www.biospherical.com/nsf/nsfrsrch/nsfhome.htm

[6] http://uvb.nrel.colostate.edu/UVB/home_page.html

[7] http://www.nist.gov/srdp

[8] The 1995 North American Interagency Intercomparision of Ultraviolet Monitoring Spectroradiometers, Early, E., Thompson, A., and Johnson, C., Journal of Research of the National Institute of Standards and Technology, Vol. 103, No. 1, National Institute of Standards and Technology, Gaithersburg, MD (January-February 1998).

[9] http://www.ciks.nist.gov/uvnetwork.htm

Advances in Laboratory Experimentation

Chapter 6

Use of Uniformly Distributed Concentrated Sunlight for Highly Accelerated Testing of Coatings

Gary Jorgensen[1], Carl Bingham[1], David King[1], Al Lewandowski[1], Judy Netter[1], Kent Terwilliger[1], and Karlis Adamsons[2]

[1]National Renewable Energy Laboratory, 1617 Cole Boulevard, Golden, CO 80401
[2]Marshall Research and Development Laboratory, DuPont Performance Coatings, 3401 Grays Ferry Avenue, Philadelphia, PA 19146

NREL has developed a new ultraviolet (UV) light concentrator that allows material samples to be subjected to uniform intensity levels of 50-100X solar UV at closely controlled sample exposure temperatures. In collaboration with industry, representative coating systems have been exposed without introducing unrealistic degradation mechanisms. Furthermore, correlations have been derived between these highly accelerated test conditions and results obtained at 1-2 suns. Such information is used to predict the degradation of materials in real-world applications. These predictions are compared with measured in-service performance losses to validate the approach. This allows valuable information to be obtained in greatly reduced timeframes, which can provide tremendous competitive advantage in the commercial marketplace.

As the performance and durability of new materials improve, the needs of the coatings industry for facilities that provide realistic expectations for in-use exposure testing of their products become increasingly stringent. An urgency exists to greatly extend the service lifetime requirements of decorative and protective coatings for the automotive industry. These businesses simply cannot afford to wait for prolonged periods of time to directly measure product lifetimes or to risk, without substantiating data, providing the warranties demanded by consumers. For example, there is tremendous competitive pressure to increase the present 5-year warranties associated with 19-month design-to-production cycle times to 10-year warranties based on a 16-month cycle time. To expedite commercialization, life projections must be made in abbreviated time frames; this necessitates using accelerated exposure test (AET) results to allow transformation from life at accelerated stress to life at in-service stress conditions. One of the harshest outdoor stresses these products must withstand is exposure to ultraviolet (UV) radiation contained in terrestrial sunlight.

The National Renewable Energy Laboratory (NREL) has previously demonstrated the ability to expose organic materials (back-metallized polymeric films) to very high levels of solar UV (1). Recently, NREL has developed a new UV concentrator system that is less complicated and more amenable to future commercialization. Using this device, material samples can be exposed to uniform intensity levels of 50-100 times (X) the terrestrial solar UV contained between 290-385 nm. To demonstrate the viability of this process, DuPont provided the characterization of appropriate response variables and a series of representative coating samples for AET at NREL's UV concentrator facility. Relevant environmental stresses were identified and incorporated into models that relate multivariate damage functions to the physical and/or chemical nature of the dominant failure mechanisms experienced by the samples. These models allowed correlations to be made between the service lifetime (SL) at accelerated stress and the SL at in-service stress conditions. AET that simulates reality (i.e., that does not introduce failure mechanisms that are not encountered in actual service) was then used to predict SL.

Experimental

UV Concentrator

NREL's UV concentrator is shown schematically in Figure 1. It consists of an array of faceted mirrors that tracks the sun in two axes and redirects sunlight back to a sample exposure chamber attached by three structural support tubes. The concentrator is designed to provide up to 100X concentration having uniform flux at high UV intensity and low visible (VIS) and near-infrared (NIR) intensity. This is achieved by coating the facets with a custom-designed 37-layer film that uses alternating high and low refractive index materials that results in high UV reflectance and low VIS/NIR reflectance.

The spectral irradiance at the sample exposure plane is presented in Figure 2. The solid line is the concentrated solar intensity as measured by an Ocean Optics SpectraScope fiberoptic spectral radiometer. The dashed curve is an American Society for Testing and Materials (ASTM) direct normal terrestrial air-mass 1.5 solar spectrum (2) multiplied by a factor of 100. Excellent agreement between the measured spectral irradiance and 100 times the ASTM standard is evident throughout the UV bandwidth (300-400 nm). A steep cutoff at about 475 nm greatly

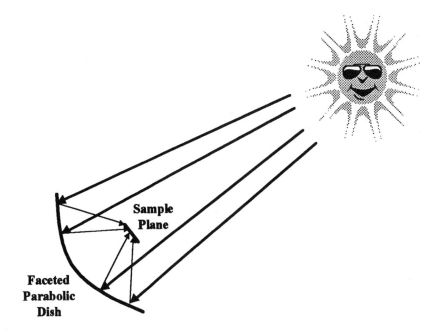

Figure 1. Schematic of UV concentrator.

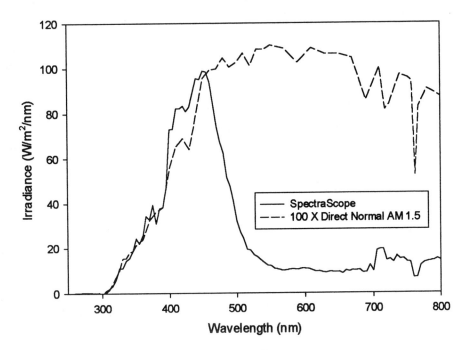

Figure 2. Spectral irradiance at sample exposure plane.

facilitates thermal control of samples during exposure. The area between the two curves in the visible and NIR portions of the spectrum represents avoided thermal loading.

The concentrator uses a close-packed array of 18 hexagonal facets, with their centers located on a near-parabolic surface. Each of the facets has a spherical curvature with a focal length of 150 cm and an outer diameter (across the vertexes) of 28 cm. The diameter of the effective aperture of the faceted dish is roughly 1.3 m. Individual facets are aimed at a single point in the target plane that is located at a fraction of the nominal focal length. This aiming strategy results in partially concentrated images of the facets and allows a wide range of nearly uniform concentration levels to be achieved by varying the number of facets and target distance. At 100X concentration the target image is a hexagon with an area of about 87 cm^2. This can conveniently accommodate sixteen samples that are each 2-cm x 2-cm in size. The various dimensions of the concentrator system can be scaled to allow increased sample exposure areas.

The flux profile was measured using NREL's BEAMCODE characterization system (a product of Coherent Inc.). A solid state array video camera is focused on a near-Lambertian target plate placed at the sample exposure plane. Grey scale images are captured to a computer frame grab board that interfaces with the BEAMCODE software package for flux distribution analysis. Results are shown in Figure 3. The shadows cast by the support tubing on the facets are clearly seen, but the impact on the uniformity is only 4% in the center and less than 2% in the shadows. These tubes also provide inlet and outlet cooling water lines and a vacuum line to the sample chamber.

Sample Exposure Chamber

The sample chamber was constructed from a 2.5-cm thick block of copper because of its excellent thermal conductivity properties. Details of the chamber are shown schematically in Figure 4. Six 0.64-cm diameter serpentine channels were machined near the top surface of the copper block to allow circulation of cooling water, thereby providing conductive cooling to samples being exposed. A grid of vacuum holes was used to hold the samples in good thermal contact with the copper substrate and to keep the samples in place during exposure while tracking. To accomplish this, 0.64-cm diameter vacuum lines were machined orthogonal to the water cooling channels in the lower half (near the bottom surface) of the copper block, along with a 0.95-cm diameter vacuum manifold channel. A grid of 0.16-cm diameter holes was drilled from the top surface to intersect the vacuum lines to deliver a vacuum pull on the backside of samples to be exposed. The copper block was also machined to accommodate fiber probes to allow real-time monitoring of spectral irradiance experienced by samples being exposed.

The ability of this sample chamber arrangement to control sample temperature during UV exposure at 100X was demonstrated. While exposed to full power "on-sun" conditions (100X), the capacity of chilling water delivered to the chamber was

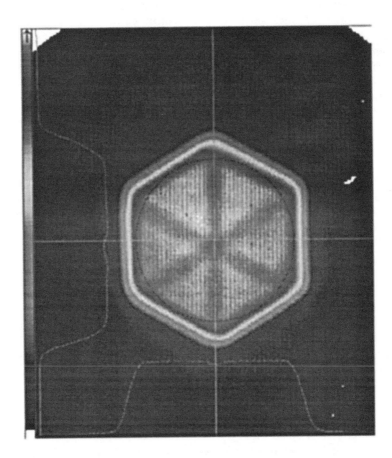

Figure 3. Distribution of UV flux at sample exposure plane.

capable of maintaining the surface of the copper block at 12°C. Test samples had the construction: bilayer paint coating / metal substrate, where the bilayer coating was a clearcoat over either a black or white pigmented paint basecoat and the metal substrate was 0.1-cm thick steel. The surface temperature of exposed samples was measured using an infrared (IR) camera (Mikron Instrument Company, Inc., Thermo Tracer TH1100) and a hand held surface thermocouple probe (Omega HPS-HT-K-12-SMP-M). The surface probe was used to calibrate the emissivity setting of the IR camera. A value of 0.92 gave good agreement between the two temperature measurements for both black and white painted samples. At 100X exposure with maximum cooling, surface temperatures ranged between 18°C and 26°C for white and black samples, respectively. With the chilling water turned off, sample temperatures reached 38°C (white) to 52°C (black). These maximum temperatures were within the targets/set-points specified by DuPont.

Measurement of UV Exposure

During exposure testing of candidate coating samples, UV irradiance was continually monitored by two pyranometers: an EKO MS-210W UV-B pyranometer (sensitive to light between 285-315 nm) and an Eppley TUVR pyranometer (sensitive to light between 290-385 nm). These instruments were integrally mounted to the UV concentrator so that they tracked the sun during material exposures. They were equipped with occulting tubes to allow only the direct portion of the solar spectrum (i.e., that part of the solar spectrum imaged by the UV concentrator) to be measured. Both pyranometers were periodically calibrated by using an Optronic OL-754 spectral radiometer that measured only the direct solar irradiance. These calibrations indicated agreement with the pyranometers to within 10%. From the one-sun direct irradiance data and knowledge of the concentrating properties of the UV concentrator, all the information needed to fully characterize the UV exposure of tested samples was provided.

Sample Exposures

Four sets of samples were provided for accelerated light exposure testing at NREL's UV concentrator facility. DuPont had previously accumulated considerable testing acumen with these sample types exposed in an Atlas Ci-65 WeatherOmeter® (WOM). This chamber uses an artificial xenon arc light source with borosilicate inner and outer filters. The typical spectral irradiance experienced by the samples is shown in Figure 5; a very close match (i.e., 1X) to an air-mass 1.5 terrestrial solar spectrum is evident throughout the UV region. During Ci-65 exposure, samples were subjected to conditions specified in Society of Automotive Engineers Standard J-1960 (*3*). A 3-hour cycle was used during which the light was on for 2 hours (40 minutes of light; 20 minutes of light with front specimen water spray; and 60

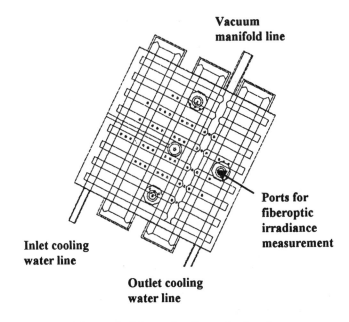

Figure 4. Schematic of sample chamber.

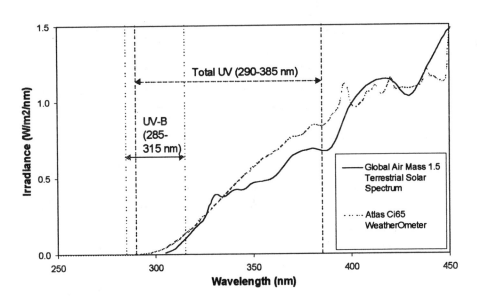

Figure 5. Spectral irradiance of Ci-65 exposure chamber vs. global air-mass 1.5 terrestrial solar spectrum.

108

minutes of light) with a sample temperature of 70°C, and the light was off for one hour with a sample temperature of 38°C.

The sample sets consisted of 16 types of bilayer paint systems (clear coat/colored basecoat) on steel substrates. Samples were 2-cm x 2-cm in size and 0.1-cm thick. The acrylic polyol clearcoats were fortified with various levels of UV absorbers (UVA) and hindered amine light stabilizers (HALS) as indicated in Table I.

Table I. Coating Samples Tested

DuPont Coating System	Coating Description	UVA % (100%= Standard)	HALS % (100% = Standard)
1.1	Acrylic Polyol / Melamine Crosslinked	100	100
1.2	Acrylic Polyol / Melamine Crosslinked	100	75
1.3	Acrylic Polyol / Melamine Crosslinked	100	50
1.4	Acrylic Polyol / Melamine Crosslinked	100	25
5.4	Acrylic Polyol / Melamine Crosslinked	100	0
5.3	Acrylic Polyol / Melamine Crosslinked	80	0
5.2	Acrylic Polyol / Melamine Crosslinked	65	0
5.1	Acrylic Polyol / Melamine Crosslinked	16	0
7.3	Acrylic Polyol / Melamine & Silane Crosslinked	100	100
7.1	Acrylic Polyol / Melamine & Silane Crosslinked	100	0
7.2	Acrylic Polyol / Melamine & Silane Crosslinked	0	100
7.4	Acrylic Polyol / Melamine & Silane Crosslinked	0	0
Comm-OEM[a]	Acrylic Polyol / Melamine Crosslinked	N/A	N/A
Exp #1[b]	Acrylic Polyol / Modified Isocyanate Crosslinked	N/A	N/A
Comm-Ref[c]	Acrylic Polyol / Isocyanate Crosslinked	100	100
Exp #2[b]	Acrylic Polyol / Isocyanate Crosslinked	N/A	N/A

[a]Commercial OEM product
[b]Experimental product
[c]Commercial refinish product

At NREL, two experiments were performed, one each at 50 suns and 100 suns UV exposure. A complete set of sixteen samples was exposed each time. Both of these experiments were continued until a cumulative UV-B dose equivalent to 2-3 years outdoor exposure in Miami, FL, was reached. During UV concentrator exposure, humidity was not controlled (ambient levels were ~30%) and sample temperatures were targeted to be between 40-50°C; the actual exposure temperatures are given in Table II, along with other details from the accelerated sunlight and Ci-65 tests. To simulate moisture conditions experienced during Ci-65 exposures, samples were exposed in a Q-Panel Lab Products QUV chamber (4) whenever they were not on-sun. The cycle profile was 4 hours with the lights on but blocked (i.e., the samples saw no additional light) at 40°C and 87% relative humidity, and 4 hours with the lights off at 40°C and 97% relative humidity (to allow water to condense on the surface of the samples). At the end of each experiment, NREL and DuPont performed various analytical characterizations. These results allowed us to demonstrate a correlation between our 50X and 100X exposures and the 1X exposures in the Ci-65 chamber.

Table II. Sample Exposure Conditions

Exposure Chamber	Sunlight Intensity	Cumulative UV-B (MJ/m^2)	Cumulative Total UV (MJ/m^2)	Temperature $(°C)$
Ci-65	1X	16.80	372.4	70±2
UV Concentrator	50X	11.25	306.0	42±5
UV Concentrator	100X	12.77	396.5	52±5

Analysis

To obtain correlations between in-use and accelerated exposure results, a suitable material-specific damage function model must be found that accurately relates changes in an appropriate response variable to relevant applied environmental stresses. The response variable can be either microscopic (e.g., changes in chemical structure) or macroscopic (e.g., loss of gloss), but ideally should be easily measured and directly related to an important property of the material being tested. For the types of coatings being investigated, a number of useful response variables have been suggested (5).

Coatings are known to be susceptible to degradation caused by cumulative light dosage (D) exposure (6):

$$D(t) \sim \int_0^t L(t)\,dt \tag{1}$$

where $L(t)$ is the time-dependent incident spectral irradiance, $I(\lambda,t)$, convoluted with the absorption spectra of the material being exposed, $\alpha(\lambda,t)$, and the quantum efficiency of the absorbed photons to propagate reactions that are harmful to the coating, $\phi(\lambda,t)$, integrated over an appropriate bandwidth (defined by λ_{min} and λ_{max}) throughout which light-induced damage occurs:

$$L(t) = \int_{\lambda_{min}}^{\lambda_{max}} I(\lambda, t) \, \alpha(\lambda, t) \, \phi(\lambda, t) \, d\lambda \qquad (2)$$

In previous work ($1, 7$) with organic materials (back-metallized polymeric films), we have obtained useful results by approximating the absorption spectra and quantum efficiency as constants in eq 2 and defining:

$$I_{UV}(t) = \int_{\lambda_{min}}^{\lambda_{max}} I(\lambda, t) \, d\lambda \qquad (3)$$

with $\lambda_{min} = 285$ nm and $\lambda_{max} = 315$ nm (for UV-B) or $\lambda_{min} = 290$ nm and $\lambda_{max} = 385$ nm (for total UV). For constant (controlled) irradiance, this leads to an approximate generalized cumulative dosage model in which the loss in performance, ΔP, (change in response variable) with time is proportional to a power law expression of the ultraviolet irradiance I_{UV} ($8, 9$):

$$\frac{\Delta P}{\Delta t} \sim (I_{UV})^n \qquad (4)$$

To account for thermal effects, an Arrhenius term can be included and the change in performance after the ith time interval is:

$$\Delta P_i = A (I_{UV})^n \, \Delta t_i \, e^{-E/kT} \qquad (5)$$

where T is the temperature (K) experienced by samples during exposure, k is Boltzmann's constant, and E is an activation energy. For constant accelerated stresses, I_{UV} and T are known; this allows eq 5 to be fit to measured values of ΔP_i and subsequent determination of the coefficients A, E, and n. For variable real-world stresses, the time dependent form eq 5 must be used:

$$\Delta P(t) = A \int_0^t [I_{UV}(t)]^n \, e^{-E/kT(t)} \, dt \qquad (6)$$

Having determined the relevant coefficients from AET's performed at constant stresses, eq 6 can be used to compute a predicted loss in performance after some time t where the relevant stresses are monitored; these predicted values can then be compared with actual measured values.

Results

The various coating systems were characterized by transmission mode FTIR analysis as a function of cumulative UV dose exposure. A skiving technique was used in sampling the top ~3-4 microns of a given specimen. This technique is not readily suitable to obtaining thinner cross sections; attempts often result in shredding. Low temperature sectioning procedures introduce additional complications in terms of sample handling. The skiving approach was used to plane a thin layer of material from the surface of each clearcoat. In previous work, this has proven to be quite acceptable to allow coating degradation mechanisms and kinetics to be studied. Samples were characterized in transmission mode using a Nicolet Magna IR Model 760 spectrophotometer equipped with a Nicolet Nic-Plan IR microscope. Typical results are shown in Figure 6. Spectra were normalized to the same absorbance scale. Although changes in the various bands of interest are small, they can be easily discerned using Nicolet OMNIC software. The key to this process is to have accurate baseline correction measurements prior to determination of envelope areas.

Samples were also analyzed with atomic force microscopy (AFM) before and after exposure. The AFM images of the exposed and unexposed samples are quite different; at a 2μ x 2μ scan the weathered sample appears to have much more micro surface topography than does the unexposed sample. The macro roughness of the coating is preserved, as evidenced by large irregularities and holes in the coating. The visible changes in the topography resulting from exposure are represented by an increase in the root mean square surface roughness from 33 to about 50 angstroms. This increase in roughness is likely representative of surface material that ablates away as the polymer system decomposes as a function of the accelerated weathering process.

The chemical changes, detected with FTIR, that occur in the organic films as a function of accelerated exposure testing are representative of bulk rather than surface photochemical reactions. This is true because the FTIR sampling depth into the coating is on the order of microns, which is much deeper than the surface roughness of the films. Because no analysis technique was used to sample and determine surface chemistry on the order of 5 nm, it is difficult to ascertain the true chemistry precisely at the surface. Although XPS allows the binding energy envelopes for oxygen or carbon (or other appropriate elements) to be monitored in the top 4 nm of the surface, deconvolution of the XPS spectra and assignment of specific functionality has been difficult. TOF SIMS has been tried with isotope substitution (O-18 for O-16) with some success, but results are difficult to interpret and quantify. The need for improved surface-specific techniques to follow photo-oxidation/hydrolysis reactions remains evident. However, we have found that the FTIR results are consistent with typical changes associated with weathering these materials at much lower acceleration factors.

Based on previous experience with these types of materials (*10,11*), a specific photo-oxidation index was chosen as the appropriate response variable. This index is calculated from FTIR data as the [OH,NH,COOH] envelope peak area (between 3200-3600 cm^{-1}) normalized for thickness by the [CH] envelope peak area (between 2800-3000 cm^{-1}). ΔP is then defined as the change (difference) between this ratio at any time t and the unweathered (t=0) value. Any number of degradation processes may contribute to ΔP (photo-oxidation, hydrolysis, etc.). Close inspection of the relevant peaks of the FTIR spectra revealed that hydrolysis did not significantly contribute to ΔP for the coatings tested. If hydrolysis was important, a generalized Eyring form of eq 5 could have been used to account for the effect of moisture (7). For samples exposed in the Ci-65 and the UV concentrator, measured values of ΔP_i are available for various levels of cumulative dose and sample temperature exposures

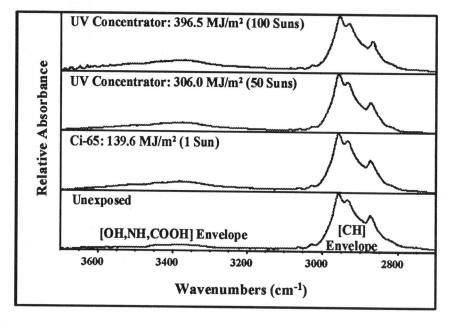

Figure 6. Relative FTIR absorbance of coating system 5.4 as a function of cumulative UV dose.

(70°C at 1X in the Ci-65 and 42°C at 50X and 52°C at 100X in the UV concentrator). Eq 5 was fit to these data to determine the model coefficients. Representative results for system 5.4 are presented in Figure 7. Here, the predicted change in photo-oxidation index is plotted vs. measured values for samples subjected to three levels of temperature and light intensity. Additional data at 50X and 100X are desirable but require future longer-term exposures. The validity of the model is indicated by the fact that a straight line having a near-unity slope can represent all of the data. Similar results were obtained for other coating systems and are summarized in Table III.

Table III. Coefficients Derived for Representative Bilayer Coating Samples

DuPont System	A	n	E (kcal/mole-K)
1.2	6.513	0.667	3.878
5.4	750.92	0.635	6.781
7.1	370.51	0.706	6.523

Values of activation energies (E) derived (4-7 kcal/mole-K) are reasonable for photo-thermally driven degradation mechanisms. The values of n (~0.67) imply that exposure to 50-100X light intensities had a net effect of only 15-25X, suggesting that some shielding or rate limiting reactions occur that do not allow all photons to participate in degradation.

To fully substantiate our contention that exposure of coatings at highly accelerated levels of light can produce useful and realistic results, it would be desirable to have measured values of ΔP_i for samples of these coatings exposed outdoors where the time dependent stress variables are known. Then, eq 6 could be used to predict ΔP_i for comparison with measured data. Unfortunately, such data are not available. However, NREL has a durability testing program that includes outdoor exposure (with carefully monitored radiometric and meteorological conditions) for another organic-based type of material, namely, samples of bulk transparent polymers.

Two types of sheet (0.32-cm thick) glazing materials have been tested at a variety of outdoor exposure test (OET) sites; these include polyvinyl chloride (PVC) and a UV-stabilized polycarbonate (PC). These materials have also been exposed (12) in an Atlas Ci-5000 WOM having a UV intensity of about 2X (at 60°C for both the PC and PVC) compared to typical outdoor terrestrial levels, and at 50X (42°C for the PC and 25°C for the PVC) and 100X (48°C for the PC and 35°C for the PVC) in our UV concentrator. The response variable was chosen to be hemispherical transmittance between 400-500 nm because, in general, that is the spectral region most sensitive to stress exposure induced loss in performance (Figure 8). The same damage functions expressed in eqs 5 and 6 were assumed. Data from the Ci-5000 and the UV concentrator exposures of the polymer glazings were used to fit eq 5 and to obtain the model coefficients corresponding to these materials; the results are

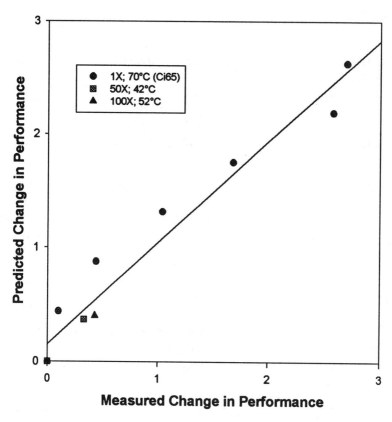

Figure 7 Measured vs. predicted [OH,NH,COOH] / [CH] for coating system 5.4.

given in Table IV. As with the DuPont coatings, the derived activation energies are not unreasonable. The value of n for PVC was similar to that found for the DuPont coating systems. For the UV-stabilized PC sample, a value of n=1 suggests that exposure of this material follows strict reciprocity even up to 100X; all incident photons fully contribute to degradation reactions that proceed at twice the rate undergone at 50X exposure and 50 times the rate experienced at 2X exposure.

Table IV. Coefficients Derived for Representative Clear Polymer Sheet Samples

Polymer Sheet	*A*	*n*	*E (kcal/mole-K)*
Polyvinyl Chloride	2892	0.669	8.440
UV-Stabilized Polycarbonate	5.497	1.093	6.688

Using the coefficients from Table IV and time-monitored values of sample temperature and UV irradiance, the loss in performance was predicted using eq 6 for both PVC and PC as exposed outdoors in Golden, CO, and Phoenix, AZ. Predicted values were then compared with actual measured data for these materials exposed at these sites. The results are presented in Figure 9. Time-dependent changes in weathering variables produce the irregular shapes of the predicted curves. Excellent agreement is evident between the measured and predicted data, thereby validating the ability to expose samples at very high light levels, our approach to data analysis (using accelerated test results to obtain model coefficients, and then the use of these coefficients to predict time-variable real-world degradation), and the assumed damage function model.

Conclusions

Correlation between highly accelerated levels (50-100X) of UV light intensity and in-use levels (~1X) has a number of significant implications. First, these materials can be tested at ultra-accelerated intensities without introducing unrealistic degradation mechanisms. This is an impressive result because the conventional wisdom has been that organic coatings could not be realistically and confidently tested at more than about 10 suns because of difficulties associated with adequately controlling sample temperature. Consequently, very abbreviated testing times can be substituted for long-time exposures at low intensity levels. This will allow much shorter development cycle times for new products; manufacturers will not be forced to wait months or years to ascertain if prospective coating systems will exhibit adequate durability. This will provide a vital competitive advantage to such manufacturers and will result in greatly improved new products.

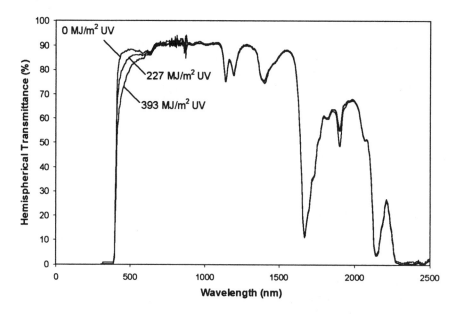

Figure 8. Spectral change in transmittance with UV exposure for UV-stabilized PC.

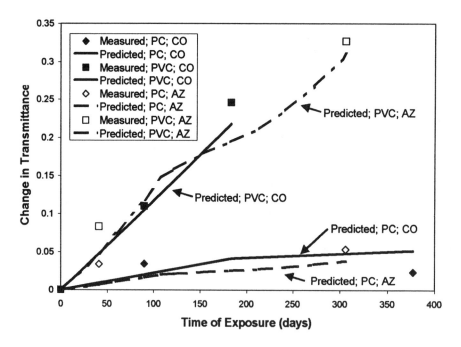

Figure 9. Measured vs. predicted change in hemispherical reflectance between 400-500 nm for two polymeric materials at two outdoor exposure sites.

Acknowledgements

NREL's UV concentrator was designed, fabricated, installed and tested as part of a cooperative project with NPO Astrophyisca, Moscow, Russia. Development was supported through the U. S. Department of Energy's Initiatives for Proliferation Prevention Program. Samples and useful discussions were provided by Basil Gregorovich at DuPont. AFM characterization was performed at NREL by Helio Moutinho. This paper is dedicated to the memory of David King who was an innovative and dedicated scientist; his insights, enthusiasm, and friendship will be greatly missed by his co-authors and all others who knew him.

Literature Cited

1. Jorgensen, G.; Bingham, C.; Netter, J.; Goggin, R.; Lewandowski, A. In *Service Life Prediction of Organic Coatings, A Systems Approach*; Bauer, D. R. and Martin, J. W., Eds.; ACS Symposium Series 722; American Chemical Society, Oxford University Press: Washington, DC, 1999; pp 170-185.
2. *Annual Book of ASTM Standards*; Standard E892-87 (Reapproved 1992); American Society for Testing and Materials: Philadelphia, PA, 1993; Vol. 12.02, pp 487-494.
3. Standard SAE J1960; The Society of Automotive Engineers, Inc.: Warrendale, PA, 1989.
4. *Annual Book of ASTM Standards*; Standard G53-91; American Society for Testing and Materials: Philadelphia, PA, 1993; Vol. 6.01, pp 1045-1050.
5. Bauer, D. R. *J. Coatings Tech.* **1994**, 66, 57-65.
6. Martin, J. W.; Saunders, S. C.; Floyd, F. L.; Weinberg, J. P. *Methodologies for Predicting the Service Life of Coatings Systems*, NIST Building Science Series 172, National Institute of Standards and Technology: Gaithersburg, MD, 1994.
7. Jorgensen, G.; Kim, H-M.; Wendelin, T.; In *Durability Testing of Nonmetallic Materials*; Herling, R. J., Ed.; ASTM STP 1294; American Society for Testing and Materials: Scranton, PA, 1996; pp 121-135.
8. Martin, J. W. *Durability of Building Materials* **1982**, 1, 175-194.
9. Carlsson, B. *Solar Materials Research and Development; Survey of Service Life Prediction Methods for Materials in Solar Heating and Cooling*; International Energy Agency Solar Heating and Cooling Program, Report # BFR-D-16-1989 (DE90 748556), ISBN 91-540-5063-4; Swedish Council for Building Research: Stockholm, Sweden, 1989; p 27.
10. Adamsons, K.; Lloyd, K.; Stika, K.; Swartzfager, D.; Walls, D.; Wood, B. In *Interfacial Aspects of Multicomponent Polymer Materials*; Lohse, D. J.; Russell,

T. P.; Sperling, L. H., Eds.; Proceedings of an American Chemical Symposium held in Orlando, Florida, August 25-30, 1996; Perseus Books: Cambridge, England, 1998; pp 279-300.

11. Adamsons, K.; Blackman, G.; Gregorovich, B.; Lin, L.; Matheson, R. *Progress in Organic Coatings* **1998**, 34, 64-74.

12. *Annual Book of ASTM Standards*; Standard G26-92; American Society for Testing and Materials: Philadelphia, PA, 1993; Vol. 6.01, pp 1033-1041.

Chapter 7

Laboratory Apparatus and Cumulative Damage Model for Linking Field and Laboratory Exposure Results

Jonathan W. Martin, Tinh Nguyen, Eric Byrd, Brian Dickens, and Ned Embree

National Institute of Standards and Technology, 100 Bureau Drive, Stop 8621, Gaithersburg, MD 20899-8621

A laboratory apparatus and a cumulative damage model are described for linking field and laboratory photodegradation results for polymeric materials. The apparatus is capable of independently and precisely controlling or monitoring the three primary weathering factors in both space and time. These factors are temperature, relative humidity, and spectral ultraviolet radiation. Linkage between field and laboratory results will be made through measurements of dose, dosage, and material damage. These measurements are input into the model to estimate the apparent spectral quantum yield and the total effective dosage for a model acrylic melamine coating. The model coating was exposed at twelve different spectral wavebands over a wide range of temperatures and relative humidities. Variables affecting the accuracy of the measurements are discussed.

Introduction

This paper describes a laboratory apparatus and a cumulative damage model, called the total effective dosage model, for use in linking field and laboratory photodegradation results for polymeric materials. This model has gained widespread acceptance in medical and biological studies since the mid-1900s (1,2). The total effective dosage model has a firm basis in the principles of photochemistry, including the Lambert-Beer law, the photochemical law, the reciprocity law and the spectral-independence-of-damage assumption, often called the law of additivity (3). This model differs from generally accepted field exposure metrics such as total exposure time (4), total solar irradiance (5,6) and total-UV irradiance (7) in that the spectrum of incident radiation is broken up into regions, each of which is treated independently and each region is weighted as to its photolytic effectiveness. For most polymeric materials, quanta at wavelengths of solar ultraviolet closer to 290 nm tend to have a higher photolytic efficiency than do quanta at longer wavelengths. Broadband exposure metrics like total solar irradiance and total-UV irradiance assume that the photolytic effectiveness of all quanta in the radiant spectrum is constant.

Total Effective Dosage Model

The number of photons absorbed by a material, the dosage, is given by

$$D_{dosage}(t) = \int_0^t \int_{\lambda_{min}}^{\lambda_{max}} E_o(\lambda,t)\left(1 - 10^{A(\lambda,t)}\right) d\lambda \, dt$$

[1]

where

λ_{min} and λ_{max} = minimum and maximum photolytically effective UV-visible wavelengths (units: nm),

$A(\lambda,t)$ = absorbance of sample at specified UV-visible wavelength and at time t, (units: dimensionless) [note, for polymers containing UV stabilizers or absorbers the absorbance term can be partitioned to account for the spectral absorbance from each component],

$E_o(\lambda,t)$ = incident spectral UV-visible radiation dose to which a polymeric material is exposed to at time t (units: J cm^{-2}),

t = elapsed time (units: s).

The total effective dosage, D_{total}, is the total number of absorbed photons that effectively contribute to the photodegradation of a material during an exposure period. It has the form

$$D_{total}(t) = \int_0^t \int_{\lambda_{min}}^{\lambda_{max}} E_o(\lambda,t)\left(1 - 10^{-A(\lambda,t)}\right)\phi(\lambda) \, d\lambda \, dt$$

[2]

where

$D_{total}(t)$ = Total effective dosage (units: J cm^{-2}) and

$\phi(\lambda)$ = spectral quantum yield, the damage at wavelength λ relative to a reference wavelength (dimensionless).

Photolytic damage to a polymeric material has been empirically related to dosage, D_{dosage}, by a damage function. Commonly published damage functions include a linear response

$$\Gamma = a\, D_{dosage}$$

a power law response

$$\Gamma = a\, (D_{dosage})^b$$

and an exponential response

$$\Gamma = a\, \exp(b\, D_{dosage})$$

where a and b are empirical constants and where damage is any quantitative, performance characteristic that is considered to be critical to a material's field performance.

The apparent quantum yield, $\phi(\lambda)$, in Equation 2 is estimated from the initial slope of the material damage, Γ, versus dosage, $D_{dosage}(t)$, curve for a material exposed at wavelength λ. Once the apparent spectral quantum yield has been estimated, then this value can be substituted back into Equation 2 after which the total effective dosage, D_{total}, can be determined. Although this can be accomplished, the estimated total effective dosage are only realistic when the damage versus dosage curves are linear.

Experiment

Materials and Specimen Preparation:

Coating

Bauer, Gerlock and others (8,9) have extensively studied the photochemistry of the model acrylic melamine coating selected for this study. The coating is made from a mixture of a hydroxy-terminated acrylic polymer and a partially-alkylolated amine crosslinking agent in a mass ratio of 70:30. The acrylic polymer contains 68% by mass normal butylmethacrylate, 30% hydroxy ethylacrylate, and 2% acrylic acid and is supplied as a mixture 75% by mass acrylic polymer and 25% 2-heptanone. The crosslinking agent used is Cytex Industries Cymel 325[1], which is a mixture containing 80% by mass melamine formaldehyde resin and 20% isobutanol solvent. The glass transition temperature of the cured coating as determined by dynamic mechanical analysis was 45° C ± 3° C for our material.

Calcium fluoride (CaF_2) disks having dimensions of 100 mm diameter and 9 mm thick were selected for the laboratory substrates. CaF_2 is transparent for wavelengths between 0.13 µm to 11.5 µm (10). For the field exposure experiments, silicon and quartz substrates were selected due to the high cost of the CaF_2 disks. Silicon is transparent for wavelengths between 1.2 µm to 15 µm and is a good substrate for studying photodegradation via FTIR analysis in the transmission mode. Quartz is transparent from 0.12 µm to 4.5 µm and, therefore, it is a suitable substrate for studying changes in spectral UV absorbance. All three substrates have excellent moisture and temperature resistances.

Solutions of acrylic polymer and melamine crosslinking agent were mixed at the manufacturer's suggested ratio, degassed, spread onto the substrates, and spin cast at 2000 rpm for 30 seconds. The coatings were then cured at 130 °C for 20 minutes. This schedule completely cured the coatings as verified by Fourier transform infrared spectroscopy (FTIR). The average and standard deviation of the cured coating thicknesses were 10.4 μm ± 0.5 μm as estimated by a prism coupling technique (11).

Apparatus

Solar Simulators

Two Oriel Instruments solar simulators[1] were procured for this study. A schematic of the simulator is shown in Figure 1. Each solar simulator's optical system included a 1000 W xenon arc lamp, a parabolic mirror, a dichroic mirror, an optical integrator, a cold mirror, a 36 cm diameter Fresnel collimating lens, and a light intensity controller (not shown in Figure 1).

Ultraviolet radiation emitted from the xenon arc is reflected from the parabolic mirror to the dichroic mirror. Visible and infrared portions of the xenon arc spectrum pass through the dichroic mirror and are absorbed by a heat sink. The ultraviolet portion of the xenon arc spectrum is reflected off the dichroic mirror through the optical integrator. The optical integrator homogenizes the beam and the Fresnel lens collimates the beam. For our setup, the heat-induced irradiance of the specimens was less than 2° C above the ambient temperature.

Exposure Cells

The laboratory exposure cells were designed to simultaneously expose different sections of the same film to 12 well-defined, spectral radiation bandwidths. A complete description of the exposure cell and the exposure cell arrangement under the two solar simulators has been given in Reference 11. This section briefly describes the design of an exposure cell and the arrangement of the solar cells under each solar simulator.

The exposure cells were arranged as shown in Figure 2. Exposure cells 1 through 4 were irradiated under the left-hand solar simulator while exposure cells 5 through 8 were irradiated under the right-hand solar simulator.

---

1 Certain commercial instruments and materials are identified in this paper to adequately describe the experimental procedure. In no case does such identification imply recommendation or endorsement by the National Institute of Standards and Technology nor does it imply that the instrument or materials are necessarily the best available for the purpose.

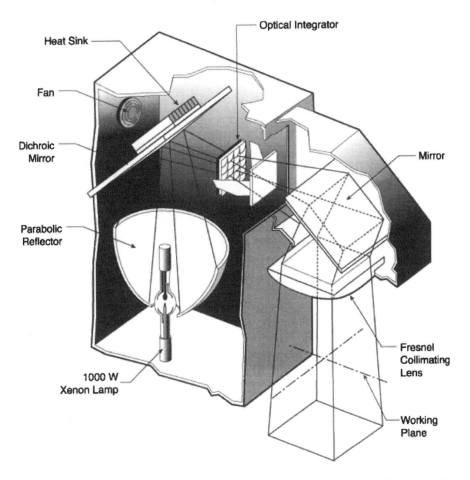

Figure 1. Schematic of solar simulator showing the arrangement of its optimal components.

124

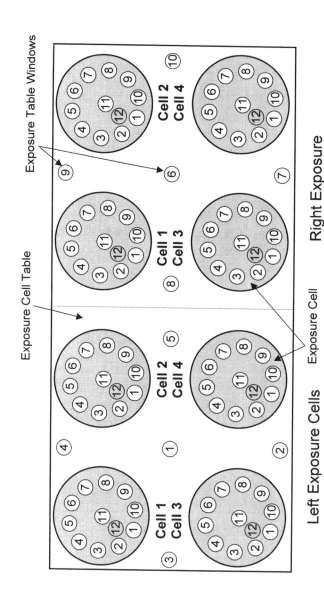

Figure 2. Arrangement of the eight exposure cells on the exposure table and the arrangement of the 12 windows within each exposure cell. The 12th window is only in the specimen disk and is used in studying non-photolytic degradation.

A cross sectional schematic of an exposure cell is shown in Figure 3. Each exposure cell has a layered design that included a filter disk, a quartz disk, a CaF_2 specimen disk and an encasement to hold the disks. The filter disk contained 11 windows: 10 on the disk perimeter and the 11th in center (the 12th window is displayed in Figure 2, but this window was only open in the specimen disk). Each window is 16 mm in diameter. A slot machined into the top of each window supports an interference filter and the bottom is beveled with a knife-edge to minimize grazing angle reflections from the interior sides of the window.

Each of the 10 perimeter windows was outfitted with a narrow bandwidth interference filter covering a different segment of xenon arc radiant spectrum from 290 nm to 550 nm. Nominally identical interference filters were used in all eight-exposure cells. The first eight interference filters had full-width-half-maximum (FWHM) values between 2 nm and 10 nm and covered the range between 290 nm and 340 nm. The nominal center wavelengths for these filters were 290 nm, 294 nm, 300 nm, 306 nm, 312 nm, 318 nm, 326 nm, and 336 nm. The two remaining filters had FWHM values greater than 10 nm and had center wavelengths of 354 nm and 450 nm. The 290-nm center wavelength interference filter was selected because 290 nm is the lowest radiation reaching the Earth's surface (12). The center wavelengths were selected to cover the range of suspected photolytic activity. The FWHM for each filter was selected in an effort to degrade the coating at the same rate. This selection process was based on three a priori assumptions: 1) the photolytic effectiveness (i.e., spectral quantum yield) monotonically decreases with increasing wavelength; 2) between 290 nm and 500 nm the apparent spectral quantum yield decreases by approximately three orders of magnitude (13); and 3) the spectral output of the xenon arc lamp approximates the solar radiant flux.

The center window (i.e., the 11th window) was termed the "full UV-radiation" window because, in the first study, it was not outfitted with a filter and, thus, transmitted all of the spectral radiant flux emitted from the solar simulator. In subsequent studies, a short pass filter (Schott filter KG-1, KG-2, KG-3, or KG-5) was inserted into this window to remove radiation below 290 nm.

The CaF_2 specimen disk was sandwiched between two aluminum plates (see Figure 3). Each plate contained 12 windows. The aluminum plates protected the CaF_2 from damage during handling while the windows allowed a portion of the film to be exposed to a specific wavelength treatment. The top aluminum plate was positioned about 1 mm above the coating while the bottom plate was in direct contact with the lower surface of the CaF_2 disk. All 12 windows in the top and bottom plates had the same dimensions and so aligned that radiation could exit the exposure cell without back reflection. The 12[th window] was termed the "no-UV" or "dark window" and was used in assessing non-photolytically induced degradation.

The exposure cell was machined to tight tolerances so that a small positive pressure could be maintained within the exposure cell chamber. The quartz disk acted as the top of the exposure cell chamber and protected the interference filters from the degrading effects of moisture while transmitting essentially the entire incident UV radiation. The CaF_2 disk acted as the bottom of the chamber while the encasement acted as the sides.

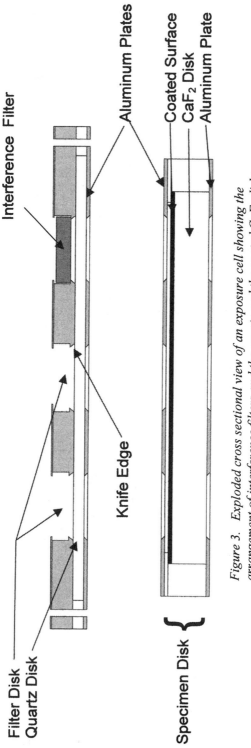

Interference Filter

Aluminum Plates

Coated Surface
CaF$_2$ Disk
Aluminum Plate

Knife Edge

Filter Disk
Quartz Disk

Specimen Disk

Figure 3. Exploded cross sectional view of an exposure cell showing the arrangement of interference filters and the quartz and the coated CaF$_2$ disks.

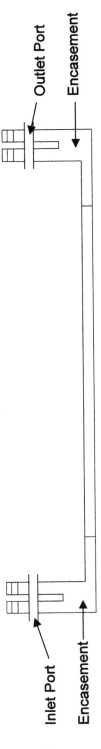

Figure 3. *Continued.*

Inlet and outlet ports were drilled into the exposure cell chamber letting air having a specified temperature and relative humidity (RH) in and out of the chamber. The conditioned air was generated in a humidity-temperature generator and pumped into each exposure cell. The temperature within the exposure chamber was monitored via a thermocouple positioned inside the chamber. The RH of the air flowing into the exposure cells was monitored by rerouting a portion of the air flowing into the chamber through a chilled mirror hygrometer for five minutes out of every 20 minutes. Temperature and RH measurements within each exposure cell were made on a continuous basis throughout the duration of the experiment.

Humidity-Temperature Generators

Two identical temperature-humidity generators were designed and fabricated to provide air of known RH and temperature to the four exposure cells positioned under each solar simulator. Each temperature-humidity generator was capable of providing conditioned air at one temperature and up to four different relative humidities. The RH of the air pumped into each exposure cell could be independently controlled and maintained to within ± 3% between 0% and 95% RH; while the temperature could be maintained at any value from a few degrees Celsius above room temperature to 70° C ± 1° C.

Measurements:

UV-Visible and Infrared Transmittance Measurements

Dosage, D_{dosage}, not time or dose, is the exposure metric used in these experiments. Dosage is the total number of photons absorbed by a pure material whereas the dose is the number of incident photons. To accurately estimate the dose and dosage, the optical properties of all of the components identified in Figures 4 had to be measured and account taken of 1) the spatial, temporal and spectral variability in the radiant flux of the light source and 2) the spectral transmittances of the interference filters, quartz plate, and coated CaF_2 disks for all eight exposure cells. These measurements were made prior to the start of the experiment and at every inspection period.

Measurements of the optical properties of the system components were highly automated (11). Automation was achieved through three devices: 1) a robotic arm connected to a HP 8452A UV-visible spectrophotometer via a fiber optic cable, 2) a 150 mm diameter PIKE automated ring inspection device inserted into the optical beam path of the HP 8452A UV-visible spectrophotometer, and 3) a second 150 mm diameter PIKE automated ring inspection device inserted into the optical beam path of a Perkin Elmer 1760X FTIR spectrophotometer.

The HP 8452A spectrophotometer was calibrated against NIST's standard reference xenon arc lamp at least twice a year. This calibration curve allowed the

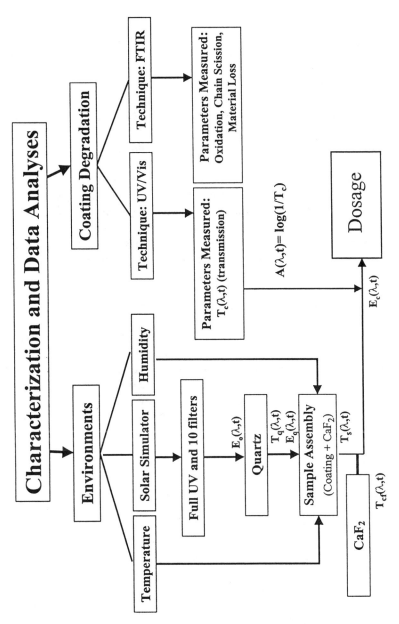

Figure 4. A flowchart displaying the various measurements for making dosage and damage assessments during each inspection.

conversion of the UV-visible signal into photon counts. About midway through the experiments, it became apparent that even a small imprecision in the spectral UV measurements could greatly affect the estimated dosage. To improve the accuracy of these measurements, the UV-visible spectrometer was recalibrated during each inspection against standard UV-visible spectrometer transmittance density filters using NIST Standard Reference Material (SRM 930d) while the FTIR spectrometer was calibrated using a 38.1 μm thick polystyrene film.

Dosage and Apparent Spectral Quantum Yield Calculation Computer Program

A software program [14] was written to estimate dose, dosage, and material damage given the input from the optical components shown in Figure 4. That is, dose and dosage were estimated from the UV-visible spectra of the lamps, the transmittances of the interference filters, quartz disk, and coated CaF_2 disks and from the UV-visible absorbance of the coating specimens. Some of the optical values changed with time. For example, the spectral radiant flux emitted by the xenon arc lamp changed non-uniformly with time as the light source aged, even though the light source was equipped with a light intensity controller. The experimental procedure involved monitoring changes in UV-visible spectra during each inspection. Once the output of the lamp and the transmittance of the interference filters, quartz disk, and coated CaF_2 disk were known, the spectral dose incident on a specimen could be estimated. Knowing the absorbance of the uncoated and coated calcium fluoride disk, the transmittance of the specimen could be determined. Radiant energy incident on the specimen that did not emerge from the other side of the specimen was presumed to have been absorbed by the specimen and was used in estimating dosage. Besides calculating spectral dose and dosage, the software package is capable of analyzing infrared spectra, plotting the damage function, estimating the apparent spectral quantum yield, and estimating the total effective dosage.

Experimental Design:

Specimens were exposed to 16 combinations of temperature and RH at each of 12 different spectral bandwidths. The nominal RH's were 20%, 40%, 70% and 90% while the nominal temperatures were 30° C, 40° C, 50° C, and 60° C. An additional experiment was conducted at 50° C and at RH close to 0%.

All eight exposure cells were simultaneously irradiated. The four exposure cells exposed under the right solar simulator (designed the right exposure in Figure 2) were held at a different temperature than the exposure cells exposed under the left solar simulator. The temperatures usually differed by 10° C. The temperature of each set of four exposure cells irradiated under the same simulator was maintained at approximately 1° C to 2° C above the specified exposure cell temperature. This was achieved by surrounding the exposure cells with 8 mm thick poly (methyl methacrylate) walls and heating the interior space. Increasing the temperature around the exposure cells was necessary to prevent the formation of condensation of water on the top surface of the coating within an exposure cell for the high humidity exposures.

Results and Discussion

Environmental Characterization

Spatial Uniformity, Collimation and Temporal Stability of UV-Visible Light Source
    Results on the spatial uniformity, collimation, and temporal stability of the UV-visible light source have been graphically and tabularly presented in Reference 11. In general, the spectral radiant flux differed from one solar simulator to another and differed from one lamp to another.

    One example is displayed in Figures 5a and 5b for the spatial radiant flux distributions at 380 nm of the right and left solar simulators. The mean and standard deviation for radiant fluxes at 380 nm for the two solar simulators were 0.13 W/m^2 ± 0.014 W/m^2 and 0.19 W/m^2 ± 0.026 W/m^2, respectively. The radiant flux from the right solar simulator was always 50 % less than the radiant flux from the left simulator; even when the same xenon arc source was used in both simulators. These spatial irradiance differences were measured at 110 locations under the two simulators at each inspection time using the robotic arm positioned underneath the exposure cell table. This data was used in the dose and dosage software program to calculate the incident radiation at every window for each exposure cell.

    The spectral radiant flux of the xenon arc light sources also changed non-uniformly with time. For example, after 1500 h of lamp operation, the spectral radiant flux decreased with respect to initial values at 290 nm, 326 nm and 450 nm by 25%, 50%, and 38%, respectively. These changes demonstrate the need for frequent monitoring of the spectral radiant flux emitted by the light source.

    Beam collimation and scattering of the radiation were measured in both solar simulators. Measurements using the fiber optic probe positioned at the end of the robotic arm and connected to the UV-visible spectrometer. The degree of collimation was ascertained by measuring the radiant flux below the center points of several randomly selected windows. Measurements were made while displacing the fiber optic probe in 10 mm increments until the sensor was 100 mm from the initial position. Radiant flux measurements were plotted against vertical distance and regression analysis was performed. The slope of the regression curve was not significantly different from zero suggesting that the radiant output from each solar simulator was highly collimated. The absence of stray light incident on the covered

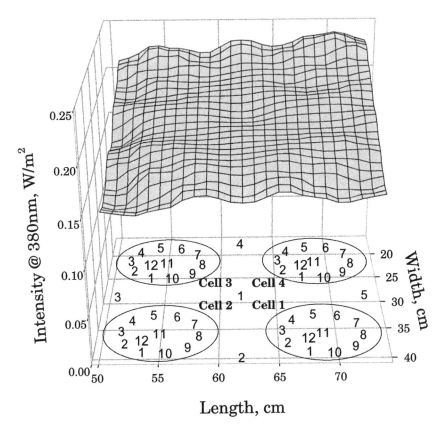

Figure 5. Spatial irradiance uniformity of (a) the right and (b) the left solar simulators

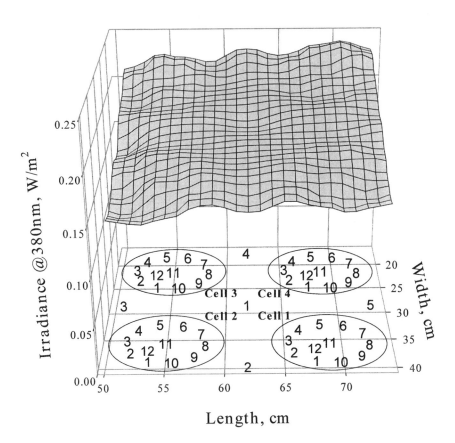

Figure 5. *Continued*

12th window was verified by positioning the fiber optic probe immediately beneath this window. No UV-visible radiation was detected.

Temporal Stability of Interference Filters, Quartz Disk and Calcium Fluoride Disk
The CaF_2 did not exhibit any loss of transmission over time. The interference filters were also photolytically and thermally stable except for the 450-nm filter; the loss of transmission for this filter ranged from 2% to 7% after six months of exposure. The quartz plates showed a small transmission loss over the course of exposure. All of these changes were accounted for in the software program used for calculating dose and dosage.

Temporal Control of Temperature and Relative Humidity Within the Exposure Cells
Both temperatures and relative humidities were well controlled. The standard deviations of the four RH levels studied (20 %, 40 %, 70 % and 90 %) ranged between 3 % and 7 % and the standard deviations of the four temperatures (30 °C, 40 °C, 50 °C, and 60 °C) ranged between 0.4° C to 1° C (11). With further refinement of the temperature-humidity generators, relative humidity was controllable to within ± 2%.

Coating Degradation

Photodegradation results for the acrylic melamine coating are presented in this section to demonstrate the approach for measuring material damage and calculating dosage. Figure 6a and 6b show FTIR spectra of the coating before and after aging for 2040 h under full UV at 50° C and close to 0 % RH, and Figure 6c is the difference spectrum between the aged and unaged samples. The major FTIR bands of interest in the cured, unaged acrylic melamine coating are the absorptions near 3530 cm⁻¹, due to OH stretching of the acrylic resin, near 3380 cm⁻¹, due to OH/NH stretching of the melamine resin (15), and at 1085 cm⁻¹ and 1015 cm⁻¹, due to C-O stretching. The band near 1555 cm⁻¹ has been assigned to the contribution of three different groups: triazine ring, CN, and CH_2 (16). Although not visible in Figure 6a, two important bands that are frequently used for curing and degradation analyses of melamine-based coatings are the bands at 910 cm⁻¹ (OCH_3) and at 815 cm⁻¹ (the triazine ring out-of-plane deformation). From Figure 6a, the cured film still contained a substantial amount of unreacted OH groups of the acrylic resin, and unreacted methylol or imino and OCH_3 groups of the melamine crosslinking agent. The presence of these groups in the cured films can affect both the photolysis and hydrolysis of acrylic melamine coatings (9,17,18).

The difference spectrum given in Figure 6c indicates that the coating has substantially degraded after 2040 h exposure. This is evidenced by the decreases in intensity in the 2750 cm⁻¹ - 3050 cm⁻¹ and 1600 - 1000 cm⁻¹ regions and increases in the intensity in the 3100 cm⁻¹ - 3400

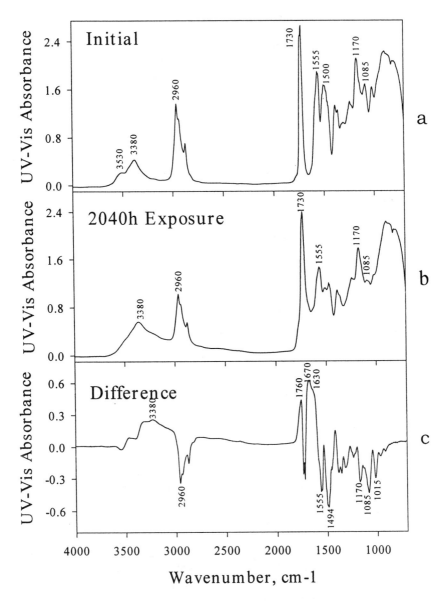

Figure 6. FTIR spectra of acrylic melamine coating (a) unaged, (b) after exposure for 2040 h in full UV, 50° C and close to 0% RH, and (c) difference spectrum (b minus a).

cm^{-1} regions and of the bands near 1760 cm^{-1}, 1670 cm^{-1}, and 1630 cm^{-1}. Analysis of amide model compounds and D$_2$O treatment of the UV-degraded specimens suggest that the band at 1670 cm^{-1} is due to the C=O of an amide and that the band at 1630 cm^{-1} is due to the NH bending of primary amines or primary amides. These assignments are consistent with Lemaire's scheme of photodegradation for acrylic melamine coatings (19) and with literature assignments of functional groups of amine and amide compounds (20). Complete assignments of the cured coating before and after exposure to the environments are given in Reference 18.

In this study, we used the amide C=O band at 1670 cm^{-1} and the C-O band at 1085 cm^{-1} to follow the oxidation and crosslink chain scission, respectively. Bauer and coworkers have utilized the integrated area in the 1550 cm^{-1} - 1750 cm^{-1} region (8), and later the intensity of the 1630 cm^{-1} band (9), to follow the photo-oxidation of acrylic melamine coatings exposed to UV conditions. Figures 7a and 7b depict FTIR intensity changes of the 1670 cm^{-1} and 1085 cm^{-1} bands, respectively, as a function of time for different spectral UV-visible conditions at 50° C and 0% RH. The changes in these band intensities were due to photolysis because post-curing effects have been removed. The photodegradation of this coating exposed to the full UV condition (inset) was much greater than that of any of the ten filters. Under full UV, the rates of both photo-oxidation and crosslink chain scission were almost constant with exposure time after an initial slow down. Further, under these exposure conditions, nearly 25 % of the C-O groups were lost after three months of exposure.

Total Effective Dosage and Apparent Spectral Quantum Yield

Figures 8a and 8b display photo-oxidation (1670 cm^{-1}) and crosslink chain scission (1085 cm^{-1}) as a function of dosage for specimens exposed to full UV, several filters and at 50 °C and close to 0 %RH. The dosage was calculated using Equation 1. Specimens exposed under the 294 nm, 300 nm, and 306 nm filters exhibited the least damage per unit dosage while specimens exposed under the 326 nm filter exhibited the greatest damage per unit dosage, despite the fact that the highest dosage was recorded at the 354 nm filter.

The apparent spectral quantum yield is taken as the initial slope of the damage/dosage curve. The use of this initial slope avoids complications from multiple degradation processes and from potential shielding of the coating material by degradation products absorbing radiation. In the process of determining this slope, a fourth-order polynomial function was fitted to the damage/dosage curve. In this first implementation of analyzing this curve, the slope at the 10 % of the dosage was used as the point at which to estimate the apparent quantum yield. One example of such fitting is shown in Figure 9 for the 1085 cm^{-1} and 1670 cm^{-1} bands, where the symbols are the experimental data and the solid lines are the fitted curves. The initial part of the damage/dosage curves at 1085 cm^{-1} exhibits considerable scatter due to

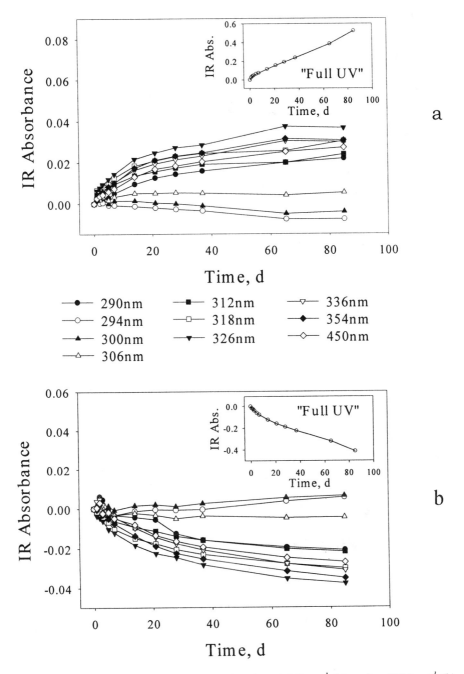

Figure 7. FTIR intensity changes of the bands at 1670 cm⁻¹ (a) and at 1085 cm⁻¹ (b) versus exposure time. Exposure conditions were 50° C and close to 0% RH.

138

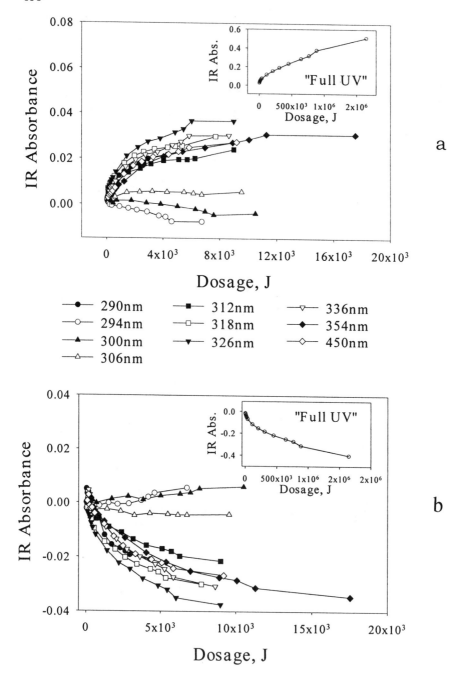

Figure 8. Oxidation (a) and crosslink chain scission (b) as a function of dosage for an acrylic melamine coating exposed to 50° C and close to 0% RH to full UV and under the 10 interference filters.

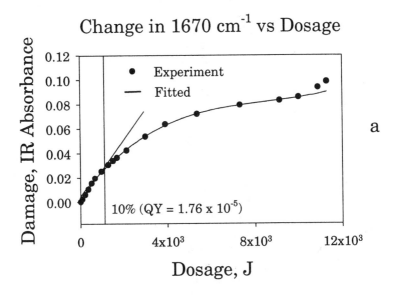

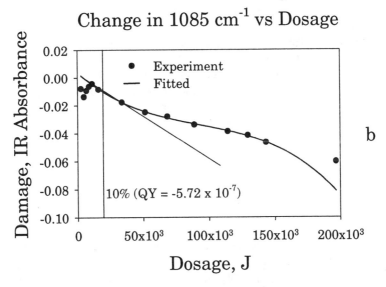

Figure 9. Experimental and fitting FTIR degradation data as a function of dosage for 1670 cm⁻¹ and 1085 cm⁻¹ bands. The initial slopes of these curves are used for obtaining the apparent spectral quantum yield.

imprecision in the measurements of very small changes at early exposures. Nevertheless, the polynomial function appears to fit most of the data well. Differentiating the polynomial gives the apparent quantum yield at a given dosage. Since the damage/dosage relationship is not linear, the apparent spectral quantum yield falls off with dosage, presumably due to increasing shielding of the coating by degradation products and to the consumption of weak links in the material.

Figure 10 presents typical apparent spectral quantum yield curves for the oxidation and crosslink chain scission of coatings exposed at 50 ° C and close to 0 %RH under full UV and the ten interference filters. The apparent quantum yield at 290 nm was the highest. In general, the apparent quantum yields in all UV exposure conditions for this coating were small, as is typical for solid polymers. Gupta et al. (21) have reported apparent quantum yield values (number of scissioned molecules per number of quanta absorbed by the polymer) in the range of 0.003-0.007 ± 0.002 for polycarbonates exposed to 300 nm to 400 nm wavelength radiation at temperatures 0° C and 55° C and at RH between 0 % and 100 %; whereas the apparent quantum yield of the same material irradiated in solution has been reported to be 0.18 (22). Dan and Guillet (23) observed that quantum yield of the chain scission in glassy polymers increases rapidly to the value obtained for the same polymer in solution as the temperature is increased to a temperature above the glass transition temperature.

Estimates of the total effective dosage and apparent spectral quantum yield were found to be very sensitive to a number of experimental variables, including the initial UV-visible absorbance of the coatings, formation of coating degradation products, anomalies in the spectral UV absorption of the substrate, and UV-visible and FTIR measurement errors. The effects of these variables on the dosage and spectral apparent quantum yield are reported in Reference 24. Protocols have been developed and experimental errors had been essentially eliminated through careful and frequent re-calibration of the UV-visible spectrophotometer using NIST's Standard UV-visible spectrometer transmittance density filters and polystyrene films to monitor the performance of the FTIR spectrometer before taking the spectra of any samples.

Conclusions

In this paper, a total effective dosage model has been presented that may have potential for linking field and laboratory photodegradation results for polymeric materials. An experimental apparatus and the experimental protocols for estimating the model coefficients have also been described.

The total effective dosage model has a basis in the principles of photochemistry and has had extensive application in the biological community. The primary inputs into the model are dose, dosage and material damage. Given this input, the apparent spectral quantum yield and the total effective dosage for a study material can be estimated.

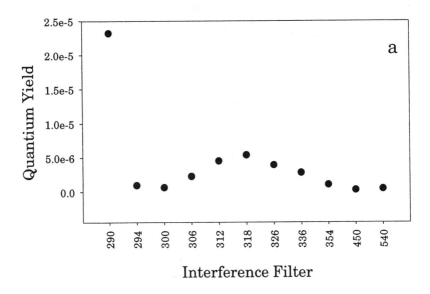

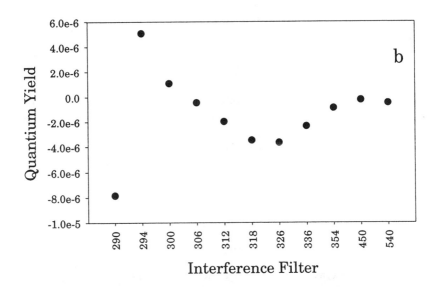

Figure 10. Quantum yields of the oxidation (a) and crosslink chain scission (b) of an acrylic melamine coating exposed to 50° C and close to 0% RH for full UV and 10 different wavelengths.

The exposure apparatus was built so that the primary exposure factors could be independently and accurately controlled and monitored within very tight bounds over both time and space. These factors include spectral UV radiation, temperature, and RH. Collimated UV radiation was produced by two 1000 W xenon arc solar simulators. Narrow band interference filters were used to segment the xenon arc spectrum into narrow bandwidths. Spectral damage to the coating was measured using Fourier transform infrared spectroscopy. Damage to the coating was attributed to hydrolysis, photolysis, and moisture-enhanced photolysis.

Experiments were conducted on the model coating covered a wide range of temperature and RH exposure conditions. A full 4x4 factorial experiment was conducted at four temperatures (30° C, 40° C, 50° C and 60° C) and four relative humidities (20 %, 40%, 70 %, and 90 %) and at each of twelve spectral wavebands. An extra experiment was conducted at 50° C and 0 % RH.

It was quickly determined that estimates of the total effective dosage and apparent spectral quantum yield were sensitive to a number of experimental variables including the formation of coating degradation products and UV-visible and FTIR measurement errors. To improve the precision of the measurements, both the UV-visible and the FTIR spectrometers were recalibrated during every inspection period. The spectral data was input into a computer program to estimate the dosage, the apparent spectral quantum yield, and the total effective dosage. This program provided near real-time updating of the total effective dosage model parameters by combining data from the current inspection period with historical data stored in the computer from previous inspections.

The total effective dosage model appears to be a good model for linking field and laboratory exposure results because the exposure apparatus built for this experiment greatly reduces temporal and spatial variations in the intensity of the three primary weathering factors – spectral ultraviolet radiation, temperature and relative humidity--it also allows separation of these effects. It was also shown that the experimental protocol for measuring or estimating the parameters of the total effective dosage model could be automated and that the outcome of these experiments for the study acrylic melamine coating are consistent with published results.

Acknowledgements:

The research reported here is part of a government/industry/university consortium on Service Life Prediction at NIST. Companies involved in this consortium include AKZO Nobel, ATOFINA, Atlas Electric Devices Inc., Dow Chemical, Dupont Automotives, Duron Inc., Eastman Chemicals, Millennium Inorganic Chemicals, PPG, and Sherwin Williams Co. Federal Highway Administration and Wright Patterson AFB provided additional funds for this research.

References

1.   A.L. Norins, in F. Urbach (Ed.), The Biological Effects of Ultraviolet Radiation, Pergamon Press, N.Y., 1969, p. 605.
2.   R. Setlow, Advanced Biological and Medical Physiology 5: 37 (1957).
3.   R.M. Sayre, R.L Olson, and M.A. Everett, M.A, Journal of Investigative Dermatology 46: 240 (1966).
4.   W.J.Clapson and J.A. Schaeffer, J.A., Journal of Industrial and Chemical Engineering, 26(9): 956 (1934).
5.   G.Z. Zerlaut and M. L. Ellinger, J. Oil Colour Chem. Ass., 64, 387(1981)
6.   G.A. Zerlaut, Accelerated and Outdoor Durability Testing of Organic Materials, W. D. Ketola and D. Grossman (Eds.), ASTM STP 1202, American Society for Testing and Materials, Philadelphia, PA, 1993, p. 3.
7.   D.M. Grossman, Accelerated and Outdoor Durability Testing of Organic Materials, W. D. Ketola and D. Grossman (Eds.), ASTM STP 1202, American Society for Testing and Materials, Philadelphia, PA, 1993, p. 68.
8.   D.R. Bauer, J.L. Gerlock, and D.F. Mielewski, Polym. Deg. Stab, 36,9 (1992).
9.   D. Bauer and D.F. Mielewski, Polym. Deg. Stab., 40, 349 (1993).
10.   J. D. Rancourt, Optical thin Films, User's Handbook, McGraw-Hill, NY, 1987, Chapter 6, p. 183.
11.   J.W. Martin, E. Byrd, E. Embree, and T. Nguyen, A Collimated, Spectral Ultraviolet Radiation, Temperature, and Relative Humidity Controlled Exposure Chamber, in review.
12.   K. Klein and B. Goldberg, Proc. Int. Solar Energy Soc. Conference, 1978, Vol. 1, p. 400.
13.   G. Kampf, K. Sommer, and Zirngiebl, Prog. Org. Coat., 19, 69 (1991).
14.   B. Dickens and E. Byrd, to be published as a NIST technical note.
15.   D. R. Bauer and R.A. Dickie, J.Appl. Polym. Sci., 18, 2014 (1980).
16.   P.J.Larkin, M.P. Makowski, and N.B. Colthup, Vibrational Spectroscopy 17(1998) 53.
17.   D.R. Bauer, J. Appl. Polym. Sci., 27, 3651 (1982).
18.   T. Nguyen, J. Martin, E. Byrd, and E. Embree, submitted for publication in the Journal of Coatings Technology.
19.   J. Lemaire and N. Siampiringue, Service Life Prediction of Coatings and Materials, J. Martin, L. Floyd, D. Bauer, (Eds.), ACS Symposium Series 722, American Chemical Society, 1998, p. 246.
20.   N.B. Colthup, L.H. Daly, and S.E. Wiberley, Introduction to Infrared and Raman Spectroscopy, 3 rd Ed., Academic Press, N.Y., 1990, p. 439.
21.   A. Gupta, A. Rembaum, and J. Moacanin, Macromolecules, 11, 1285 (1978).
22.   A. Gupta, A. R.Liang, J. Moacanin, R. Goldbeck, and D. Kliger, Macromolecules, 13, 262 (1980).
23.   E. Dan and J.E. Guillet, Macromolecules, 6, 230 (1973).
24.   J. Martin, T. Nguyen, E. Byrd, B. Dickens, and E. Embree, submitted for publication in Polymer Degradation and Stability.

Chapter 8

Integrating Sphere Sources for UV Exposure: A Novel Approach to the Artificial UV Weathering of Coatings, Plastics, and Composites

Joannie W. Chin, W. Eric Byrd, Edward Embree, and Jonathan Martin

Building Materials Division, National Institute of Standards and Technology, 100 Bureau Drive, Stop 8621, Gaithersburg, MD 20899-8621

The primary method for obtaining laboratory weathering data for a wide range of commercial polymer products including coatings, textiles, elastomers, plastics and polymeric composites is through the use of ultraviolet radiation exposure chambers (UV chambers). Although numerous improvements have been made in the design of UV chambers over the last 80 years, the repeatability and reproducibility of the exposure results from these chambers have remained elusive. This lack of reproducibility and repeatability is attributed to systematic errors in their design, operation, and control which, in turn, have prevented comparisons of the performance of materials exposed in the same environment, comparisons of the performance of the same material exposed in different laboratories, and the comparison of field and laboratory results. This paper describes an innovative UV chamber design having a basis in integrating sphere technology that greatly reduces the magnitude of these errors, as well as provides additional experimental capabilities.

Introduction

Ultraviolet radiation exposure chambers (UV chambers) are the primary means for generating laboratory weathering data for a wide range of commercial products including coatings, elastomers, plastics and polymeric composites (collectively termed *construction materials*) (*1*). Over the years, numerous technical improvements have been implemented in the design, construction, and control of these UV chambers. However, the repeatability and reproducibility of the exposure results obtained from these chambers have remained elusive (*2,3,4,5,6*).

The repeatability and reproducibility of exposure results are affected not only by variability in material response, but also by *experimental* and *systematic* errors. Experimental errors and material variability, which are present in all experiments, are random in nature and can be compensated for through the use of proper experimental designs.

Systematic errors, on the other hand, are uncompensated, non-random sources of error which bias experimental results. Common sources of systematic errors associated with existing UV chambers include human/machine interactions, high specimen temperatures, non-uniform irradiance over the surface of a specimen, and temporal changes in exposure conditions. These errors can be minimized by standardizing test procedures, making changes in existing exposure equipment, or circumventing them through the use of alternate UV chamber designs. In this paper, known sources of systematic errors are discussed and an innovative UV chamber, based on integrating sphere technology, which may be capable of reducing the magnitude of these systematic errors, is presented. In addition, it will be shown that the use of an integrating sphere-based UV chamber allows for greater experimental capability in terms of testing conditions.

Current Laboratory UV Weathering Instrumentation

Commercially available UV chambers began to appear circa 1920. Atlas Electric Devices[†] introduced its carbon arc "Fade-O-Meter" in 1918; several years later, Nelson (*7*) published preliminary exposure results for a UV chamber

---

[†] Certain trade names and company products are mentioned in the text or identified in an illustration in order to adequately specify the experimental procedure and equipment used. In no case does such an identification imply recommendation or endorsement by the National Institute of Standards and Technology, nor does it imply that the products are necessarily the best available for the purpose.

design using a mercury arc lamp in 1922, and Buttolph (8) patented several modifications to Nelson's UV chamber in 1924.   UV chambers containing fluorescent lamps were introduced at a later date.

Numerous modifications in the early UV chamber designs have been made over the last 80 years.   These modifications have been aimed at improving the repeatability and reproducibility of exposure results and include:

- The identification of a more temporally stable ultraviolet light source
- The identification of spectral radiant power distribution which more closely approximates the maximum solar ultraviolet radiant power (9,10,11,12), and
- The re-design of the exposure racks within a chamber to improve the spatial irradiance uniformity over the dimensions of a specimen and among specimens.

Specific examples of improvements include:

- Identification of cut-off filter combinations to remove radiation below 290 nm
- The introduction of photopic sensors and feedback-control devices for minimizing temporal changes in the radiant power (13,14,15), and
- The introduction of the three-tier exposure racks in xenon arc UV chambers.

These changes have greatly reduced the variability in exposure results, but they have not fully resolved issues related to the lack of repeatability or reproducibility (5,6).

Factors Affecting Reproducibility and Repeatability of Current UV Chambers

Common sources of systematic error found in current UV chambers include human/machine interactions, unnatural exposure conditions, non-uniform irradiance over the dimensions of a specimen and among specimens, the inability to accurately and precisely measure ultraviolet radiation dose, and temporal changes in exposure conditions. Of these sources of error, unnatural exposure conditions, non-uniform irradiance and temporal changes in exposure conditions have been targeted by NIST researchers as those which could be mitigated by the use of a novel integrating sphere-based UV chamber.

Unnatural Exposure Conditions

Xenon arc lamps are almost always operated at a higher current density than mercury arc lamps because of the smaller cross-sectional area of the xenon atom relative to that of the mercury atom (16). This high current density makes the xenon arc a very "hot source" (17,18); that is, xenon arcs emit a substantial amount of energy in the visible and infrared regions. As such, the temperature of exposed specimens has been observed to approach 60 °C (19,20).

Improved temperature control of UV chambers can be achieved by removing the primary source of thermal energy, visible and infrared radiation, while maintaining the photolytically effective ultraviolet radiation. This can be accomplished by introducing a heat controlling optical element (e.g., a dichroic mirror) between the light source and the specimen. Dichroic mirrors or filters are designed to transmit (or reflect) ultraviolet radiation through (or off of) the dichroic mirror and onto the specimens while reflecting (or transmitting) the visible and infrared portions to a heat sink.

Ultraviolet radiation with a wavelength less than 290 nm does not reach the earth's surface (21,22); thus the presence of radiation with wavelengths below 290 nm may induce "unnatural" chemistry in materials. Efforts have been made by UV chamber manufacturers to eliminate these wavelengths by carefully researching light sources and utilizing cut-off filters. Although not a perfect match, certain grades of fluorescent lamps or xenon arc lamps equipped with borosilicate/borosilicate filters appear to be a close approximation to the solar spectrum (12).

Temporal Variation

The radiant power output from an arc source is known to be unstable over time. This temporal instability can be attributed to both equipment variables and changes in the light source as it ages (light source plus the filters).

The major equipment variable affecting radiant power output is a change in the electrical current density over time. The higher the current density, the higher the temperature of the plasma and, thus, the greater the radiant power. Also, the higher the current density, the higher the proportion of the total radiant power emitted in the ultraviolet region (23,24). Thus, temporal stability of a lamp depends heavily on the stability of the lamp power supply.

The optical properties of the arc lamps, filters (in the case of xenon arc lamps), and phosphors (in the case of fluorescent lamps) change with time through an aging process. The tungsten electrodes operate near the melting temperature of tungsten when the lamp is on, thus causing tungsten to be sputtered off the electrodes and deposited onto the interior walls of the glass cylinder (*16,25*). These deposits reduce the spectral transmittance properties and eventually cause the glass envelope to devitrify and crack (*23*). In commercial xenon arc chambers, efforts have been made to minimize this effect by placing a metallic sleeve around the anode.

The glass enclosure surrounding the arc also ages. This can occur whenever the arc plasma touches the enclosure, causing it to melt and to deposit silicon onto the interior surface of the glass. The quartz and other glass filters surrounding a xenon arc also age through a process called solarization (*26*). Solarization is the reduction in transmittance of a filter resulting from the exposure to short wavelength UV radiation and to high temperatures.

Spatial Irradiance Uniformity

Ensuring spatial irradiance uniformity over the dimensions of a specimen and between specimens is a prime consideration in designing any optical system. Spatial uniformity is needed to determine the spectral ultraviolet radiation dosage received by a specimen. Spectral ultraviolet radiation dosage must be known in order to compare the performance of materials exposed in the laboratory and those exposed in the field (*27,28*).

Spatial irradiance uniformity is difficult to attain in current UV chambers due, in part, to the larger surface area over which uniformity must be controlled. Factors affecting irradiance uniformity include reflectance from the specimen and walls of the chamber and physical limitations imposed by the optical system (e.g., the geometry of the light source and the dimensions of the specimens). The remainder of this paper addresses a proposed solution to the problem of spatial irradiance uniformity in conventional UV chambers, through the use of an integrating sphere as a UV source.

Integrating Sphere Theory

The theory of integrating spheres, as well as their uses in a wide variety of applications, is well-established (*29,30,31*). An integrating sphere is a hollow spherical chamber with an inner surface coated with a diffuse reflecting, or *Lambertian*, coating. Light entering an integrating sphere undergoes multiple diffuse reflections at the interior surface, resulting in an uniform field of light within the sphere. This collected light can then serve as a means of measurement or as a source of uniform illumination. This latter function will be utilized in the novel UV weathering device.

Integrating sphere theory has its origin in the theory of radiation exchange between diffuse surfaces (*32*). Although the general theory can be complex, the symmetry of the sphere simplifies the analysis. Consider the exchange of radiation between two differential elements of diffuse Lambertian surfaces A_1 and A_2, with areas dA_1 and dA_2 as shown in Figure 1. The fraction of the total flux leaving A_1 (Φ_{A_1}) and arriving at A_2 (Φ_{A_2}) is given by:

$$\frac{\Phi_{A_2}}{\Phi_{A_1}} = dF_{1-2} = \frac{\cos\theta_1 \cos\theta_2}{\pi S^2} dA_2 \tag{1}$$

where dF_{1-2} is the exchange factor. If these two surface elements are contained inside a diffuse sphere surface, and $S = 2R\cos\theta_1 = 2R\cos\theta_2$, as shown in Figure 2, then equation 1 becomes:

$$dF_{1-2} = \frac{dA_2}{4\pi R^2} \tag{2}$$

Equation 2 is independent of the locations of the two elements and as well as the distance between the elements. This result is significant because it states that the fraction of flux received by A_2 is the same for every other point on the sphere surface.

If the differential areas dA_1 and dA_2 become finite areas A_1 and A_2, then Equation 2 becomes:

$$dF_{1-2} = \frac{1}{4\pi R^2} \int_{A_2} dA_2 = \frac{A_2}{4\pi R^2} \tag{3}$$

Equation 3 is also independent of dA_1, allowing it to be expressed as:

$$dF_{1-2} = \frac{A_2}{4\pi R^2} = \frac{A_2}{A_s} \tag{4}$$

where A_s is the total surface area of the sphere. Thus, the fraction of energy received by A_2 is equal to the fraction of surface area that it takes up within the sphere.

The physical significance of the previous analysis is that every point on a sphere is equally illuminated by reflections from every other point. This leads to the conclusion that theoretically, not only should the output intensity be uniform across the plane of an exit port, but every exit port should have the same spectral output as every other exit port. This property of integrating spheres will be

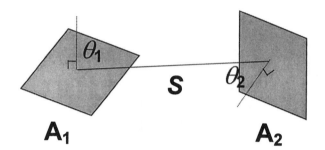

Figure 1. *Exchange of radiation between two diffuse surface elements*

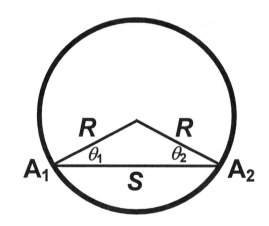

Figure 2. *Exchange of radiation in a spherical enclosure.*

exploited to produce a artificial UV weathering device with improved spatial irradiance uniformity. Thus, it can be seen that the use of an integrating sphere as a uniform UV source can potentially resolve one of the major problems in current UV chamber design; that is, non-uniform irradiance across the dimensions of a specimen and from specimen to specimen.

The throughput, at a given port, of an integrating sphere with multiple ports is given by:

$$\frac{\Phi_o}{\Phi_i} = \frac{\rho f_e}{1 - \rho(1 - f_{tot})} \tag{5}$$

where Φ_i is the input flux, Φ_o is the output flux, ρ is the average sphere wall reflectance, f_e is the fraction of the sphere surface area taken up by the exit port of interest, and f_{tot} is the total fraction of the sphere surface area taken up by all of the exit ports. This equation assumes that no portion of the input flux is directly incident on any of the exit ports. It is generally recommended that f_{tot} be less than 5% if sphere output uniformity is critical. If a known input flux is known, the output flux of an actual sphere can be calculated via equation 5.

In order for the above-mentioned equations to be correct and for the successful use of an integrating sphere as an optical device, it is critical that the sphere interior surface scatter light in a perfectly Lambertian fashion over the wavelength region of interest. If the scattering is non-Lambertian, then the basic assumptions for the standard integrating sphere analysis are violated. The effects of non-Lambertian reflectance on integrating sphere measurements are presented by Hanssen (33).

In the wavelength regions of interest, namely 290 nm to 400 nm, materials which provide the greatest degree of Lambertian reflectance are based on barium sulfate powder or polytetrafluoro-ethylene (PTFE). The reflectance of pressed PTFE powder has been studied by Weidner and Hsia (34), and was measured to be > 0.98 in the region from 250 nm to 2000 nm. In recent years, a solid, machinable PTFE material has been developed with the high reflectance of the powder, but with greater durability and resistance to temperature, moisture and corrosive chemicals. PTFE maintains its reflectance indefinitely under normal laboratory conditions if not contaminated and does not need to be repacked or recoated (35). In the event that reflectance does decrease over time due to surface contamination, the material can be sanded or vacuum baked (36) to regenerate its original reflectance.

Integrating Sphere Based UV Chamber

NIST researchers are currently developing a novel integrating sphere-based UV chamber, and have procured a 2 m diameter integrating sphere. A photograph and schematic representation of the NIST 2 m sphere are shown in Figures 3 and 4, respectively. The sphere construction is based on a modular panel design, which allows individual panels to be removed for modification or repair. The interior of the sphere is lined with PTFE panels, and currently contains thirty-two 11.2 cm diameter ports, and a 61 cm diameter top port to accommodate a high intensity UV light source.

The proposed lamp system for the 2 m sphere is a microwave-powered, electrodeless lamp system with an output which is rich in the region between 290 nm and 400 nm. A multiple lamp system will be utilized, which has been calculated to provide an output greater than 50 equivalent suns (where a "sun" is defined as the integrated irradiance between 305 nm and 400 nm taken from the direct normal spectral solar irradiance distribution in ASTM G159). A multiple lamp system, in contrast to a one lamp system, allows the response of a material to various irradiance levels to be evaluated, thus providing opportunities to test the law of reciprocity.

Dichroic reflectors in the lamp housings will remove (80 to 90) % of the infrared and visible output from the lamps. In the event that wavelengths > 400 nm are predicted to be instrumental in the photodegradation of certain materials, the dichroic reflectors can be replaced with conventional reflectors, thus retaining a greater proportion of the visible spectrum. Figure 5 shows the output spectrum for the lamp system with the dichroic reflectors in place.

As was discussed earlier, due to the fact that every point on the sphere surface is theoretically equivalent to every other point, the monitoring of UV spectral intensity is quite simple. For instance, a fiber optic probe could be inserted at an arbitrary point in the sphere wall (away from the "first strike" region of the light source) and connected to a spectroradiometer to provide a measure of the spectral radiance within the sphere. Through photofeedback processing, lamp power can be continually adjusted to compensate for temporal instabilities in, and diminishment of lamp output, over time.

This design appears to be capable of mitigating most, if not all, of the systematic errors from unnatural exposure conditions and spatial irradiance non-uniformity discussed earlier. Prior to the design and fabrication of the 2 m integrating sphere, a prototype 50.8 cm sphere was utilized to determine the suitability of integrating spheres for use in artificial weathering devices. Spectral UV measurements were made in the center of two 12.7 cm diameter apertures of this sphere, a schematic of which is shown in Figure 6. Spectral output from the two exit ports for a 1000 W xenon arc source are overlaid in Figure 7. It should be pointed out that no discernible difference in the spectral output is apparent.

Figure 3. NIST 2 m integrating sphere.

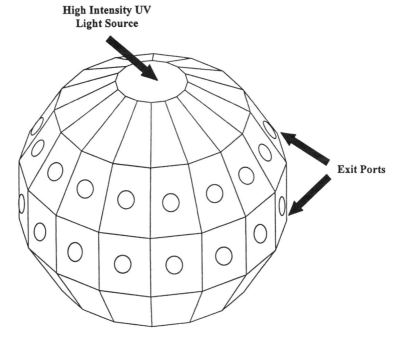

Figure 4. Schematic of 2 m integrating sphere.

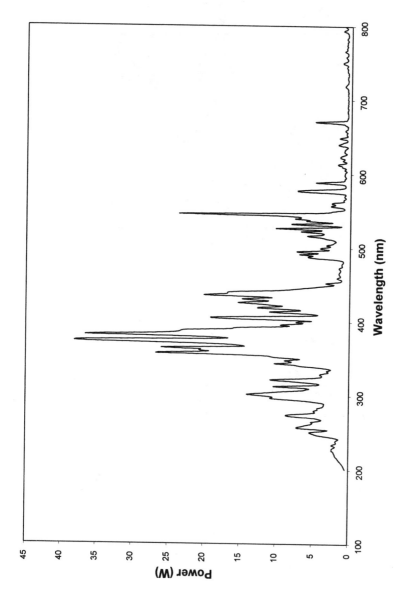

Figure 5: Spectral output for lamp system used in 2 m integrating sphere

Right exit port

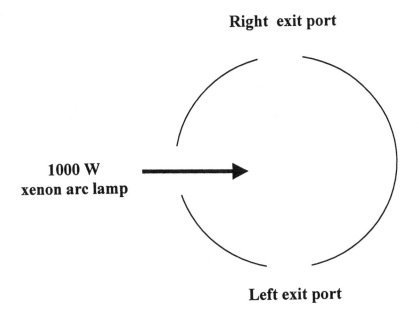

1000 W xenon arc lamp

Left exit port

Figure 6. Schematic of 50.8 cm integrating sphere.

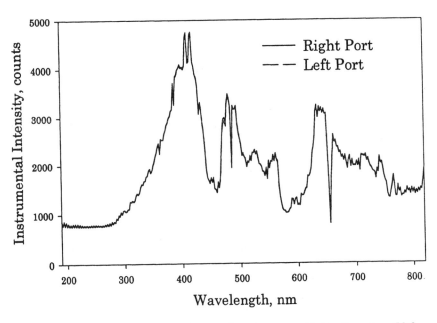

Figure 7. Comparison of output from right exit port and left exit port on 50.8 cm integrating sphere.

In order to study the environmental durability of materials used in building and construction applications, it would be advantageous to uniformly irradiate specimens under a variety of conditions. This could be accomplished by equipping each port with a specimen chamber in which temperature, relative humidity, mechanical loading and other factors can be independently controlled.

Because each chamber is independent of the others, a multiplicity of environmental conditions can be evaluated in a given experiment. While the UV irradiance would be identical between the ports, narrow band-pass filters or neutral density filters can be installed at the exit port to study the effect of a narrow wavelength region or to adjust the intensity, respectively. For instance, it would be possible to expose a specimen at one exit port to 60 °C, 95 % relative humidity, polychromatic light; whereas, at another exit port, the specimen could be exposed to 50 °C, 95 % relative humidity, and 290 nm radiation. The capability to apply mechanical stresses to the specimens while they are undergoing UV exposure can also be achieved, as shown in Figure 8. Other unique exposure environments can also be created, including freeze/thaw cycling and acid rain.

It will be necessary to situate the specimen chamber at some distance from the integrating sphere exit port to accommodate specimens which need to undergo mechanical loading, due to the physical space taken up by the loading frame and specimen grips. In order to convey the highly uniform radiation from the sphere exit port to the above-described specimen chambers without loss of uniformity and with minimal loss in intensity, non-imaging optical devices will be utilized. Such non-imaging optical devices are also known as compound parabolic concentrators, Winston cones, and cone concentrators, and are considered to be more efficient than conventional image-forming optics in concentrating and collecting light. These devices date back to the 1960's and were once used for solar collection; a detailed treatment of this subject is given by Welford and Winston (37). A typical cone concentrator is shown in Figure 9. In the application at hand, they will be used to collimate the diffuse output from the sphere exit port to within 20° and transfer it to the specimen surface.

Summary

UV chambers play an important role in comparing or predicting the performance of construction materials and determining the effect of different weathering factors on the performance of a construction material. Although significant modifications have been made in current UV chamber designs, controlling the systematic errors and thus the repeatability and reproducibility of these chambers has remained elusive. An integrating sphere UV chamber design

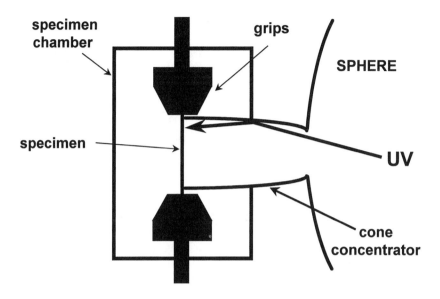

Figure 8. Proposed specimen chamber with capability for specimen loading.

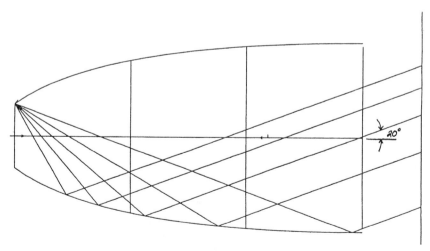

Figure 9. Schematic of typical cone concentrator.

is being developed which appears to be capable of mitigating known sources of systematic errors. Assuming that the sphere is properly designed and that the UV lamps are correctly integrated, uniform irradiance across the dimensions of a specimen and from specimen to specimen is assured by integrating sphere physics.

The improvements offered by an integrating sphere UV chamber over conventional weathering instrumentation are summarized as follows:

- Simplicity in design and easy accessibility to the light source, exposure cells and specimens.
  The light source, exposure cells, and specimens are located on the exterior of the integrating sphere and, thus, they are readily accessible even while an experiment is in progress.

- Removal of visible and infrared radiation from the radiant flux
  High specimen exposure temperatures can be minimized by removing most of the visible and infrared portions of the radiant energy flux emitted by the light source prior to entering the sphere. By removing most of the visible and infrared radiation emitted by the light, the temperature within the chamber can easily be maintained at slightly above room temperature.

- Removal of radiation below 290 nm.
  By positioning a cut-off filter after the dichroic mirror, radiation below 290 nm can be removed from the radiant flux. Moreover, interference or cut-off filters can be positioned in front of each specimen so that each specimen can be uniquely irradiated using any combination of wavelengths.

- Spatial irradiance uniformity.
  As a uniform radiation source, an integrating sphere is capable of minimizing errors due to non-uniform irradiance over the dimensions of a specimen and from specimen-to-specimen. Spatial irradiance uniformity does not depend on the light source, the age of the light source, or batch to batch variability, but, instead, it is controlled by the physics of integrating spheres.

- Temporal irradiance monitoring and control.
  Temporal changes in the spectral radiant intensity of the light source can not be controlled. Through the use of the integrating sphere, however, temporal variations can be easily spectroradiometrically monitored. This is possible because the radiant power within the integrating sphere remains uniform and, thus, changes in the radiant power can be easily monitored.

- Experimental Flexibility
Finally, use of an integrating sphere provides the opportunity to simultaneously and independently expose a multiplicity of specimens each to its own exposure environment. This can be achieved by positioning specimens in individual exposure cells and uniformly irradiating the specimens by projecting the output from the exit ports with the use of non-imaging optical devices. The environmental and operating conditions within each exposure cell can be uniquely selected.

Acknowledgments

This project was conducted under the auspices of the NIST Coatings Service Life Prediction Consortium, which includes as its industrial members Atlas Electric Devices, Dow Chemical, DuPont Automotive Coatings, Duron Inc., and PPG. Federal agency participants include the Federal Highway Administration (FHWA), Forest Products Laboratory, National Renewable Energy Laboratory (NREL), and Wright Patterson Laboratories. A patent has been applied for the described research. Special thanks are given to Dr. Shuang-Ling Chong for FHWA's generous support for this research

Literature Cited

1. Ellis, C.; Wells, A.A. *The Chemical Action of Ultraviolet Rays*, Reinhold Publishing Company: New York, NY, 1941.
2. Nowacki, L.J. *Official Digest - Federation of Societies for Coating Technology* 1962, *4*, p. 1191.
3. Nowacki, L.J. *Official Digest – Federation of Societies for Coating Technology* 1965, *37*, p.1371.
4. Association of Automobile Industries, *J. Coatings Technology* 1986, *58*, p. 57.
5. Fischer, R.M.; Ketola, W.D.; Murray, W.P. *Progress in Organic Coatings* 1991, *19*, p. 165.
6. Fischer, R.M. in *Accelerated and Outdoor Durability Testing of Organic Materials, ASTM STP 1202;* Ketola, W.D.; Grossman, D.,Eds.; American Society for Testing and Materials: Philadelphia, PA, 1994; p. 112.
7. Nelson, H.A. *Proc. of the ASTM* 1922, *33*, p. 485.
8. Buttolph, L.J. United States Patent 1,818,68, 1924.
9. Hirst, H.R. *Journal of the Society of Dyers and Colourists* 1925, *41*, p. 347.

160

10. Weightman, H.E. *Rubber Age* **1928**, *23*, p. 75.
11. Hirt, R.C.; Schmitt, R.G.; Searle, N.D.; Sullivan, P. *Journal of the Optcal Society of* America **1960**, *50*, p. 706.
12. Searle, N. in *Accelerated and Outdoor Durability Testing of Organic Materials, ASTM STP 1202;* Ketola, W.D.; Grossman, D., Eds.; American Society for Testing and Materials: Philadelphia, PA, 1994; p. 52.
13. Caldwell, M.M. Gold, W.G. Harris, G.; Ashurst, C.W. *Photochemistry and Photobiology* **1983**, *37*, p. 479.
14. Kockott, D. *Die Angewandte Macromolekulare Chemie* **1985**, *137*, p. 1.
15. Fedor, G.R.; Brennan, P.J. in *Accelerated and Outdoor Durability Testing of Organic Materials, ASTM STP 1202*, Warren D. Ketola and Douglas Grossman, Eds., American Society for Testing and Materials, Philadelphia, 1994, p. 199.
16. Philips, R. *Sources and Applications of Ultraviolet Radiation,* Academic Press, New York, NY, 1983.
17. Thouret, W.E. *Illuminating Engineering* **1960**, *55*, p. 295.
18. Schäfer, V. *Applied Polymer Symposia* **1967**, *4*, p. 111.
19. Martin, K.G.; Campbell, P.G; Wright, J.R. *Proceedings of the American Society for Testing and Materials* **1965**, *65*, p. 809.
20. Clark, J.E.; Harrison, C.W. *Applied Polymer Symposia* **1967**, *4*, p. 97.
21. Barker R.E. *Photochemistry and Photobiology* **1968**, *7*, p. 275.
22. Klein, W.H.; Goldberg B. *Proceedings of the International Solar EnergySociety Conference*, New Delhi, India, 1978, vol.1; p. 400.
23. Goncz, J.H. *ISA Transactions* **1966**, *5*, p. 28.
24. Goncz, J.H. *Journal of the Optical Society of America* **1966**, *56*, p. 87.
25. Hirt, R.C.; Searle, N.D. *Applied Polymer Symposia* **1967**, *4*, p. 61.
26. Coblentz, W.W.; Long, M.B.; Kahler, H. *Bureau of Standards, Scientific Paper* **1919-1920**, *1*, p. 1.
27. Martin, J.W. *Progress in Organic Coatings* **1993**, *23*, p. 49.
28. Martin, J.W.; Saunders, S.C.;Floyd, F.L.; Wineburg, J.P. *Federation Series on Coatings Technology*, Federation of Societies for Coatings Technology: Blue Bell, PA, 1996.
29. Edwards, D.K.; Gier, J.T.; Nelson, K.E.; Ruddick, R.D. *Journal of the Optical Society of America* **1961**, *51*, p. 1279.
30. Goebel, D.G. *Applied Optics* **1967**, *6*, p. 125.
31. Carr, K.F. *Surface Coatings International* **1997**, *10*, p.490.
32. Carr, K.F. *Surface Coatings International* **1997**, *8*, p. 380.
33. Hanssen, L.F. *Applied Optics* **1996**, *35*, p. 3597.
34. Weidner V.R.; Hsia J.J. *J. Opt. Soc. Am.* **1981**, *71*, p. 856.
35. Storm, S.L.; Springsteen A. *Spectroscopy* **1998**, *13*, p. 9.
36. *NASA Tech Briefs* January 1996, p. 64.
37. Welford, W.T.; Winston R. *High Collection Non-imaging Optics*; Academic Press: New York, NY, 1989.

Advances in Analytical Measurements

Chapter 9

The Role of Fundamental Mechanistic Studies in Practical Service Life Prediction

David R. Bauer[1] and Karlis Adamsons[2]

[1]Ford Motor Company, Research Laboratory, MD3182, 2101 Village Road,
Dearborn, MI 48121
[2]Marshall Research and Development Laboratory, DuPont Automotive,
3401 Grays Ferry Avenue, Philadelphia, PA 19146

Chemical and physical measurements of coating degradation and stabilization have played a key role in gaining a fundamental understanding of coating failures. Such fundamental mechanistic studies are playing an increasingly important role in coating development and practical service life prediction. For the coating supplier, fundamental studies provide insights into how to develop new resins and select new additives to produce formulations with improved performance. The fundamental approach can drastically shorten coating development time by providing quick direction as to trends in performance. For the coating user, fundamental studies have improved test protocols, provided direction in setting performance goals, and helped ensure robust application procedures. The end result is reducing the risk of in-service failure and improving the performance/cost ratio.

Introduction

Over the past 20 years, a significant effort has been made to develop an understanding of the critical factors that control coating failures. Fundamental studies of chemical and physical changes have provided much insight into how and why coatings fail [1]. For example, studies of ultraviolet light absorber (UVA) loss have led to an understanding of basecoat/clearcoat delamination [2]. Other studies including ESR measurements of HALS nitroxide concentration, FTIR measurements of crosslink scission and oxidation, and NMR measurements of polymer end-group concentrations have provided insights into why some coating formulations weather faster than others [3-7]. As we will show, such fundamental studies are finding increasing practical applications for both coating suppliers as well as users. Coating

162   © 2002 American Chemical Society

suppliers are using the fundamental studies to develop improved coatings and to reduce product development time. Coating users are using the studies to select more appropriate accelerated tests and to align performance specifications with customer satisfaction targets. In this paper, we first describe the most common coating weathering failures and discuss limitations of conventional weatherability testing. We then describe a number of the fundamental techniques that have been used to study coating failure and discuss how the knowledge gained from those studies has been used to improve both coating development and service life prediction capabilities.

Automotive Weathering Failures and Conventional Testing

There are two basic types of automotive coating technologies that have been used over the past several years, monocoats and basecoat/clearcoats. Monocoats contain (highly dispersed) pigment particles in the top coating layer to achieve the desired color. As monocoats weather, the polymers in the monocoat erode away exposing pigment particles, which leads to a loss of gloss and ultimately chalking. Pigments and colorants may also degrade leading to color change or fade. Gloss loss and fade occur gradually during service life. These phenomena all occur at the very top layer of the coating and can be repaired by waxing and/or polishing. In basecoat/clearcoats, a clear polymer layer protects the pigmented basecoat. Gloss loss and color fade tend to be much slower in basecoat/clearcoats versus monocoats. Basecoat/clearcoats can ultimately fail by abrupt cracking or delamination of the clearcoat from the basecoat. Cracking and delamination require repainting, an expensive process. Cracking can also occur in monocoats, however, by the time cracking occurs, the coating is so dull that it would likely be considered to have already failed. It is also possible for the topcoat system (monocoat or basecoat/clearcoat) to peel off of the primer. This can occur if there is sufficient light transmitted through the topcoat and if the primer is sufficiently sensitive to that light. Such delaminations would also require repainting to restore appearance. Both coating types are also susceptible to damage by acid rain. In the case of monocoats, both etching and color changes can occur. In the case of basecoat/clearcoats, clearcoat etching can be observed.

Traditionally, automotive paint/coating systems have been exposed outdoors in USA climates such as those found in Florida and Arizona. Florida represents a climate that is hot and humid, while Arizona represents a climate that is hot and dry. A specific location in Florida (Jacksonville) is used to evaluate the effects of acidic depositions on paint films.

Although such climates are useful in providing a good approximate worse case of what to expect elsewhere in the continental USA, many years are required (especially for basecoat/clearcoats) before appearance changes are apparent. As a result, a variety of accelerated exposure conditions are used to evaluate coatings. In order to induce failures in as short a time as possible, accelerated tests often include exposure conditions that cannot occur in service. For example, the radiation used (i.e., QUV lamps, Xenon arc lamps, carbon arc, etc.) does not accurately represent solar radiation encountered at the earth's surface, particularly in the UV range. Certain coating chemistries are particularly sensitive to UV light below 295 nm and tests that employ such light can give results that are not consistent with in-service exposures. Even though accelerated test results have been widely used, the test results are usually considered to be tentative pending confirmation by outdoor results. There have been numerous cases where accelerated tests have rejected durable coatings or passed coatings that fail in service. In general, the more realistic the accelerated test, the longer the exposure that is required to evaluate performance leading to longer development times.Techniques Used to Study Coating Degradation

A wide variety of techniques have been applied to the question of coating degradation and stabilization. The most important techniques have been FT-IR and Raman measurements of coating compositional changes, ESR measurements of free radical formation and HALS stabilization, UV/VIS measurements of coating transmission and UVA retention, Chromatographic methods for additive analysis, and ToF-SIMS measurements of oxidation.

Although weathering is usually thought of as a surface phenomenon, it is often important to be able to map chemical changes as a function of depth in various coating layers. Many of the above techniques have been adapted to provide this depth resolution. A summary of commonly used techniques is provided to frame the discussion of application to service life prediction. More detailed descriptions and recent results are described elsewhere in this volume. It is important to note that many coating failures result from a combination of chemical change and mechanical stress. It is critical to relate the chemical changes to changes in mechanical properties. Photooxidation and changes in crosslink structure, for example, lead to changes in the energy to propagate fractures in the coating, which is critical for crack resistance.

FT-IR Analysis

Various IR techniques have evolved and been applied in studying coatings [8-17]. Techniques including transmittance, attenuated total reflectance (ATR), diffuse reflectance (DRIFT), and photoacoustic (PAS) mode experiments have been used. Their selection depends on the nature of the sample, specific locus of interest, and technique hardware/software/experience available. Note that the sampling approach used will often determine the volume element, and thus surface or depth specificity, achievable.

The simplest transmission mode IR measurement on coatings involves preparing a coating of the appropriate thickness (~5-7 μm) on an IR transparent substrate. For exposure studies, a silicon wafer is a good substrate. The clearcoated silicon wafers can be anchored onto metal holders whose position in the IR spectrophotometer-sampling compartment can be reproduced to permit monitoring the same area (cross-section) of the sample following each exposure period.

Commercial coating samples are optically too thick (clearcoats are generally 40-50 μm and full automotive paint systems are > 100 μm) or applied on non-IR transmitting substrates for direct measurement by transmission IR. Surface skiving or microtoming are sampling techniques useful in obtaining samples in the ~2-7 μm thickness range from automotive paint/coating materials. Microtoming also provides the possibility of depth profiling. For example, depth analysis can be done using IR microscopy of thin layer cross-sections, which must be supported flat and securely on an X-Y micro-translating stage. Section handling requires considerable skill and training. For example, the small cross-sections must be immobilized and supported flat.

Depth analysis by IR-microscopy is more easily accomplished by microtoming ~4-10 μm thick coplanar sections. A depth profile is generated by sequential analysis of these sections. Sections on this thickness scale for automotive systems can be quickly prepared using room-temperature slab-microtomy. Note that longer-term outdoor or accelerated aging of automotive systems can cause them to become fragile and difficult to section. Use of very sharp cutting blades (i.e., carbide or diamond) and optimizing the cutting angle can minimize this problem. Slightly thicker sections can be cut to help maintain integrity of sections during handling. In addition, many slab-microtomes can operate under cryo-conditions, which often permit sectioning of weaker, degraded paint/coating layers without shredding. Overall, this approach is equally convenient for sectioning a single layer or the entire multi-layered system. Normally this combination of sample preparation and IR analysis is found effective in determining the general chemistry as a function of depth [11].

ATR obtains IR spectra by passing IR light through an internal reflection element (IRE) in contact with the surface of a sample. The evanescent wave that penetrates into sample results in spectra that probe the surface and near surface of that sample. Subtle differences in surface/near-surface composition as a function of depth can be obtained using different ATR internal reflection elements (IRE's), which provide a means to vary the IR beam penetration depth over ~0.5-3 μm. The IRE material used (i.e., typically zinc selenide [ZnSe], germanium [Ge] or diamond), as well as angles of beam incidence, dictates the penetration depth. In addition, ATR measurements can usually be made with minimal sample preparation. Diamond is the hardest of the IRE materials and is often used in analysis of materials that are quite hard or that possess irregular surfaces. Under sufficient contact pressure the material opposite the diamond will deform to provide uniform surface contact, optimizing the quality of the resultant ATR spectrum. In this context, automotive paint/coating materials are easily deformed by the diamond, resulting in high quality spectra.

DRIFT surface/near-surface sampling is efficiently done using a sampling technique known as "Si-Carb". This type of sampling employs surface polishing, where a silicon carbide sandpaper (600-1200 grit) disk is used with controlled rotary motion to abrade the surface of a paint/coating system. The surface specificity of this technique is generally limited to the ~5-10 μm depth range, depending on the time and pressure applied in polishing. Analysis is performed on the surface of the sandpaper disk, where a new disk is used to obtain a background spectrum. Reproducible surface abrading can be obtained using a Dremel ™ Moto-Tool (or similar) drill and corresponding drill stand. Many of these drills permit variable rotation speed ranging from 5,000-30,000 rpm.

Photoacoustic spectroscopy (PAS) is a technique that directly measures the absorbance spectrum of a polymer-based paint/coating system layer without requiring a direct radiation reflection or transmission measurement. When modulated, mid-infrared radiation is absorbed by the sample, the material heats and cools in response to the modulated infrared energy impinging on it. Thereafter, the heating and cooling is converted into a pressure wave that directly impacts on the coupling gas (usually He) in contact with the sample. Thermal expansion and contraction of the gas is then detected by an acoustic detector (i.e., sensitive microphone) and transduced into the photoacoustic signal. Note that the acoustic detector replaces the standard infrared radiation detector in the FT-IR spectrophotometer. Contributions to the signal come from each region of the sample in which infrared radiation is absorbed. Of course, significant contributions require that the thermal wave amplitude has not effectively decayed prior to crossing the sample/coupling-gas boundary The sampling depth at a given wavenumber can be varied in PAS measurements [18-22] by changing the mirror velocity. A slower mirror velocity effectively allows for a greater contribution from lower layers by increasing the thermal wave decay length. Conversely, a faster mirror velocity minimizes the thermal contribution from deeper layers. Experiments such as these where the mirror velocity is constant are categorized as linear scan. Also, the sampling depth increases with decreasing wavenumber by more than a factor of three going from 4000 to 400 cm[-1] across a mid-IR range spectrum. For a detailed discussion of the principles and instrumentation required in performing the PAS experiments one can refer to the work of McClelland et. al. [23].

More recently, another approach has been evolving for doing PAS experiments. Using a "step scan" IR spectrophotometer the mirror is moved incrementally in steps [14, 24-26], rather than continuously with a constant velocity (linear scan). The step scan approach eliminates the time-dependence of the interferogram and temporal Fourier frequencies that are created. Specifically, phase modulation step scan experiments are conducted wherein the movable mirror is moved to a given position and then one of the other mirrors is oscillated back and forth at a fixed frequency to modulate the impinging IR radiation. In addition, there are two important phase modulation parameters: frequency and amplitude. The phase modulation frequency refers to the number of oscillations per second (i.e., Hz). The phase modulation amplitude measures the range (i.e., distance) of the mirror oscillation. This is usually indicated in terms of the laser wavelength (i.e., HeNe λ). Ultimately, step scan capability offers an efficient means of getting spectra whose depth information does not depend on wavenumber. PAS has been identified as a useful tool for near-surface analysis of paint/coating systems as well as various types of fabrics, fibers and films [13-18, 24, 27-35].

The PAS and DRIFT techniques have been compared in the context of near-surface analysis techniques [28]. A variety of polymer-based films/coatings, fibers and fabrics were studied. It was observed that DRIFT showed an enhancement in band intensities of near-surface species, however PAS appeared to have a significantly smaller sampling depth. In all cases PAS was found to be a quick and broadly applicable tool for polymeric materials.

Raman Analysis

Raman spectroscopy provides information on paint/coating system materials that is both similar and complimentary to that obtained from IR spectroscopy [14, 24, 36-44]. An extensive review of Raman spectroscopy in study of a broad range of polymeric materials (including paint/coating systems) is available [42]. The two techniques provide detailed information on molecular structure. Both monitor interactions between source radiation and matter that result in vibrational (energy level) transitions characteristic of inherent molecular substructures. Raman is an inelastic photon scattering process, usually involving narrow-band source radiation from a laser. Although both are vibrational spectroscopies, they differ primarily in the impact of molecular symmetry. Raman is particularly useful in analysis of molecular structures that are more symmetrical (i.e., polymer backbone, isolated/conjugated double bond, and certain organic pigment/dye unsaturated ring substructures). IR tends to be particularly useful in analysis of polymer side chain functional group, cross-linker, and additive substructures that are asymmetrical. This highlights the complimentary nature of the Raman and IR techniques.

The Raman signal is relatively weak. It has been reported that about one Raman photon is obtained for every 10^8 photons interacting with a sample. Also, background fluorescence observed with many materials can be quite strong, making it impractical to obtain any useful Raman signal [40]. Many developments have occurred in recent years including more sensitive CCD-type detectors, filters that have high fluorescence rejection efficiency, and laser sources in the near-IR range. Laser sources with frequencies in the near-IR range allow the user to avoid much of the fluorescence typical of many industrial paint/coating materials [40, 45]. Note that many colorants (i.e., pigments and dyes) result in characteristic high levels of fluorescence. These new laser sources are now making Raman analyses of such materials practical in a modern industrial laboratory.

Raman microscopy has evolved over recent years, permitting practical analysis of very small spot areas [8, 46-48]. Areas of ~2 µm diameter, or slightly less, are possible with commercial instrumentation. Essentially, Raman microscopy is conducted by integrating an optical microscope to a Raman spectrometer with a laser for excitation. A Renishaw spectrometer/imaging microscope equipped with an X-Y micro-translating stage has been successfully used in surface and cross-section chemical line profiling and area mapping. A 50X- or 100X-objective can be used to provide a laser spot size of ~2 µm and ~1 µm, respectively. By comparison, IR gives a diffraction limited spot size of ~10 µm, with practical analysis in the ~20-30 µm range. The smaller spot sizes possible in Raman microscopy allow detailed chemical mapping of surfaces. This is useful if there is component segregation, or domain formation, on a scale larger than the spot size and within the translation capability of the microscope stage [49]. Chemically selective imaging offers a means of highlighting specific components or substructure density across a surface [14, 24, 50-52]. In addition, the high spatial resolution permits effective chemical composition mapping in failure (i.e., defect) analyses and cross-sectional depth profiling [8, 44, 46, 53]

Total internal reflection Raman spectroscopy was developed as a tool for analysis of polymeric coatings [54] . Analogous to the IR ATR technique (described earlier), an internal reflection element (i.e., sapphire) is used in direct contact with the sample surface. This technique is capable of characterizing surface coating layers as thin as ~0.05 μm.

Raman was involved in various multi-technique studies of organic polymeric paint/coating systems [37, 42]. Urban and co-workers used various IR techniques, such as ATR and step-scan PAS, to complement the Raman analysis [14, 50] . Advantage was taken of the complementary nature of the chemical data obtained from these techniques. One group combined dynamic mechanical analysis (DMA), dilatometry, microscopy and Raman analysis [55] . In this case, correlations were determined between chemical composition, physical/mechanical properties, and the appearance. Another group explored a variety of applications of Raman in forensics studies [56].

ToF-SIMS Analysis

Secondary ion mass spectrometry (SIMS) is a surface specific analysis tool that has been successfully applied in the characterization of paint/coating systems [57-63]. Interfaces obtained through delamination or sectioning techniques have also been successfully studied [8, 57, 64-67]. Adhesion to pretreated (i.e., plasma, chemical) surfaces has been investigated [63], including polymer/polymer and polymer/metal interfaces. SIMS uses an ion beam, often in the KeV energy range, to bombard a surface causing sputtered emission of ions, neutrals and electrons. Secondary ions, both positive and negative, are then analyzed with a mass spectrometer. The combination of a pulsed ion source with a Time-of-Flight mass analyzer results in a powerful analytical tool characterized by high sensitivity, high mass resolution, and high surface specificity. This technique is commonly referred to as ToF-SIMS. During a typical experiment only the top monolayers are sampled, thus the high surface specificity. Conditions using lower primary ion currents can be optimized to provide spectra containing a significant abundance of molecular ions or larger substructure ion fragments. This can provide extensive composition and structural information about the outermost polymeric surface. Higher primary ion currents cause higher levels of molecular fragmentation, typically results in very complex spectra, which can be difficult to interpret. Surface isotopic tracer studies have been reported in the context of monitoring chemical modification or degradation [8, 68]. Herein the isotopic species can often be conveniently monitored to determine surface reaction mechanisms and kinetics.

Interfaces between paint/coating layers have been chemically characterized by either delamination or sectioning (co-planar to the surface) or by high-resolution techniques looking directly at cross-sections of the system [64-67]. A major application of ToF-SIMS analysis to coating characterization has been depth profiling by analysis of microtomed cross-sections. Having a resolution of ~2 μ spot size permits detailed analysis of a cross-section, either in line-scan or raster (multi-line) mode [8, 46]. Line-scan images can be created from a single pass (string) of spectra, which is quickest, or from averaging two or more passes of spectra, which takes longer, but allows for spectral averaging. Gerlock et al. developed a technique for determining the extent of photooxidation as a function of depth by photolyzing coating samples in the presence of $^{18}O_2$ [68]. These experiments have been critical in elucidating the mechanisms of various delamination failures. Other researchers have also reported results involving exposures to both $^{18}O_2$ and H_2O^{18} [8]. As a general rule-of-thumb, if the level of ^{18}O is high, then that locus of a system is more reactive and (thus) susceptible to degradation in performance. In other words, less chemical change suggests a more durable system.

The depth distribution of automotive paint/coating system light stabilizers has been determined by ToF-SIMS analysis. In this case microtomed sections co-planar to the surface were obtained and extracted by supercritical fluid extraction (SFE). Thereafter the SFE extracts were analyzed by ToF-SIMS [69, 70].

UV-Vis Analysis

The UVA concentration as a function depth into the coating can be analyzed by transmission UV/VIS-microscopy using thin layer cross-sections, which must be supported flat and securely on an X-Y micro-translating stage. This technique monitors all UV/VIS absorbing species encountered in the beam path, including UV-screeners (UVA's) and certain paint/coating system degradation products. Gerlock et al. have reported UV/VIS depth profiling with cross-sections ~10 μm thick [12]. As before, such thin layer cross-sections can be obtained by use of microtomic techniques. The small cross-sections must be handled carefully and are immobilized on a UV-transmitting quartz slide. Longer chain terminal alcohols, ranging from n-hexanol to n-decanol, are useful for this purpose. They have no UV absorption over the wavelength range of interest and effectively immobilize the sample onto the quartz surface. Masking of the incident beam to a small rectangle has been done, ~5 μm wide (along the normal depth axis) with variable length (parallel to the surface). Note that the mostly particle free clearcoat layers give the highest quality spectra relative to the other particle containing system layers. Scattering of the incident UV/VIS radiation can be quite substantial, even for these relatively thin sections. Therefore, most UV/VIS transmission-mode analyses that have been reported are on clearcoats or particle-free binders of other layers.

Transmission-mode analysis by monitoring UV absorbing extractables in paint/coating sections microtomed co-planar to the surface can also be very effective. Haacke et al. and Adamsons et al. have reported studies of clearcoats and clearcoat/basecoat bi-layers[71-76]. Commercially available 'slab' microtomes allow the user to routinely prepare ~2 cm x 2 cm sections co-planar to the surface in the ~5-10 micron thickness range. Such sectioning permits use of full (or 'real world') automotive systems, not just isolated clearcoat layers. Once sections are obtained, solvents such as CH_2Cl_2 or chloroform are effective for extracting UVA species and do not interfere with UV/VIS measurements. Either filtration or centrifigation removes solids, in order to prevent particle light scattering. The solutions are stored airtight, thus preventing solvent evaporation.

HPLC Analysis

Chromatography-based techniques such as HPLC are quite effective in analyzing depth profiles of species extracted from slab-microtomed sections. DuPont [76] has shown the utility of this approach in study of clearcoats and clearcoat/basecoat bilayers. As cited earlier, when using UV/VIS spectrophotometric analysis, a solvent appropriate to the material of interest is used to extract each section. In fact, the same solutions obtained in the previous case in study of UVA's and HALS can be used for HPLC analysis. The chromatographic signals obtained for a given sample are normalized to the amount of material extracted.

ESR Analysis

Electron Spin Resonance is useful for monitoring free radicals in materials. It has been used in coatings to measure free radical formation rates and HALS stabilization chemistry [5, 7]. In this technique, the sample is placed in a strong magnetic field that splits the energy levels of unpaired electrons (free radicals). The Zeeman splitting is monitored by probing the sample with microwave radiation. For quantitative applications such as those described here, it is necessary to develop careful sample preparation and placement procedures [77].

Applications for Coating Development

UVA Permanence: Evaluation of Accelerated Test Protocols

As noted above, UVA permanence in automotive coatings is a critical factor in basecoat/clearcoat delamination. The UV-fortification package plays a key role in long-term durability of a clearcoat and, hopefully, the lower layers and interfaces of a coating system as well. Conventional weatherability testing only reveals delamination problems after 3-5 years exposure in Florida. Accelerated weathering tests did not usually reproduce this failure mode. Thus, selection of UV-fortification packages required very long exposure times (>5 years under field conditions). The fundamental studies of UVA loss have provided techniques that can reduce the evaluation time to ~1 year or less. UVA content and distribution have been followed in materials exposed both outdoors (in various climates around the world) and under accelerated conditions (i.e., QUV, Xenon boro/boro, Cleveland Humidity Cabinet). This type of fundamental information provides product designers with important feedback in the selection of the final UV-fortification package. Selection of a UVA species includes factors such as permanence, stability, amounts required, and cost.

It is important to understand why conventional accelerated tests failed to predict UVA loss and basecoat/clearcoat delamination. Studies have been conducted to determine the UVA depth profiles for clearcoats in automotive coating systems for different exposure conditions. The depth sampling was performed using co-planar microtoming followed by extraction as described above. Depth profiles are shown in Figures 1-3 after 2000 hours exposure under standard QUV, UV-only and CHC conditions. From comparison of loss rates in these exposures with those observed in outdoor exposures, it can be concluded that exposures which use short wavelength light to accelerate degradation do not proved adequate acceleration for UVA loss relative to photooxidation and are likely to miss that failure mode.

Monitoring Clearcoats and Multi-layer Paint Systems As a Function of Exposure Time: Additive Selection

FT-IR studies of clearcoat photo-oxidation have provided numerous insights into coating weatherability that have resulted in the development of improved materials. For example, FT-IR has been used to analyze the rate of photo-oxidation in a coating with various types/levels of UV fortification. FT-IR is sensitive to relatively small changes in chemistry, which allows rapid evaluation of photo-oxidation rates under outdoor-like (accelerated) exposures. Figure 4 shows a typical acrylic polyol / melamine cross-linked type of clearcoat formulation either unfortified, fortified with UVA only (at standard loading levels), fortified with HALS only (at standard loading levels), or fully fortified. The objective was to determine the effectiveness of UVA or HALS alone, and in combination. Figure 5 shows the same acrylic polyol / melamine cross-linked clearcoat formulated at varying levels of UVA and no HALS. The objective was to determine the effectiveness of a particular UVA at various concentrations. Analogous studies have been conducted to determine the effectiveness of a particular HALS at various concentrations and to evaluate different HALS under different exposure conditions [6]. These results clearly showed that a common photo-stabilizer that performed well in harsh accelerated tests was not effective under milder conditions. This led to the development of improved HALS materials that have been incorporated into most automotive coating systems.

FT-IR has also been used to compare the rate of photo-oxidation of coatings made from polymers created by different polymerization methods [3]. The technique has identified solvents, initiators and catalysts that produce polymers having greater weatherability. The results led to modifications of conventional polymerizations to improve overall coating durability. Again, the sensitivity of FT-IR allowed for rapid analysis of relative performance under outdoor-like exposures. Clear indications of performance (and, to an extent, service lifetime prediction) can be seen in a few thousand hours of accelerated exposure allowing for rapid, reliable formulation evaluation.

ToF-SIMS is a very effective technique for evaluating the rate of photooxidation at different levels in the coating. This can lead to insights into the role of basecoat formulation on basecoat/clearcoat delamination at different levels of UVA protection.

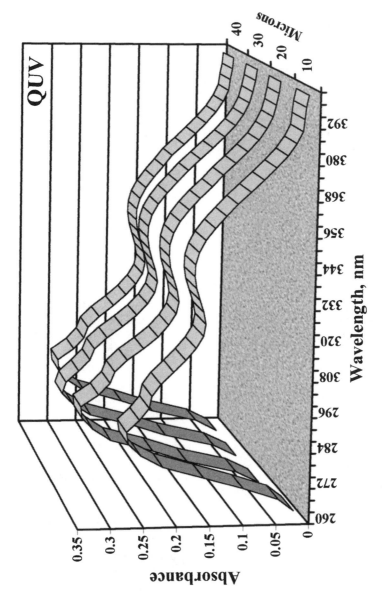

Figure 1. UVA profiles from a standard thermoset acrylic exposed 2000 h to standard QUV exposure.

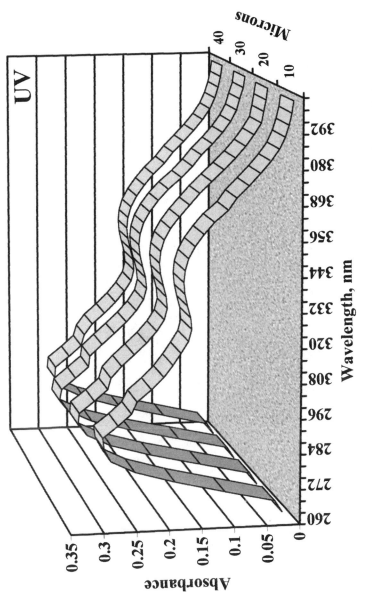

Figure 2. UVA profiles from a standard thermoset acrylic exposed 2000 h to UV light from a QUV exposure chamber with no condensing humidity cycle.

174

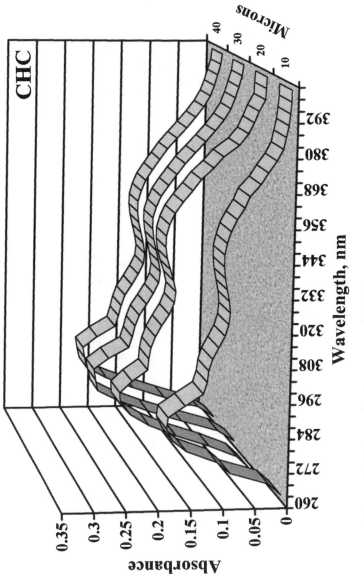

Figure 3. UVA profile from a standard thermoset acrylic exposed 2000 h to 50°C condensing humidity.

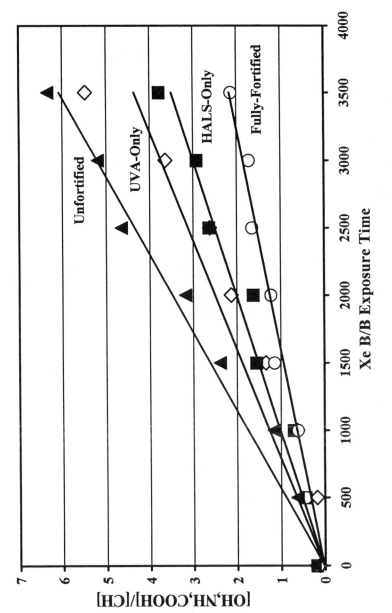

Figure 4. Increase in photooxidation with exposure to Xenon Arc (boro-boro filter) for a standard thermoset acrylic as a function of stabilizer content.

176

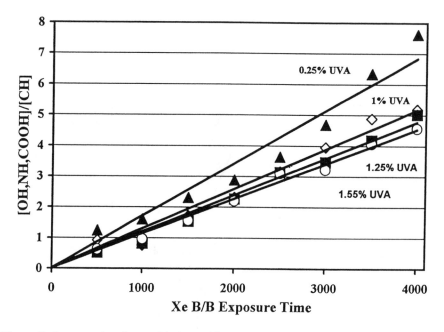

Figure 5. Increase in photooxidation with exposure to Xenon Arc (boro-boro filter) for a standard thermoset acrylic as a function of concentration of UVA.

ToF-SIMS can also be used to differentiate between photooxidation and hydrolysis as a degradation mode. Studies to track hydrolysis and water adsorption have been conducted using an exposure chamber where systems are exposed to H_2O^{18} vapor at standard temperatures, but no irradiation (dark). Monitoring the impact of O^{18} containing water vapor gives a product developer additional input on probable long-term durability in humid environments. Extensive hydrolysis at/near to a surface or interface (i.e., clearcoat//basecoat) can breakdown the mechanical integrity of the system in that region. Also, surface roughness due to material erosion will increase. Like photo-oxidation, hydrolysis can then impact mechanical performance and appearance, respectively. Knowledge of the hydrolysis resulting from specific exposure time/conditions is often critical in selecting components in a formulation.

Fundamental Measures of Scratch Performanc: Effects of Cure and Aging

Studies have been undertaken to explore the correlations between cure and mechanical performance changes in an automotive clearcoat as a function of initial cure and aging. Cure was assessed using IR analysis of the ratio of unreacted isocyanate to styrene. Chemical measurements were compared to swelling measures of cure. Mechanical property measurements include hardness, % elastic recovery after deformation, the critical coefficient of friction, and the critical normal and tangential loads to fracture. The friction and load measurements were made using a microscratch nano-indentor. The nano-indentor provides quantitative information about scratch and mar performance that is more reproducible than conventional test methods.

Results for the various measurements are reported in Table I. The initial low state of cure is evident by a large residual isocyanate signal in the IR and by a high swell ratio. The poor network formation results in a soft coating with low elastic recovery. The critical normal load to fracture is also low while the critical friction is high. This results in poor scratch and mar resistance relative to the well-cured system. The well-cured coating does not change significantly on room temperature aging. The conversion of the small amounts of residual isocyanate does not significantly change the network structure or mechanical properties. The low-cured coating gradually completes cure over about a week at room temperature leading to mechanical properties that are essentially equivalent to the well-cured coating. The implications of this result for manufacturing are clear. Under-cured coatings must be handled carefully during assembly and shipping to avoid damage while the coating is completing its cure. These studies are being extended to evaluate the effects of photooxidation and hydrolysis on mechanical property performance.

Coating User Application: Minimizing In-Service Risk of Failure

The use of chemical measurements to rapidly compare the performance of different formulations has already been discussed as a tool for rapid coating

Table I. Changes in Cure and Mechanical Properties with Time

	IR Peak Ratio NCO/Sty	Swell Ratio	Hardness (N/mm2)	% Elastic Recovery	Critical Coeff. Of Friction	Critical Normal Load (mN)	Critical Tangential Load (mN)
Initial							
Low Cure	1.43	1.77	56	17	0.9	9.62	8.69
High Cure	0.45	1.55	127	32	0.76	12.29	9.4
1 Day							
Low Cure					0.83	9.04	7.46
High Cure					0.75	12.61	9.46
7 Day							
Low Cure	0.77	1.56	110	37	0.74	12.05	8.89
High Cure	0.33	1.55	129	38	0.73	12.44	9.05
30 Day							
Low Cure	0.38	1.56	130	42			
High Cure	0.17	1.58	130	43			

development. It is often necessary to make changes to a coating formulation after it is in production. There is rarely time to "requalify" the coating using conventional long-term exposures. Chemical measurement may either be used to demonstrate that the change did not impact weathering performance, or, in the case of a weathering issue, to demonstrate that the change is at least going in the right direction. As the supplier's use of fundamental measurements becomes more routine, the need for the coating customer to duplicate such measurements will diminish. Users will be able to evaluate chemical and mechanical test results provided by suppliers to determine the wisdom of a particular contemplated change.

While comparison of the performance of two or more coatings allows one to select the "best" coating for weathering performance, it does not provide information as to whether or not the "best" was good enough or indeed, if the "worst" was also good enough. In general, there will always be a relationship between the risk of failure and the level of weathering performance of a coating. It is not a trivial matter to determine what that relationship is. While it is possible to relate the incidence of in-service failures to the level of coating performance for some coating systems and some failure modes, having to rely on the existence of in-service failures to determine the relationship, hardly a desirable state of affairs. The fundamental mechanistic studies play a critical role in developing such relationships. There will also be a relationship for a particular coating technology, between the performance of that coating and its cost. The goal of the coating user is to minimize the cost of coating the product while insuring that the risk of failure is tolerable. In the case of basecoat/clearcoat systems, this is made more difficult by the fact that the most serious failure modes occur abruptly and cannot be easily repaired. Thus, unlike gloss loss in monocoats, there is no early warning of failure and the cost of failure to the company is large. Under these circumstances, the use of fundamental measurements in service life prediction becomes a necessity rather than a luxury.

The first step in applying fundamental mechanistic information is the creation of failure models for all the critical failure modes. This is necessary since the fundamental measurements provide indirect (as opposed to direct, visual measurements) indications of performance. From basic mechanisms of failure, we establish what parameters to measure for particular failure modes. Then, since the goal is to generate absolute criteria for performance, it is critical to have a quantitative relationship between the parameters that are being measured and the absolute performance (time-to failure for all failure modes). Failure models have been discussed in detail previously [78]. We will use the failure model created for basecoat/clearcoat delamination to illustrate how fundamental measurements can be used to insure that risk of this particular failure is minimized in service. The failure model for this particular failure is given in Eq. 1

$$TF = EX \ T_{BC} + EX \cdot \frac{A_o}{k} L \tag{1}$$

where TF is the time-to-failure distribution function, EX is the photooxidation harshness distribution function relative to Florida, T_{BC} is the time to delamination for unprotected basecoat in Florida exposure, EX' is the distribution function for UVA loss relative to Florida, A_o is the initial absorbance per unit thickness of UVA, k is the rate of loss of UVA in Florida, and L is the minimum clearcoat thickness distribution function. The first point that has to be made is that many of the critical parameters are in fact distribution functions. What we are really calculating is the percent of likely failures at a given time in a given market for a given coating. For example, we may set a criterion that a coating system must satisfy 90% of all customers in a given market after 10 years in service. This means that at most 10% of customers at 10 years in service will perceive any failure. For any catastrophic failure (delamination, cracking), we can set particular performance expectations (e.g., no more than 5% for either delamination or cracking after 10 years in service).

One critical point in the application of fundamental mechanistic studies to practical service life prediction is that the failure model that is developed (i.e., Eq. 1) required a large amount of fundamental measurements to derive. From the basic mechanism of delamination to the dependence of the rate of UVA loss on environmental variables, fundamental studies were essential to the development of the failure model.

Based on Eq. 1, for this failure mode, it is clear that there are three material parameters and one application parameter that have to be measured. The application parameter, minimum clearcoat film thickness, has to be measured on actual vehicles. Of the material parameters, one can be determined from initial properties (the total clearcoat absorbance). The others, basecoat stability and UVA loss rate, have to be measured by exposure. The stability of the basecoat can be measured either by exposing UVA free systems until delamination occurs or by using ToF-SIMS to monitor photooxidation at the basecoat/clearcoat interface. The rate of loss of UVA can be determined from relatively short exposures based on validated kinetic models of UVA loss. By parameterizing the failure mode, it is possible to greatly shorten the time required for evaluation without using accelerating exposure conditions. For example, demonstration of performance against this failure mode could require exposures as long as 10 years outdoors. UVA loss rates can be measured in a year or two. Basecoat stability can be determined in 2-4 years. Thus, 10 year performance can be anticipated in at most 4 years of exposure. Even shorter times (~0.5 years) are possible using accelerated tests whose acceleration factor has been validated using the appropriate chemical measurements.

Using this approach requires that quantitative failure models be developed for all failure modes of interest. In a previous paper, several failure models were proposed [78]. One failure for which there was no good failure model was failure by cracking. Recent work presented elsewhere in this volume describes the use of fracture energy as a parameter to correlate with cracking failures. With the addition of fracture energy measurements to the critical parameters, it is possible to anticipate all known basecoat/clearcoat-weathering failures using fundamental measurements.

Another aspect where fundamental studies play a critical role is in developing relationships between in-service performance and performance in outdoor and laboratory exposures. Fundamentally, developing these relationships involves determining the dependence of critical performance variables on environmental variables. For most coating failure modes, the critical variables are either photooxidation or loss of UVA. As noted in Eq. 1, these two processes have different dependences on environmental factors. The rate of UVA loss has been found to depend mostly on UV light dose [79]. The rate of photooxidation is a complex function of light intensity, wavelength distribution, temperature, and humidity [79]. By measuring the rate of photooxidation under different controlled experimental conditions, it is possible to develop models for the dependence of photooxidation on environmental variables for different materials. When combined with environmental data from different regions or data from laboratory tests, it is possible to construct in-service exposure distribution functions or compute laboratory acceleration factors relative to standard Florida exposures [79, 80]. This allows the user to set specific performance targets in terms of years of performance in standard outdoor exposure tests for given market regions. These targets then get converted to performance targets for specific parameters or combination of parameters (e.g., basecoat stability and UVA loss). Studies of acceleration factors in laboratory weathering tests help select more appropriate accelerated tests and test conditions as well as to provide a means to understand and control test variability. Of particular importance is the identification of exposure conditions that lead to poor predictions of in-service performance. For example, the fundamental studies have clearly shown that exposures that include UV light shorter than 290 nm tend to distort the degradation chemistry and the balance of degradation and stabilization reactions leading to highly variable acceleration factors. As a result of these changes, different exposure conditions are now being specified.

Conclusion

It should be clear that fundamental studies have played and will continue to play and increasingly important role in coating development and implementation. In addition to shortening development time and establishing more realistic standards of performance, the fundamental measurements have a major advantage in that they provide the opportunity for a shared knowledge base between the supplier and user. Fundamental performance data generated by the supplier can be used by the end user and does not need to be redone. This also can shorten time to implementation. In the past, a supplier would develop a coating using internally developed methods. Validation at the user level would essentially start from scratch on submission. By agreeing on fundamental criteria up front, the fundamental data generated during the development process can be used to support validation.

References

1. Bauer, D. R. *J. Coat. Technol.* **1994,** *66* (no. 835), 57-65.
2. Gerlock, J. L., Smith, C. A., Nunez, E. M., Cooper, V. A., Liscombe, P., Cummings, D. R., Dusbiber, T. G., in *Polymer Durability*, Clough, R. L., Billingham, N. C., Gillen, K. T., Eds. Advances in Chemistry Series No. 249, American Chemical Society, Washington, DC, 1996, 335-347.
3. Gerlock, J.L., Bauer, D. R., Briggs, L. M., Hudgens, J. K. *Prog. Org. Coat.* **1987,** *15*, 197-208.
4. Gerlock, J. L, Mielewski, D. F., Bauer, D. R., Carduner, K. R. *Macromolecules* **1988,** *21*, 1604-1607.
5. Bauer, D. R., Gerlock, J. L., Mielewski, D. F., Paputa Peck, M. C., Carter III, R. O. *Polym. Deg. Stab.* **1990,** *27*, 271.
6. Bauer, D. R., Gerlock, J. L., Mielewski, D. F. *Polym. Deg. Stab.* **1990,** *28*, 115-129.
7. Bauer, D. R., Gerlock, J. L., Mielewski, D. F., Paputa Peck, M. C., Carter III, R. O. *Polym. Deg. Stab.* **1990,** *28*, 39-51.
8. Adamsons, K., Stika, K., Swartzfager, D., Walls, D., Wood, B., Lloyd, K.; *Polym. Mater. Sci. Eng.* **1996,** *75*, 482-483.
9. Kagawa, K.; *Hyomen Gijutsu* **1999,** *50(5)*, 431-436.
10. Shoda, K., Ishida, H.; *Bosei Kanri* **1989,** *33(4)*, 106-113.
11. Haacke, G., Brinen, J.S., Larkin, P.J.; *J. Coat. Technol* **1995,** *67*(843), 29-34.
12. Gerlock, J.L., Smith, C.A., Cooper, V.A., Dusbiber, T.G., Weber, W.H.; *Polym. Degrad. Stab.*, **1998,** *62*, 225-234.
13. Hodson, J., Lander, J.A.; *Polymer* **1987,** *28*(2), 251-6.
14. Urban, M.W.; *Polym. Mater. Sci. Eng.* **1998,** *78*, 18-19.
15. Carter III, R.O., Paputa Peck, M.C., Bauer, D.R.; *Polym. Degrad. Stab.* **1989,** *23*, 121-134.
16. Bauer, D.R., Paputa Peck, M.C., Carter III, R.O.; *J. Coat. Technol.* **1987,** *59(755)*, 103-109.
17. Carter, R.O., Bauer, D.R; *Polym. Mater. Sci. Eng.* **1987,** *57*, 875-879.
18. Takaoka, K.; *Shikizai Kyokaishi* **1986,** *59*, 347-356.
19. Muraishi, S.; *Bunko Kenkyu* **1984,** *33*, 269-270.
20. Bruecher, K.H., Perkampus, H.H.; *Fresenius Z. Anal. Chem.* **1986,** *325*, 676-682.
21. Ochiai, S.; *Toso Kogaku* **1985,** *20*, 192-195.
22. Dolby, P.A., McIntyre, R.; *Polymer* **1991,** *32*, 586-9.
23. McClelland, J.F., Jones, R.W., Luo, S., Seaverson, L.M.; A Practical Guide to FT-IR Photoacoustic Spectroscopy; Proper Sample Handling with Today's IR Instruments; Edited by P. Coleman, CRC Press.
24. Urban, M.W.; *ACS Abstracts, 215-th ACS National Meeting, Dallas*, March 29-April 2, PMSE-011 (1998).
25. Gregoriou, V.G., Hapanowicz, R.; *Spectroscopy* **1997,** *12*, 37-41.
26. Wahls, M.W.C., Leyte, J.C.; *Appl. Spectrosc.* **1998,** *52*, 123-127.
27. Delprat, P., Gardette, J.L.; *Polymer* **1993,** *34*, 933-7.

28. Yang, C.Q.; *Appl. Spectrosc.* **1991**, *45*, 102-8.
29. Jasse, B.; *J. Macromol. Sci., Chem.* **1989**, *A26*, 43-67.
30. Urban, M.W.; *Polym. Mater. Sci. Eng.* **1989**, *61*, 132-6.
31. Urban, M.W.; *J. Coat. Technol.* **1987**, *59(745)*, 29-34.
32. Vidrine, D.W.; *Polym. Prepr.* **1984**, *25(2)*, 147-148.
33. Carter, R.O., Palmer, R.A., Dittmar, R.M., Manning, C.J., Bains, S.S., Chao, J.L.; Proc. SPIE Int. Soc. Opt. Eng. 1145 (Int. Conf. Fourier Transform Spectrosc., 7-th), pgs. 362-363 (1989).
34. Gerlock, J.L., Smith, C.A., Carter III, R.O., Dearth, M.A., Korniski, T.J., Kaberline, T.J., Devries, J.E., Cooper, V.A., Dupuie, J.L., Dusbiber, T.G.; *Surf. Coat. Aust* **1997**, *34*, 14-16.
35. Gerlock, J.L., Smith, C.A., Nunez, E.M., Cooper, V.A., Liscombe, P., Cummings, D.R.; *Polym. Prepr.* **1993**, *34(2)*, 199-200.
36. Craver, C.D.; *Pract. Spectrosc.* **1977**, *1*, 933-1000.
37. Craver, C.D.; Spectroscopic Methods in Research and Analysis of Coatings and Plastics; ACS Symp. Ser., Appl. Polym. Sci. (2-nd Ed.), pgs. 703-738 (1985).
38. Ellis, G., Claybourn, M., Richards, S.E.; *Spectrochim. Acta, Part A* **1990**, *46A(2)*, 227-241.
39. Smyrl, N.R., Howell, R.L., Doyle, Jr.,M., Oswald, J.C.; *Pract. Spectrosc.* **1988**, *6*, 211-228.
40. Claybourne, M., Agbenyega, J.K., Hendra, P.J., Ellis, G.; *Adv. Chem. Ser.* **1993**, *236*, 443-482.
41. Ikeda, T., Ohkubo, Y.; *Nippon Insatsu Gakkaishi* **1997**, *34*, 280-286.
42. Xue, G.; *Prog. Polym. Sci.* **1997**, *22*, 313-406.
43. Sun, B.; Non-Destructive Analysis of Paints and Coatings by Raman Spectroscopy; Florida State Univ. Microfilms Int., Order No. DA9544333, 110 pp, Diss. Abstr. Int. B **1995**, 56(8), 4289.
44. Takahashi, M.; *Prog. Org. Coat.* **1986**, *14*, 67-86.
45. Schrader, B., Baranovic, G., Keller, S., Sawatski, J.; *Fresenius' J. Anal. Chem.* **1994**, *349*, 4-10.
46. Adamsons, K., Litty, L., Lloyd, K., Stika, K., Swartzfager, D., Walls, D., Wood, B.; ACS Symp. Ser. 722 (Service Life Prediction of Organic Coatings), pgs. 257-287 (1999).
47. Witke, K., Brzezinka, K.W., Reich, P.; *Materialpruefung* **1997**, *39*, 316-322.
48. Koenig, J.L.; *Chem. Technol.* **1972**, *2*, 411-415.
49. Boyer, H. Jobin-Yvon; *Spectrochim. Acta, Part B,* **1984**, *39*, 1527-1532.
50. Zhao, Y., Urban, M.W.; ACS Abstracts, 215-th ACS National Meeting, Dallas, March 29 – April 2, PMSE-015 (1998).
51. Claybourne, M.; *Polym. Mater. Sci. Eng.* **1994**, *71*, 143-144.
52. Mannik, L.; *Energy Res. Abstr.* **1988**, *13(15)*, Abstr. No. 34411.
53. Dupuie, J.L., Weber, W.H., Scholl, D.J., Gerlock, J.L.; *Polym. Degrad. Stab.* **1997**, *57*, 339-348.
54. Iwamoto, R., Miya, M., Ohta, K., Mima, S.; *J. Am. Chem. Soc.* **1980**, *102*, 1212-1213.

184

55. Lange, J., Andersson, H., Manson, J.A.E., Hult, A.; *Surf. Coat. Int.* **1996,** *79,* 486-491.

56. Kuptsov, A.H.; *J. Forensic Sci.* **1994,** *39,* 305-318.

57. DeVries; *J. Mater. Eng. Perform.* **1998,** *7,* 303-311.

58. Erenrich, E.H., Leone, E.A., Sedgwick, R.D.; *Proc. Int. Waterborne, High-Solids, Powder Coat. Symp.* **1996,** *23,* 327-339.

59. de Lange, P.J., Mahy, J.W.G.; *Fresenius' J. Anal. Chem.* **1995,** *353,* 487-493.

60. Schrepp, W.; *Double Liaison – Phys., Chim. Econ. Peint. Adhes.* **1995,** *42,* 469-470.

61. Short, R.D., Ameen, A.P., Jackson, S.T., Pawson, D.J., O'Toole, L., Ward, A.J.; *Vacuum* **1993,** *44,* 1143-1160.

62. Gresham, G.L., Groenewold, G.S., Bauer, W.F., Ingram, J.C.; *Proc. SPIE Int. Soc. Opt. Eng.* **1999,** *3576,* 66-72.

63. Van Ooij, W.J., Sabata, A.; *Surf. Interface Anal.* **1992,** *19,* 101-113.

64. DeVries, J.E., Haack, L.P., Prater, T.J., Kaberline, S.L., Gerlock, J.L., Holubka, J.W., Dickie, R.A., Chakel, J.; *Prog. Org. Coat.* **1994,** *25,* 95-108.

65. DeVries, J.E., Hhaack, L.P., Prater, T.J., DeBolt, M. , Gerlock, J.L., Holubka, J.W., Dickie, R.A.; *Polym. Mater. Sci. Eng.* **1992,** *67,* 69-70.

66. Schuetzle, D., Carter III, R.O., DeVries, J.E.; *Adv. Org. Coat. Sci. Technol.* **1989,** *Ser. 11,* 229-245.

67. Aulton, S.; *Radcure Coat. Inks: Appl. Perform.,* **1996,** Paper 12.

68. Gerlock, J.L., Prater, T.J., Kaberline, S.L., deVries, J.E.; *Polym. Degrad. Stab.* **1995,** *47,* 405-411.

69. Andrawes, F.F., Valcarcel, T., Haacke, G., Brinen, J.; *Anal. Chem.* **1998,** *70,* 3762-3765.

70. Haacke, G., Andrawes, F.F., Campbell, B.H.; *J. Coat. Technol.* **1996,** *68(855),* 57-62.

71. Haacke, G., Longordo, E., Brinen, J.S., Andrawes, F.F., Campbell, B.H.; *J. Coat. Technol.* **1999,** *71,* 87-94.

72. Haacke, G., Longordo, E., Brinen, J.S., Andrawes, F.F., Campbell, B.H.; *Proc. Int. Waterborne, High-Solids, Powder Coat. Symp.,* **1998,** *25,* 433-443.

73. Haacke, G., Longordo, E., Andrawes, F.F., Campbell, B.H.; *Prog. Org. Coat.* **1998,** *34,* 75-83.

74. Haacke, G., Longordo, E., Andrawes, F.F., Campbell, B.H.; *Proc. Int. Conf. Org. Coat.: Waterborne, High Solids, Powder Coat.* **1997,** *23,* 171-185.

75. Haacke, G.; ACS Abstracts, 215-th ACS National Meeting, Dallas, TX, March 29 – April 2, PMSE-195 (1998).

76. Adamsons, K.; Short Course Presentation at The Atlas School for Natural and Accelerated Weathering (ASNAW); April 28-30, Miami, Florida, 1999.

77. Gerlock, J. L., Mielewski, D. F., Bauer, D. R.; *Polym. Deg. Stab.* **1988,** *20,* 123-134.

78. Bauer, D. R.., in "Service Life Prediction of Organic Coatings: A Systems Approach", Bauer, D. R., Martin, J. W., Eds. Oxford University Press (ACS Symp. Ser. 722), **1999** pp. 378-395.

79. Bauer, D. R.; *Polym. Deg. Stab.* **2000,** *69,* 297-306.

80. Bauer, D. R., *Polym. Deg. Stab.* **2000,** *69,* 307-316.

Chapter 10

Chemical Depth Profiling of Automotive Coating Systems Using IR, UV-VIS, and HPLC Methods

Karlis Adamsons

Marshall Research and Development Laboratory, DuPont Performance Coatings,
3401 Grays Ferry Avenue, Philadelphia, PA 19146

Today's high performance automotive coating systems require very specific surface, interface and depth profile characteristics to create products capable of providing customers (i.e., vehicle owners/operators) with desired appearance characteristics and service life. This is the case for both original equipment manufacturer (OEM) and Refinish targeted systems and associated components. The various system components (i.e., clearcoat / basecoat / primer / electrocoat on zinc phosphate treated steel substrate or clearcoat / basecoat / primer on adhesion promoter treated plastic substrate) must be durable, compatible and integrate with each other to produce systems that are commercially viable. This article focuses on the measurement technologies, including IR, UV-VIS, and HPLC based analyses, that have been applied in our laboratories on surface/near-surface and depth profiling study of automotive coating systems.

The relationship of chemical functionality, material distribution, and network architecture to product performance (i.e., chemical resistance, mechanical performance and appearance characteristics) can be best understood by knowledge of system composition as a function of locus. The measurement technologies detailed herein are broadly available, although the sampling techniques involve specialized expertise and equipment. Each of these analytical tools will be covered from the applications viewpoint. Hopefully, this will provide researchers insight into application of these tools in addressing their own characterization needs and technical managers with the incentive to invest in such analytical capability.

Introduction

A wide variety of sampling techniques and measurement technologies have been used for surface, interface, and depth profiling analysis in the study of automotive OEM and Refinish coating systems [1,2,11,15,16,19,26,38,45,55,60]. The ability to obtain detailed chemical composition, physical property, and mechanical performance information for a given layer, or on a multi-layered system, will be increasingly important in designing future systems. The primary advantage for these types of information is that change(s) can be identified and monitored well in advance of appearance change(s) being detectable. Appearance changes are usually detectable only after a period of time sufficient for damage to accumulate and (in turn) manifest as macroscopic events (i.e., gloss loss, delamination, cracking, and erosion). The induction time is longest for appearance changes to be monitored, but considerably shorter for chemical, physical and mechanical changes. The ability to follow changes earlier and to correlate them with time-to-failure offers coating system developers a much sought after means of service life prediction. Coating system consumers are demanding products that have long-term durability with respect to appearance, while also maintaining their resistance to chemical and mechanical damage. Development of these systems requires that one can readily determine the chemical composition and associated properties following initial system cure, as well as outdoor/accelerated exposure time/conditions, and to do so in a competitively short timeframe.

Automotive coating suppliers and automobile manufacturers have long relied on appearance-based measurement technologies to monitor finish condition and durability. Measurements commonly used include gloss, distinctness of image, orange peel, and haze. The primary limitation in using appearance-based measurements to determine coating system durability is in the time required for significant (i.e., measurable and reproducible) changes to take place. Years may be required for significant appearance changes to take place in many outdoor environments. Chemical, physical and mechanical changes can often be reliably monitored well before appearance-based changes are sufficient to be measured. The demand for decreased cycle time in the development of new coating products is (effectively) forcing the coatings industry to identify methods to take advantage of these other property changes in predicting durability or service life.

A better understanding of the chemistry, including factors like component concentration heterogeneity, segregation and stratification effects, interlayer mixing, and/or migration, is becoming essential to creation of current/future high performance products. This understanding permits controlled, customer-guided evolution of products with tailored chemical composition, physical properties, mechanical performance and appearance characteristics. Often a multi-technique approach, using routine as well as state-of-the-art technologies, is necessary to adequately characterize and (thus) understand these systems [1,2,16,22,26,38,56]. The techniques described herein can provide complementary types of chemical information. This, in turn, can result in a more detailed view of chemical composition and component distribution, particularly as a function of factors such as system preparation, application, and environmental exposure.

Levels of complexity inherent in the design of coating systems (*Figure 1*) can be targeted with appropriate measurement technologies. Analysis of the general chemistry present in a coating layer, but non-locus specific, is done to provide basic information on the overall chemical composition. Analysis of chemical detail with respect to morphology is done to establish the locus or distribution of chemistry. Here morphology is taken in a three-dimensional context, not just surface topology. Analysis of chemistry with respect to properties such as acid etch resistance, scratch & mar resistance, surface/bulk hardness, and solvent transport, is done to determine if desired product characteristics have been achieved. Analysis of overall performance usually includes a combination of chemical, physical, and mechanical factors. Knowledge of the correlations between these factors, even if somewhat incomplete, has been found very useful in design of state-of-the-art coating systems for our automotive finishes customers.

The ability to realistically approximate the durability or service life performance of a coating system is becoming increasingly important [63,64]. Measurement technologies are continuously being identified and applied for this purpose. The more accurate the service life predictions, the better in terms of risk management for the product supplier. They allow one to significantly decrease the time required to bring a product to the marketplace and provide confidence as to a product's (probable) long-term performance. Reducing the cycle time in product development gives one a critical advantage in today's competitive marketplace. It is well known that the "window of opportunity" in the automotive finishes marketplace is getting ever shorter for specialized, high-performance coating system products. Success or failure in realizing such marketplace opportunities can rapidly determine the success or failure of the supplier's business. From the standpoint of the direct product consumers (i.e., automobile/truck manufacturers and refinish shops), such predictions on long-term performance provide confidence and opportunities for risk management as well. Ultimately, their success or failure in the marketplace is dictated by the degree of customer (i.e., vehicle owner/operator) satisfaction.

The primary focus of this review is on chemical analysis at a given coating system locus, however studies involving correlations between chemistry and other types of properties will be included as appropriate. Specific correlations between chemical composition and appearance, physical/mechanical properties, or morphology have been identified. These correlations, when well founded, help product developers quickly determine performance under field or accelerated exposure time/conditions. It is also believed that such correlations offer developers a much more intimate and detailed understanding of system architecture. This also provides developers with the ability to optimize their product to suit a coating product consumer's needs.

Multi-layered automotive systems are applied on either metal or plastic substrate as shown in *Figure 2*. Clearcoat/basecoat (CC/BC) systems, which are widely used on both zinc/phosphate treated steel or on plastic substrates have been used to demonstrate the utility of the various measurement technologies. Common system dimensions are as follows: clearcoat is ~40-50 μ (microns), basecoat is 30-50 μ, primer is ~20-30 μ, and electrocoat is ~20-30 μ. Note that each of these system layers is on the order of 20 to 50 μ in thickness. Effective composition profiling of these systems as a function of depth requires techniques with either superb spatial and depth resolution and/or specialized sample preparation. Applications reported in this review are from the automotive industry, although it is expected that these techniques can be successfully used in detailed examination of other coating systems. Many applications are possible in areas as diverse as building materials, coil coatings, Teflon® cookware, maintenance finishes, food/pharmaceutical packaging, barrier/transport layers, electronic components and solar cells.

Surface / Near-Surface Characterization: Chemistry

Surface specific chemical characterization methods applied in study of automotive coating systems involve various types of probe beams (i.e., photons, electrons, ions), analysis beams (i.e., photons, electrons, ions), sampling depths, detection limits, spatial resolution (including lateral and height), and chemical information (i.e., elemental, functional groups, oxidation states). This article focuses on the various measurement technologies, specifically IR and UV-VIS, and associated sampling techniques routinely used in these studies. A brief review of these methods, as well as Raman, ToF-SIMS, and ESR analysis, is available in this volume under the title *The Role of Fundamental Mechanistic Studies in Practical Service Life Prediction*.

Infrared Analyses (Surface/Near-Surface or Thin Layer Volume Element). Infrared (IR) analysis has had a long and successful history in surface/near-surface chemical composition and component distribution analysis of automotive coating systems [39,44,47,48,49,50]. Various sampling techniques and IR measurement technologies have evolved and/or been applied in study of such systems. Methods including attenuated total reflectance (ATR) [26,27,40,59], diffuse reflectance (DRIFT) [50], photoacoustic (PAS) [14,24,26], and transmittance [1,2,38] mode experiments have been performed. Their selection depends on the nature of the sample, specific locus of interest, domain size and technique hardware/software/experience available. Note that the sampling technique used determines the volume element, and thus surface or depth specificity, achievable.

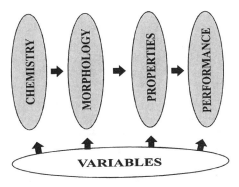

Figure 1. Levels of complexity in design of automotive coating systems. Chemistry is most fundamental level. Performance (customer perceived) is the most complex level.

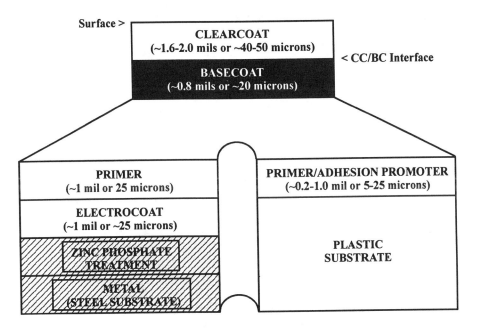

Figure 2. Cross-section of typical automotive coating systems: left branch illustrates sequence of layers over a steel substrate; right branch illustrates sequence of layers over a plastic substrate. Typical layer thickness is given.

ATR or transmission mode IR analyses using surface scraped, skived, or microtomed sections are generally limited to a depth resolution of ~0.5-5 μm. These techniques have been successfully applied using IR spectrophotometers configured with a standard sampling compartment or equipped with an IR-microscope and X/Y- translating stage. The primary advantage of using an IR-microscope equipped spectrophotometer is in the ability to obtain locus specific chemical information on relatively small samples. Experiments have been reported using IR-microscopes on samples as small as 30-100 μm in diameter [1,2]. Often there is a trade-off between IR spectral signal/noise (S/N), resolution, and acquisition time. Also, efficient handling of the smaller samples on a microscope stage requires additional skill and training.

Subtle differences in surface/near-surface composition as a function of depth can be obtained using different ATR internal reflection elements (IRE's), which provide a means to vary the IR beam penetration depth over the ~0.5-4 μm range [43,47]. The IRE material used (i.e., typically ZnSe, Ge or diamond) in direct contact with the sample, as well as angles of beam incidence/detection, determine the penetration depth. Another benefit is that ATR measurements can usually be made with minimal, if any, additional sample preparation. Diamond is the hardest of the IRE materials and is often used in analysis of materials that are quite hard or that possess irregular or roughened surfaces. If sufficient contact pressure is applied, then the material opposite the diamond will deform to provide uniform surface contact, optimizing the quality of the resultant ATR spectrum. *Figure 3* illustrates use of ATR analysis to monitor clearcoat degradation as a function of Xe B/B exposure time in the ~2700-3600 cm^{-1} range. An important advancement in recent years has been the design and application of IRE's such as ZnSe that have a thin layer of diamond on their surface [66] to increase their hardness and, as a result, virtually eliminating IRE surface deformation. Here the IRE effectively acts as if it is ZnSe, but with a much more durable (i.e., diamond) surface. A study detailing hindered amine light stabilizer (HALS) reduction in an acrylic clearcoat (i.e., acrylic polyol, melamine cross-linked) that was Xe B/B exposed in shown in *Figure 4*.

Transmission mode IR experiments can often provide very high quality spectra. However, there are practical limits in the thickness of the sample being analyzed due to beam attenuation. Typically, one can avoid spectral distortions (i.e., primary absorption artifacts and/or scattering) by using sample thickness under 5-7 μm. Surface skiving (micro-planing) or micro-toming are sampling techniques useful in obtaining samples in the ~2-7 μm (minimum) thickness range from automotive paint/coating materials [1,2] when cut at room temperature. Normally automotive finishes, including all of the component layers, are substantially thicker. Note that many automotive clearcoats are in the ~40-50 μm thickness range (*Figure 2*). Such sampling techniques can allow surface/near-surface characterization of these materials, as well as depth profiling through sequential sectioning.

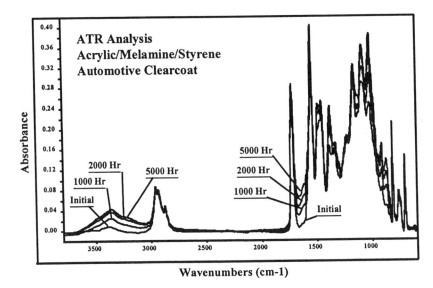

Figure 3. ATR analysis illustrating acrylic/melamine/styrene clearcoat degradation as a function of Xenon borosilicate/borosilicate exposure time/conditions in 600-3800 cm-1 range. Exposure according to SAE J-1960 Jun 89 TM. Exposure timeframe: 0-5000 Hrs.

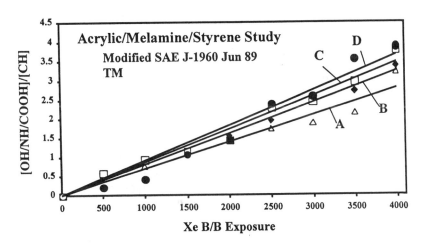

Figure 4. ATR study of hindered amine light stabilizer (HALS) reduction series for an acrylic/melamine/styrene based clearcoat. System A (▲ series) = 0.83% HALS. System B (◆ series) = 0.60% HALS. System C (■ series) = 0.40% HALS. System D (● series) = 0.20% HALS. Spectra obtained on Nicolet Nexus 470 FT-IR ESP spectrophotometer equipped with the Nicolet Smart DuraSampl/R ATR system. Photo-oxidation Index vs Xenon borosilicate/borosilicate (SAE J-1960 Jun 89 TM) exposure time/conditions.

Ford Scientific popularized a type of transmission mode IR experiment for study of automotive clearcoats as a function of accelerated exposure time/conditions [12,16]. Here a uniform, thin layer of clearcoat was applied onto the surface of a silicon wafer, targeting an ~5-7 μm thickness following selected (usually standard) cure conditions (i.e., time/temperature). A cross-section view is shown in *Figure 5*. The clearcoat application can be easily done by several approaches, including spin-coating, draw-downs, or solvent dilution with controlled drop spreading. This thickness of the clearcoat gave high-quality transmission mode spectra. These clearcoated silicon wafers were thereafter anchored onto metal holders whose position in the IR spectrophotometer sampling compartment could be highly reproduced. This was done to permit monitoring the same area (cross-section) of the sample following each exposure period. Although it was realized that such a clearcoat study only approximated the behavior of a clearcoat in an actual automotive system, normally containing clearcoat/basecoat/primer/electrocoat/steel or clearcoat/basecoat/primer/plastic multi-layers, it was considered useful in rating/ranking of clearcoats. Clearcoat performance as a function of photo-oxidation and hydrolysis was determined from plots of [OH, NH, COOH]/[CH] area ratios (i.e., photo-oxidation index) versus exposure time as shown in *Figures 6*. The [OH, NH, COOH] value includes the hydroxy, amine and carboxylic acid envelop (peak areas) in the 2200-3700 cm^{-1} range. The [CH] value includes the methylenic envelop (peak areas) in the 2700-3000 cm^{-1} range. In addition, [CH] area versus exposure time was used to follow clearcoat mass loss. This is considered valid as long as the sample area (i.e., locus) analyzed is reproduced after each exposure period. Commonly, exposures were conducted in an Atlas Weatherometer according to SAE J-1960 Jun 89 TM protocol, using a Xenon lamp with an inner/outer borosilicate filter set to approximate natural outdoor conditions in the ~280-360 nm range. Subsequently, Ford Scientific researchers had determined that PAS analyses of full (or multilayer) automotive systems were a more appropriate indicator of real world performance [13,14,26,27,32].

Ford Scientific has determined that PAS may be ideally suited for monitoring chemical changes associated with photo-oxidation and hydrolysis for automotive systems, specifically in clearcoat layers [18,21,32,33,41]. Their intent was to analyze a relatively thick volume element and not depth profiling per se. Earlier they had developed an approach (detailed above) wherein a thin clearcoat layer was applied to a (mid-IR range transmitting) silicon wafer substrate. These samples were weathered and the resultant chemical change was followed by transmission-mode IR analysis. The problem they had with this earlier approach was that it only approximated what would happen to a clearcoat in an actual automotive system. Here the clearcoat was isolated from the rest of the system (i.e., typically including basecoat, primer, and electrocoat on a zinc phosphate treated steel substrate). Since normal component or degradation species migration between layers of a system (i.e., clearcoat/basecoat bilayers are of particular concern) is eliminated, it is not certain how well the earlier experiments would predict actual system in-use (or field) performance. Also,

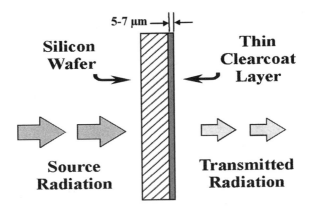

Figure 5. Cross-section view of Ford transmission mode IR analysis. The silicon wafers used have been either round (~12 mm dia.) or square (~17x17 mm2). Initial clearcoat target thickness is typically ~7 µm.

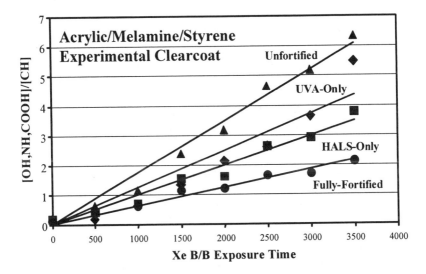

Figure 6. IR analysis of an Acrylic/Melamine/Styrene-based automotive clearcoat system: Plot of [OH,NH,COOH]/[CH] versus Xenon borosilicate/borosilicate exposure time (hours). Key to UV-screener fortification: Fully fortified (both UVA and HALS); HALS only; UVA only; and unfortified (no UVA or HALS). The clearcoats were applied on Si-wafers, exposed, and IR analysis done in transmittance mode.

the prior transmission-mode experiments limited the thickness of the clearcoat to ~5-7 μm in thickness to prevent or minimize spectral (i.e., primary absorption, scattering) distortions. Since normal clearcoat layers are ~40-50 μm thick, PAS experiments were designed to monitor chemical change in a relatively thicker volume element. They targeted an ~8-12 μm thick layer starting at the surface. *Figure 7* illustrates the point that a full (multi-layered) coating system can be routinely analyzed in these experiments. *Figure 8* shows a typical exposure time data set in the ~2700-3600 cm^{-1} range. Photo-oxidation index values, as documented in the CC/Si-wafer studies, are routinely determined to track automotive clearcoat degradation.

DRIFT surface/near-surface analysis of automotive finishes has been done using "Si-Carb" (i.e., silicon carbide sandpaper) sampling [1,2,65]. Surface specificity of this approach is in the ~5-10 μm depth range, due to the practical limitations of polishing with fine grit silicon carbide sandpaper. This experiment is described in *Figure 9*.

Selection of sampling approach and measurement technology are often dictated by equipment and expertise available. Ideally, the surface specific volume element that is targeted in a given experiment is appropriate to the chemical information being sought. Our recent experience suggests that ATR mode measurements using an IRE (i.e., ZnSe) with a thin layer of diamond at the sample contact surface (DiComp™ diamond ATR system) are fast, reproducible, and provide ranking similar to that observed with the other techniques. This tends to be our technique of choice assuming that the sampling volume element is acceptable. PAS mode experiments are also very useful, particularly if a less surface specific (thicker) volume element is needed.

Ultraviolet-Visible Analyses (Thin Layer Volume Element). T-mode analysis by a typically configured ultraviolet-visible (UV/VIS-) spectro-photometer can be done using thin layer clearcoat application over UV/VIS range transmissive substrates (i.e., quartz microscope slides). These clearcoat coated quartz slides are indexed so that the position in the sample compartment holder is reproduced. This is important in the event of non-uniform coating thickness across the slide. Clearcoat thickness in the ~6-10 μm range is used with typical (initial) loading levels of UV-screener fortification. Higher loading levels of UV-screener package components or the evolution of significant concentrations of UV-absorbing degradation products may require use of thinner coating thickness. The system chemistry, as well as the exposure time/conditions, dictates this.

Depth Profiling: Chemistry

Chemical depth profiling techniques applied in study of automotive coating systems involve many of the previously cited techniques found useful for surface/near-surface or thin layer characterization. Specific depth sampling approaches have been developed and used together with the various chemical

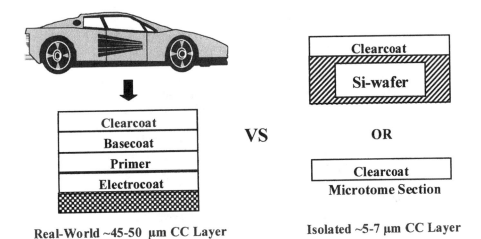

Real-World ~45-50 μm CC Layer   Isolated ~5-7 μm CC Layer

Figure 7. PAS mode (IR) analysis can be done on commercial multi-layered automotive coating systems. Specified by Ford for their coating suppliers. The sample is normally obtained as a small disk (~1 cm dia.) obtained with a mechanical punch. Transmission mode (IR) analysis typically uses a thin (~ 7 μm) layer of clearcoat on a Silicon substrate or microtome section of clearcoat obtained co-planar to the automotive system surface.

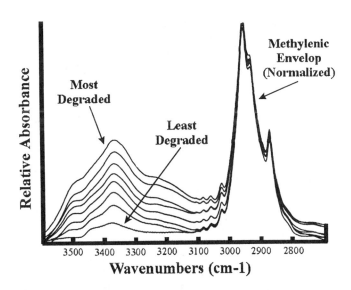

Figure 8. PAS mode (IR) spectra of an acrylic/melamine-based clearcoat subjected to QUV FS-40 exposure over a period of ~4000 Hrs. Spectra shown cover only the ~2700-3600 cm-1 range used in the standard determination of photo-oxidation index values.

analysis methods to obtain actual chemical depth profiles. Common approaches include cross-sectioning of a given layer or multi-layered system, sequential microtoming or skiving co-planar to the surface, or Si-Carb polishing through a sample. Direct analysis of the material obtained from depth sampling can be done using IR, Raman, ESCA, ToF-SIMS, and UV-VIS techniques, to name those that are more common referenced. Direct analysis primarily uses the spectroscopic techniques in chemical depth profiling. Indirect analysis involving solvent extraction can also be done, primarily to monitor additive and degradation product locus/concentration as a function of depth. In the case of indirect analysis of solvent extracts, spectroscopic, spectrometric, as well as chromatographic techniques have been successfully used in chemical depth profiling.

The main reasons for chemical depth profiling are to determine general chemical composition or component distribution as a function of depth. In practice, this can range from tracking a single species (i.e., additive or degradation product), a molecular sub-structure (i.e., repeating unit in a polymer chain, cross-linker, or specific functional group), stratification (i.e., gradient formation), segregation (i.e., component de-mixing), general composition to component mixing, or even migration across interfaces. Targeted areas of research involving depth profiling of automotive coating systems are many: degree/type of crosslinking after curing [17,52]; depth dependence of cross-link density versus glass transition temperature [9,10]; UV-screener locus, migration or degradation as a function of exposure time/conditions [16,62]; surface/near-surface versus bulk chemical composition [1,2,39,51]; composition at/near interfaces [20,22,42,54,58]; tracking photo-oxidation and/or hydrolysis as a function of exposure time/conditions [1,2,15,19,20,22,35,36,37,46]; mixing of components between system layers [20]; segregation of pigment particles at/near interfaces [44]; interaction between pigment particles and binder [57]; defect formation at a given locus [47]; monitoring adsorption of moisture; or solvent trapping.

Infrared Analyses (Depth Profiling). Depth profiling of automotive coating systems has been reported using various IR-based techniques, including transmittance- (T-), ATR-, DRIFT-, and PAS-mode analyses [1,2,3,6,25,26]. Due to the spatial resolution inherent in these techniques, ranging from ~10-1000 μm, analysis of typical automotive system layers by cross-section is generally not practical. However, a variety of depth-sampling approaches have been developed, making many of these techniques quite practical. Also, the wide availability of IR-based techniques in many laboratories, and their versatility, makes them attractive as tools for routine measurements. The intent in this section is to highlight the IR depth profiling techniques that are commonly available, including the various associated depth sampling approaches reported.

T-mode analysis by IR-microscopy can be done using thin layer cross-sections, which must be supported flat and securely on an X-Y micro-translating stage [12], while using appropriate optical masking. These thin layer cross-sections can be obtained by use of microtomy techniques, however section handling during the measurements is not trivial, and requires considerable skill and training. Often, small cross-sections must be immobilized and supported flat. In the case of a typical clearcoat of ~40-50 μm thickness, the spatial resolution limits the number of sequential steps for the clearcoat to ~4-6 in the normal direction. Clearcoat layers, which tend to be optically particle free, give the highest quality spectra relative to the other particle (i.e., pigment, mica flake, metal flake, latex) containing system layers. Scattering of the incident IR radiation by particles (or particle aggregates) in other layers can be quite substantial, even for these relatively thin sections. Therefore, most IR T-mode analyses that have been reported are on clearcoats or particle-free binders of other layers. The particle-free binders are often prepared as models in study of binder chemistry degradation.

T-mode analysis by IR-microscopy is more readily accomplished by microtoming ~4-10 μm thick sections co-planar to the surface. A depth profile is generated by sequential analysis of these sections [12]. Sections from automotive systems can be quickly prepared using room-temperature slab-microtomy. A convenient size of flat panel section for slab-microtoming is ~1"x1". Caution should be exercised in preparation and handling of automotive system sections due to their fragile nature. Longer-term outdoor or accelerated aging of automotive systems often causes them to become somewhat difficult to section and mechanically fragile. Use of very sharp cutting blades (i.e., carbide or diamond) and optimizing the cutting angle can minimize this problem. Slightly thicker sections can be prepared from exposed samples to help maintain integrity of sections during handling. In addition, many slab-microtomes can operate under cryo-conditions, which often permit sectioning of weaker, degraded coating layers without tearing or shredding. Overall, this approach is equally convenient for sectioning a single coating layer (i.e., clearcoat or monocoat), a bi-layer (i.e., clearcoat/basecoat), or the entire multi-layered coating system (i.e., clearcoat/basecoat/primer/electrocoat). Normally this combination of sample preparation and T-mode IR analysis is found effective in determining the general chemistry as a function of depth. Stratification, segregation or migration of components and changes in cross-linking density across a given coating layer, as well as interlayer mixing of components, can often be established. *Figure 10* gives a clearcoat depth profile of the general chemistry. The transmission-mode analyses were done on a Nicolet Magna-IR 760 FT-IR spectrophotometer equipped with a Nicolet Nic-Plan IR-microscope.

A potassium bromide (KBr) abrasion technique has been used to follow strata in a phenolic-modified epoxy coating [53]. Depth sampling was done by first wetting the coating with a swelling solvent (i.e., hexane, toluene or xylene) and then gently abrading the surface with KBr powder. The technique is effectively a type of polishing. The KBr/polymer mixture was pressed into a pellet and analyzed in T-mode. Repeated polishing of the same area provided a crude

198

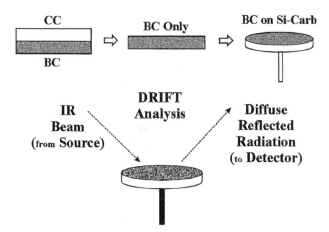

Figure 9. Si-Carb sampling for DRIFT mode (IR) analysis of an automotive basecoat. An aluminum pedestal is typically used to support the silicon carbide (Si-Carb) sandpaper pad. The post was fitted for DRIFT analysis using a SpectraTech Applied Systems FT-IR spectrophotometer equipped with a DRIFT module.

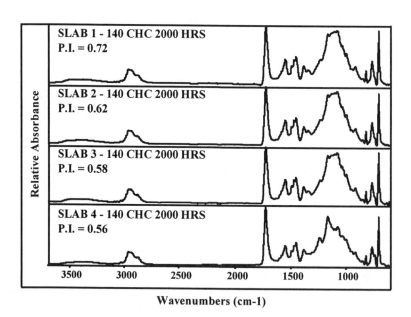

Figure 10. Cleveland humidity cabinet exposure series for an acrylic/ melamine/silane-based clearcoat layer. Transmission mode (IR) analysis of a slab microtomed series. Photo-oxidation index (P.I.) values and general IR spectra are given for each section.

means of depth profiling. The Si-Carb polishing technique (note below in this section) has been used with similar success [1,2].

ATR-mode (IR) depth profiling can be done directly from the top surface [12,40,43] or from a revealed surface. Standard internal reflection elements (IRE's), by selecting composition and angles of beam incidence, have been used to sample (top-down) to various depths. The configuration for this type of analysis is done using a typical spectrophotometer sample compartment. Normally, the absorption of polymer-based coating systems beyond ~10 μm is (in our experience) problematic, often ranging anywhere from spectral distortions to total absorption of incident radiation. Since ATR analysis is effectively a top-down measurement, the volume element being sampled includes spectral information from a given maximal depth and all material above this depth. Difference (i.e., subtraction) spectra are commonly obtained to help resolve spectral detail from a given (sub-surface) volume element. However, penetration of radiation through clearcoats is wavelength and material dependent, resulting in some uncertainty as to the actual depth element being sampled. Penetration of radiation through the other, particle containing, layers have an additional complication of light scattering and potentially stronger absorption of radiation. This type of ATR analysis is usually limited to surface/near-surface analysis. Note that polishing, grinding or microtoming away upper layers can prepare a surface from a deeper volume element [51].

ATR-mode (IR) depth profiling can be done directly from microtomed sections. Surface roughness caused by the microtome cutting edge or chatter during cutting can result in poor contact when using the standard IRE crystals, which are often ~1/2" x 3/4" in contact area or even larger. The quality of the contact in a single- or multi-bounce IRE is a major contributor to the quality of the spectrum. ATR analyses are now possible on surface areas that are ~50-100X smaller in area than traditional elements, using either an IR-microscope configured with an ATR-objective [67] or the DiComp ™ diamond ATR system [66,68,69].

Recent developments and availability of ATR-objectives that can be used with IR-microscopes have virtually eliminated the IRE/coating contact problem [67]. The size, shape, and applied pressure of the ATR-objective IRE used requires an area of ~3-4 mm in diameter that is adequately smooth to ensure good contact. Automotive coating system layers are sufficiently soft to permit excellent contact across the face of the ATR-objective IRE, even when using microtomed sections. Note that adequate pressure must be applied to optimize contact, and thus quality of the spectrum.

Another approach involves use of the DiComp ™ diamond ATR design that locates a thin diamond layer in optical contact with a focusing element such as ZnSe or KRS-5 [66,68,69]. This provides an ATR system that has a broad spectral range and high throughput. It is also corrosion resistant, abrasion-proof, and has excellent mechanical strength.

The depth resolution obtained via microtoming is thus limited by the thickness of the sections. Practical handling at room temperature dictates use of sections ~6 μm or thicker. Automotive coating sections that are thinner than ~5-6 μm tend to break apart while handling. Aging of the coatings through photo-oxidation or acid/base hydrolysis tends to mechanically weaken the coatings even more (relative to an un-exposed coating), increasing handling difficulty. A typical depth profile of an automotive clearcoat layer can be conveniently obtained using a DiComp ™ diamond ATR system to obtain the spectra. The diamond plate is in optical contact with ZeSe (as the IRE). ATR-mode analyses such as these have been routinely done on a Nicolet Nexus 470 FT-IR spectrophotometer configured with this ATR system and Nicolet's ESP ™ operating system [12].

DRIFT-mode (IR) depth profiling relies on the use of a Si-Carb polishing technique, as detailed earlier in the case of surface/near-surface chemical characterization [1,2]. This is accomplished by successive application of the Si-Carb polishing procedure. The depth achieved, or amount of material abraded, can be monitored with a profilometer or caliper depth gauge. It has been determined that a Dremel ™ drill with variable rotation speed setting control and a manually adjustable vertical drill stand can be successfully used in sequential sampling. Note that Si-Carb polishing for depth sampling requires considerable training and care to be effective in practice. Often the Si-Carb depth sampling is somewhat crude, relative to that routinely achieved by microtomy. DRIFT-mode analyses have been conveniently done on an Applied Systems / Spectra Tech FT-IR spectrophotometer equipped with a powder/Si -Carb disk sampling module [1,12].

PAS-mode (IR) "linear scan" experiments can be readily configured to allow for variable depth sampling at a given wavenumber [24,25,29,45]. These measurements can be conveniently done with many commercial FT-IR spectrophotometers by changing the mirror velocity to effectively access a given sample depth. A detailed discussion of the principles and instrumentation required in performing these experiments can be found in the work of McClelland et. al. [61]. PAS-mode (IR) "step scan" experiments have also been applied in depth profiling study of automotive finishes. These measurements require an IR spectrophotometer capable of moving the mirror incrementally.

PAS-mode (IR) measurements have been successfully used by a number of laboratories in characterizing automotive, maintenance and fiber paint/coating systems. Certain groups have focussed their efforts in study of interfacial regions, between coating layers (i.e., CC//BC) and coatings over plastic substrates [28,31]. PAS has been reported in study of particle sedimentation processes in pigmented coatings [34] and residual unsaturation in photo-cured acrylate-based formulations [30]. Although nearly all of this work is focussed in the mid-IR range, Ford Scientific has also reported applications of PAS in the UV range [23]. This work was specifically directed at evaluation of the distribution of UV-absorbing benzotriazole-based (UVA) additives in clear paint layers as a function of depth.

Thermoplastic olefins (TPO) are now commonly used in automotive bumper and interior fascia component applications. PAS techniques have been used extensively in the study of component distribution in TPO [13,14]. The primary components of TPO include a crystalline polypropylene (PP) phase and an ethylene-polypropylene rubber (EPR) phase. Using PAS they were able to determine the distribution of PP and EPR components as a function of depth.

As cited in the prior *Infrared Analyses (Surface/Near-Surface or Thin Layer Volume Element)* section, selection of sampling approach and measurement technology are often dictated by equipment and expertise available. When information at a given depth locus (i.e., CC/BC interface, CC particle/defect) is desired, PAS mode experiments can often be efficiently configured for this purpose. When a detailed chemical depth profile is desired, the combination of slab microtomy and ATR (DiComp^{TM} system) is very efficient, requiring minimal data reduction. In all of the IR-based measurement technologies the analysis tends to be top-down, rather than from a cross-section. The practical analysis spot size is the limiting factor. Technologies such as ToF-SIMS or Raman microprobe may be better suited to cross-section depth profiling due to a much smaller analysis spot size (< 2 μm dia.), but are in general significantly more expensive and/or training/experience intensive.

Ultraviolet-Visible Analyses (Depth Profiling). T-mode analysis by ultraviolet-visible (UV/VIS-) microscopy can be done using thin layer automotive system cross-sections, which must be supported flat and securely on an X-Y micro-translating stage. *Figure 11* provides a diagram of sample preparation steps for obtaining clearcoat cross-sections that can be handled relatively easily on a stage. *Figure 12* provides a diagram of the depth profiling analysis. This technique monitors all UV/VIS absorbing species encountered in the beam path, including primarily UV-screeners (UVA's), hindered amine light stabilizers (HALS) and certain paint/coating system degradation products. Note that the beam path is effectively a function of the beam cross-section and thickness of the microtomed section. The cross-section handling is essentially identical to that detailed earlier for T-mode analysis by IR-microscopy. Ford Scientific has reported UV/VIS T-mode depth profiling with cross-sections as thick as ~10 μm [16]. As before, such thin layer cross-sections can be obtained by use of standard microtomy techniques. The small cross-sections must be handled carefully and are immobilized on a UV-transmitting quartz slide. Longer chain terminal alcohols, ranging from n-hexanol to n-decanol, are useful for immobilizing the samples. They have no UV absorption over the wavelength range of interest and effectively immobilize the sample onto the quartz surface. Masking of the incident beam to a small rectangle has been done, ~5 μm wide (along the normal depth axis) with variable length (parallel to the surface). Note that the mostly particle free clearcoat layers give the highest quality spectra relative to the other

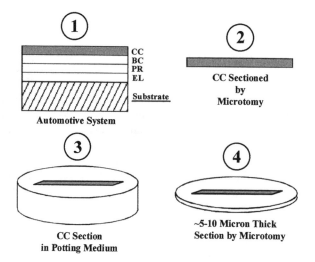

Figure 11. UV-VIS depth profiling sample preparation. This technique uses a X-Y translating stage and a UV-VIS microscope. Key: (1) Multi-layered automotive coating system; (2) Clearcoat is isolated from automotive coating system; (3) Clearcoat section is potted on edge showing cross-section; and (4) A slab microtome is used to obtain cross-section of clearcoat embedded in potting medium.

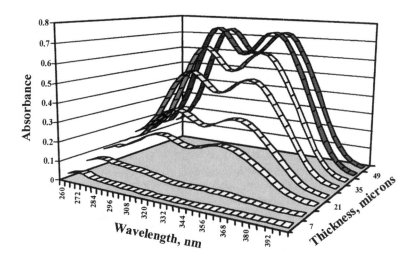

Figure 12. UVA depth profile across an automotive coating system clearcoat/basecoat bilayer. UVA is benzotriazole type. Microtome section thickness is ~7 μ's. The first 6 sections from the coating system surface are clearcoat. The next section is primarily basecoat. The final section is only basecoat. Note evidence for absorption in the 260-300 nm range, which is not due to the UVA additive, but probably degradation species.

particle containing system layers. Scattering of the incident UV/VIS radiation can be quite substantial, even for these relatively thin sections. Therefore, most UV/VIS transmission-mode analyses that have been reported are on clearcoats or particle-free binders of other layers.

T-mode analysis by monitoring UV absorbing extractables in coating sections microtomed co-planar to the surface can also be very effective. *Figure 13* provides a diagram of the analysis. Successful study of clearcoats and clearcoat/basecoat bi-layers has been reported [3,4,5,6,7,8,10,11,12]. In a number of cases the use of IR and ToF-SIMS analyses has also been reported to monitor UVA and HALS adsorption to certain pigment and ultrafine oxide particles [3,4,6,7,8]. A solvent appropriate to the material of interest to extract each section [12]. For example, extraction of UV/VIS absorbing UVA's or HALS commonly encountered in the automotive industry is readily accomplished with solvents such as methylene chloride (CH_2Cl_2) or chloroform $(CHCl_3)$. Since these solvents do not absorb in the wavelength range of interest (~260-460 nm), the extract solutions can be analyzed directly with a UV/VIS spectrophotometer. Filtration or centrifigation techniques can be used to remove solids, in order to prevent particle light scattering. The solutions are stored airtight, thus preventing loss of solvent through evaporation. The UV/VIS absorption is then normalized to the amount (i.e., weight) of material extracted. *Figure 14* shows a (primarily) benzotriazole-type UVA depth profile in a clearcoat subjected to various common (degradation) accelerating environments. This experiment allows comparison of different environments such as those found in UV, QUV and CHC exposures with respect to UVA permanence. Issues of non-bound, benzotriazole-type UVA migration, extraction and/or degradation can now be readily addressed.

High Performance Liquid Chromotography Analyses (Depth Profiling). Techniques such as high performance liquid chromatography (HPLC) or supercritical fluid chromatography (SFC) have been found quite effective in determining depth profiles of species extracted from slab-microtomed sections. DuPont has shown the utility of this approach in study of clearcoats and clearcoat/basecoat bi-layers [12]. As detailed previously, when using UV/VIS spectrophotometric analysis a solvent appropriate to the material of interest is used to extract each slab microtomed section. In HPLC analysis, the same solutions obtained in the previous case in study of UVA's and HALS can be used. Depth profiles are obtained when the chromatographic peak areas for a given species in microtomed sequence of sections are normalized to the weight of material extracted and a plotted as a function of slab locus. Issues of UVA permanence, specifically related to type and rate of chemical degradation, interlayer migration, and pigment adsorption can now be addressed.

204

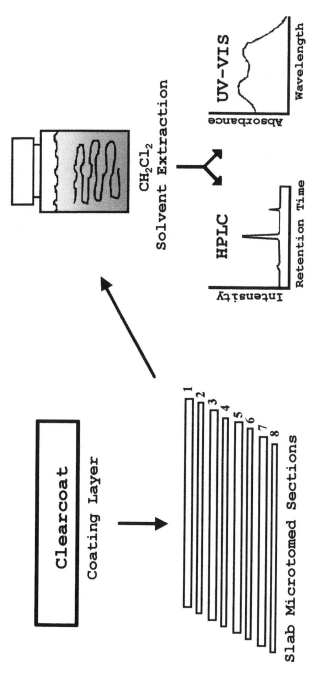

Figure 13. Flowchart for depth profiling by UV or HPLC analysis. A coating layer or multi-layer coating system is slab microtomed co-planar to the surface. Each section is weighed and UVA extracted into 5 mL solvent (usually CH$_2$Cl$_2$). Each solution must be filtered or centrifuged to give a particle free solution. UV-VIS or HPLC analysis is conducted on the solutions and the resultant signals are normalized to the weight of the extracted section.

State-of-the-Art: Overview

Chemical composition and depth profiles, including component distribution/re-distribution, in automotive coating systems can be effectively determined with modern analytical tools. This offers a particular advantage to companies that research and develop these materials. Knowledge of the chemistry at a given locus, and as a function of outdoor/accelerated exposure time/conditions, allows one to minimize product development time. This, in turn, allows a company to bring products to the marketplace as early as possible, while maintaining high confidence as to expected in-field product performance. Tools such as these also provide an advantage to direct coating customers (i.e., automobile manufacturers) using these materials. Defect identification or at least characterization is often possible. In addition, chemical information (i.e., new species formation, degradation reaction rates, and/or component distribution/re-distribution) obtained can be used to help predict service life performance. This report details analytical tools and methodology that are generally available to many laboratories and that have been applied in characterization of these systems.

Surface characterization and depth profiling tools such as those described herein allow developers of automotive coating systems an intimate look at chemical composition and component distribution. Studies are underway in various laboratories to determine correlations between chemistry and physical properties or mechanical performance, all of which tend to be early indicators of impending appearance and in-use changes. Since chemical changes can (usually) be monitored long before appearance changes are observed, tools to document chemistry and to track chemical changes have great potential in predicting service life performance. Many of the analytical tools for chemical characterization are very sensitive and species specific, permitting analyses at low concentrations in spite of often complex background matrices. Some of the tools are also ideal for identifying species and tracking changes at surfaces and interfaces, a function of their relatively high resolving power. Future developments in the sampling techniques and analytical methods will inevitably increase the efficiency of gathering chemical information, regardless of locus. Also, the time required for gathering this information will decrease. Specifically, data-basing and data-harvesting tools are being developed, which will permit rapid and effective use of this chemical information.

The focus of the applications detailed herein has been on automotive coatings or automotive coating systems, however these analytical tools are certainly not limited to the characterization of these materials. In many cases, successful applications with other polymer-based materials have been reported in the literature. Tools such as these have been used in other areas as diverse as textile fibers, flooring systems, pharmaceutical drug-delivery coatings, engineering materials, food packaging products, architectural coatings/plastics, aeronautical composites, and medical implants. However, many of these tools have been developed in recent years and, as a result, applications in many areas have not been explored.

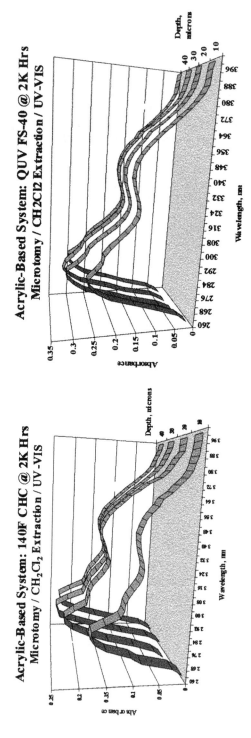

Figure 14. UV-VIS depth profile of acrylic-melamine clearcoats subjected to Cleveland humidity cabinet (CHC), Q-Panel QUV FS-40, and Q-Panel UV-Only FS-40 exposures. Prior to sectioning each sample had 2000 Hr exposure. Slab microtomy was used to obtain samples as a function of depth.

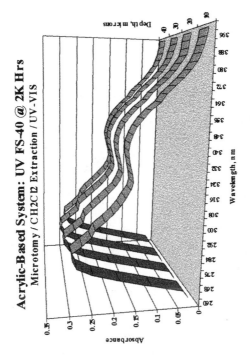

Figure 14. *Continued.*

208

References

1. Adamsons, K.; Lloyd, K.; Stika, K.; Swartzfager, D.; Walls, D.; and Wood, B.; *Interfacial Aspects of Multicomponent Polymer Materials*; Edited by Lohse et al., Plenum Press, New York, **1997**, pp. 279-300.

2. Adamsons, K.; Stika, K.; Swartzfager, D.; Walls. D.; Wood, B.; and Lloyd, K.; *Polym. Mater. Sci. Eng.*, **1996**, *75*, pp. 482-483.

3. Haacke, G; Longordo, E.; Brinen, J.S.; Andrawes, F.F.; and Campbell, B.H.; *J. Coat. Technol.*, **1999**, *71*, pp. 87-94.

4. Haacke, G.; Longordo, E.; Brinen, J.S.; Andrawes, F.F.; and Campbell, B.H.; *Proc. Int. Waterborne, High-Solids, Powder Coat. Symp.*, 25-th, **1998**, pp. 433-443.

5. Andrawes, F.F.; Valcarcel, T.; Haacke, G.; and Brinen, J.S.; *Anal. Chem.*, **1998**, *70(18)*, pp. 3762-3765.

6. Haacke, G.; Longordo, E.; Andrawes, F.F.; and Campbell, B.H.; *Prog. Org. Coat.*, **1998**, *34(1-4)*, pp. 75-83.

7. Haacke, G.; Longordo, E.; Andrawes, F.F.; and Campbell, B.H.; *Proc. Int. Conf. Org. Coat.: Waterborne, High Solids, Powder Coat. 23-rd*, **1997**, pp. 171-185.

8. Haacke, G.; *Book of Abstracts*, 215-th ACS National Meeting, Dallas, TX, March 29 – April 2, **1998**, PMSE-195.

9. Haacke, G.; *Relax. Phenom. Polym., Proc. Symp.*, **1994**, pp. 167-173.

10. Haacke, G.; Andrawes, F.F.; and Campbell, B.H.; *J. Coat. Technol.*, **1996**, *68(855)*, pp. 57-62.

11. Haacke, G.; Brinen, J.S.; and Larkin, P.J.; *J. Coat. Technol.*, **1995**, *67(843)*, pp. 29-34.

12. Adamsons, K.; *Short Course Presentation at The Atlas School for Natural and Accelerated Weathering (ASNAW)*; April 28-30, **1999**, Miami, Florida.

13. Pennington, B.D.; Ryntz, R.A.; and Urban, M.W.; *Polymer*, **1999**, *40(17)*, pp. 4795-4803.

14. Pennington, B.D.; Ryntz, R.A.; and Urban, M.W.; *Book of Abstracts*, 214-th ACS National Meeting, Las Vegas, NV, September 7-11, **1997**, PMSE-337.

15. Gerlock, J.L.; Prater, T.J.; Kaberline, S.L.; Dupuie, J.L.; Blais, E.J.; and Rardon, D.E.; *Polym. Degrad. Stab.*, **1999**, *65(1)*, pp. 37-45.

16. Gerlock, J.L.; Smith, C.A.; Cooper, V.A.; Dusbiber, T.G.; and Weber, W.H.; *Polym. Degrad. Stab.*, **1998**, *62(2)*, pp. 225-234.

17. Dupuie, J.L.; Weber, W.H.; Scholl, D.J.; and Gerlock, J.L.; *Polym. Degrad. Stab.*, **1997**, *57(3)*, pp. 339-348.

18. Gerlock, J.L.; Smith, C.A.; Carter III, R.O.; Dearth, M.A.; Korniski, T.J.; Kaberline, T.J.; DeVries, J.E.; Cooper, V.A.; Dupuie, J.L.; and Dusbiber, T.G.; *Surf. Coat. Aust.*, **1997**, *34(7)*, pp. 14-16.

19. Gerlock, J.L.; Prater, T.J.; Kaberline, S.L.; and DeVries, J.E.; *Polym. Degrad. Stab.*, **1995**, *47(3)*, pp. 405-411.

209

20. DeVries, J.E.; Haack, L.P.; Prater, T.J.; Kaberline, S.L.; Gerlock, J.L.; Holubka, J.W.; Dickie, R.A.; and Chakel, J.; *Prog. Org. Coat.*, **1994**, *25(1)*, pp. 95-108.
21. Gerlock, J.L.; Smith, C.A.; Nunez, E.M.; Cooper, V.A.; Liscombe, P.; and Cummings, D.R.; *Polym. Prepr. (Am. Chem. Soc., Div. Polym. Chem)*, **1993**, *34(2)*, pp. 199-200.
22. DeVries, J.E.; Haack, L.P.; Prater, T.J.; DeBolt, M.; Gerlock, J.L.; Holubka, J.W.; and Dickie, R.A.; *Polym. Mater. Sci. Eng.*, **1992**, *67*, pp. 69-70.
23. Carter III, R.O.; *Opt. Eng.*, **1997**, *36(2)*, pp. 326-331.
24. Takaoka, K.; *Shikizai Kyokaishi*, **1986**, *59(6)*, pp. 347-356.
25. Muraishi, S.; *Bunko Kenkyu*, **1984**, *33(4)*, pp. 269-270.
26. Urban, M.W.; *Polym. Mater. Sci. Eng.*, **1998**, *78*, pp. 18-19.
27. Urban, M.W.; *Book of Abstracts, 215-th ACS National Meeting*, Dallas, March 29-April 2, **1998**, PMSE-011.
28. McDonald, W.F.; *Int. SAMPE Symp. Exhib. 39*, **1994**, pp. 2518-29.
29. Dolby, P.A.; and McIntyre, R.; *Polymer*, **1991**, *32(4)*, pp. 586-9.
30. Factor, A.; Tilley, M.G.; and Codella, P.J.; *Appl. Spectrosc.*, **1991**, *45(1)*, pp. 135-8.
31. McDonald, W.F.; and Urban, M.W.; *J. Adhes. Sci. Technol.*, **1990**, *4(9)*, pgs. 751-64.
32. Carter III, R.O.; Paputa Peck, M.C.; and Bauer, D.R.; *Polym. Degrad. Stab.*, **1989**, *23(2)*, pp. 121-134.
33. Bauer, D.R.; Paputa Peck, M.C.; and Carter III, R.O.; *J. Coat. Technol.*, **1987**, *59(755)*, pp. 103-109.
34. Bruecher, K.H.; and Perkampus, H.H.; *Fresenius Z. Anal. Chem.*, **1986**, *325(8)*, pgs. 676-682.
35. Lemaire, J.; and Siampiringue, N.; *ACS Symp. Ser. (Service Life Prediction of Organic Coatings) 722*, **1999**, pp. 246-256.
36. Lemaire, J.; Gardette, J.L.; and Lacoste, J.; *Makromol. Chem., Macromol. Symp. 70-71 (34-th Int. Symp. on Macromol)*, **1992**, pgs. 419-431.
37. Siampiringue, N.; Leca, J.P.; and Lemaire, J.; *Eur. Polym. J.*, **1991**, *27(7)*, pp. 633-641.
38. Adamsons, K.; Litty, L.; Lloyd, K.; Stika, K.; Swartzfager, D.; Walls, D.; and Wood, B.; *ACS Symp. Ser. 722 (Service Life Prediction of Organic Coatings)*, **1999**, pp. 257-287.
39. Lamers, P.H.; Johnson, B.K.; and Tyger, W.H.; *Polym. Degrad. Stab.*, **1997**, *55(3)*, pp. 309-322.
40. Huang, J.; and Urban, M.W.; *Appl. Spectrosc.*, **1993**, *47(7)*, pp. 973-981.
41. Carter, R.O.; Palmer, R.A.; Dittmar, R.M.; Manning, C.J.; Bains, S.S.; and Chao, J.L.; *Proc. SPIE Int. Soc. Opt. Eng. 1145 (Int. Conf. Fourier Transform Spectrosc., 7-th)*, **1989**, pp. 362-363.
42. Nishioka, T.; Nishikawa, T.; and Teramae, N.; *Kobunshi Ronbunshu*, **1989**, *46(12)*, pp. 801-807.
43. Shoda, K.; and Ishida, H.; *Bosei Kanri*, **1989**, *33(4)*, pp. 106-113.
44. Takahashi, M.; *Prog. Org. Coat.*, **1986**, *14(1)*, pp. 67-86.

210

45. Ochiai, S.; *Toso Kogaku*, **1985**, *20(5)*, pp. 192-195.
46. English, A.D.; and Spinelli, H.J.; *ACS Symp. Ser. 243 (Charact. Highly Cross-Linked Polym.)*, **1984**, pp. 257-269.
47. Kagawa, K.; *Hyomen Gijutsu*, **1999**, *50(5)*, pp. 431-436.
48. Howdle, S.M.; Ramsay, J.M.; and Cooper, A.I.; *J. Polym. Sci., Part B: Polym. Phys.*, **1994**, *32(3)*, pp. 541-549.
49. Noda, K.; Yasui, T.; and Inaba, T.; *Bosei Kanri*, **1993**, *37(12)*, pp. 439-445.
50. McEwen, D.J.; and Cheever, G.D.; *J. Coat. Technol.*, **1993**, *65(819)*, pp. 35-41.
51. Le, P.C.; Zeilinger, H.; and Weigl, J.; *Wochenbl. Papierfabr.*, **1993**, *121(8)*, pp. 284-293.
52. Klygin, V.N.; Gagina, I.A.; Petrzhik, G.G.; and Simakov, Y.S.; *Lakokras. Mater. Ikh Primen.*, **1984**, *2*, pp. 30-32.
53. Pearson, L.R.; *Polym. Character., Interdisciplinary Approaches, Proc. Symp.*, **1971**, pp. 137-145.
54. Schuetzle, D.; Carter III, R.O.; and DeVries, J.E.; *Adv. Org. Coat. Sci. Technol. Ser.*, **1989**, *11*, pp. 229-245.
55. Osterhold, M.; and Armbruster, K.; *Macromol. Symp. 126 (6-th Dresden Polymer Discussion Surface Modification)*, **1997**, pp. 295-306.
56. DeVries; *J. Mater. Eng. Perform.*, **1998**, *7(3)*, pp. 303-311.
57. Ohsawa, K.; *Toso To Toryo*, **1968**, *164 (33-6)*, pp. 39-44.
58. Nishioka, T.; and Teramae, N.; *Anal. Sci. 7 (Suppl., Proc. Int. Congr. Anal. Sci., Pt. 2)*, **1991**, pp. 1633-1636.
59. Zhao, Y.; and Urban, M.W.; *Book of Abstracts, 215-th ACS National Meeting*, Dallas, March 29 – April 2, **1998**, PMSE-015.
60. Craver, C.D.; *ACS Symp. Ser., Appl. Polym. Sci. (2-nd Ed.)*, **1985**, pp. 703-738.
61. McClelland, J.F.; R.W. Jones, R.W.; Luo, S.; and Seaverson, L.M.; *A Practical Guide to FT-IR Photoacoustic Spectroscopy; Proper Sample Handling with Today's IR Instruments*; Edited by P. Coleman, CRC Press.
62. Gerlock, J.L.; Smith, C.A.; Nunez, E.M.; Cooper, V.A.; P. Liscombe, P.; Cummings, D.R.; and Dusibiber, T.G.; *Adv. Chem. Ser.*, **1996**, *249 (Polymer Durability)*, pp. 335-347.
63. Martin, J.W.; Saunders, S.C.; Floyd, F.L.; Wineburg, J.P.; Methodologies for Predicting the Service Lives of Coating Systems; *NIST/BSS-172 Report, Order No. PB95-146387*, **1994**.
64. Martin, J.W.; *ACS Symp. Ser. 722 (Service Life Prediction of Organic Coatings)*, **1999**, pp. 1-20.
65. Culler, S.R.; Diffuse Relectance Infrared Spectroscopy: Sampling Techniques for Qualitative/Quantitative Analysis of Solids, *Pract. Sampling Tech. Infrared Anal.*, CRC, Boca Raton, FL., **1993**.
66. Technical Note - ATR (Attenuated Total Reflectance): ARK HATR, Thunderdome HATR, Gemini HATR and DRIFT; Nicolet Instrument Corporation, 9901 Business Parkway, Suite H, Lanham, MD. 20706; (301)-459-2940 or (800)-237-2800; www.nicolet.com.

67. Application Note AN-9585 Infrared Microspectroscopy of Hard Samples Utilizing a Diamond ATR Objective; Nicolet Instrument Corporation, 9901 Business Parkway, Suite H, Lanham, MD. 20706; (301)-459-2940 or (800)-237-2800; www.nicolet.com.

68. Application Note - The Analysis of Metal Coatings: QC Paint Identification; SensIR Technologies, 15 Great Pasture Rd., Danbury, CT. 06810; (203)-207-9710; www.sensir.com.

69. Technical Note – Thunderdome HATR; SpectraTech, Incorporated, 2 Research Drive, Shelton, CT. 06484-8998; (203)-926-8998 or (800)-843-3847; www.spectra-tech.com.

Chapter 11

A Brief Review of Paint Weathering Research at Ford

J. L. Gerlock, C. A. Smith, V. A. Cooper[1], S. A. Kaberline, T. J. Prater,
R. O. Carter III, A. V. Kucherov[2], T. Misovski, and M. E. Nichols

Ford Research Laboratory , Ford Motor Company, MD 3182-SRL,
2101 Village Road, Dearborn, MI 48124
[1]Current address: Ford Motor Company Central Laboratory, 15000 Century Drive,
Dearborn, MI 48120
[2] Visiting scientist: Zelinsky Institute of Organic Chemistry, RAS, Moscow, Russia

A variety of spectroscopy and sample preparation techniques are described
that allow chemical composition changes to be followed for isolated
clearcoat samples, clearcoats in complete paint systems, and all coating
layers in complete paint systems as a function of exposure. The analysis
techniques provide photooxidation rate, ultraviolet light absorber
effectiveness and longevity, and hindered amine light stabilizer effectiveness
and longevity information that can be used to dramatically reduce the
possibility of introducing inferior clearcoat/basecoat paint systems in service.

INTRODUCTION

Modern automotive paint systems are both physically and chemically complex.[i]
A simple clearcoat/basecoat paint system consists of at least 4 coating layers: 1) 40-50
□μm of clearcoat over 2) 15-25 μm of basecoat, over 3) 20-35 μm primer, over 4) 20-
25 μm of electro-coat. The electro-coat (e-coat) layer provides corrosion protection.
It is electro-deposited on all metal surfaces and cured at temperatures as high as 160-
185°C. The high cure temperature also serves to harden the underlying steel substrate.

Most e-coats contain aromatic amines and tend to be light sensitive. Spray primer is applied over e-coat and cured to smooth the e-coat surface and improve adhesion of basecoat. Primers can be pigmented to match basecoat color. Basecoat is sprayed over cured primer, solvent is allowed to flash away and clearcoat is applied. Clearcoat and basecoat are cured at the same time (wet-on-wet) at temperatures as high as 140°C. Clearcoat provides gloss and serves to screen underlying coating layers from ultraviolet (UV) light. Variations on this standard paint system are also possible. An anti-chip layer can be inserted between the basecoat and primer layers at specific points on the vehicle to reduce stone damage. Double layers of clearcoat, tinted clearcoat, and double layers of differently pigmented basecoats can be used to improve appearance and/or achieve special appearance effects. Each year, these variations are combined with new color choices. And, in the future, formulations will continue to change to reduce solvent emissions, improve acid etch, scratch, chip and mar properties. Taken together, it is obvious that the automotive industry will continue to introduce a large number of "new" paint systems. From the long-term weathering performance point of view, every paint system change represents a step away from proven to non-proven weathering performance and opens the possibility that an inferior paint system will be placed in service.

The most reliable way to reduce the risk of placing inferior paint systems in service is to weather samples that accurately represent formulation and process variables outdoors at several different exposure locations for many years and observe their behavior. Unfortunately, this approach is usually not practical and essentially impossible when paint formulations and processes must evolve quickly. As a result, decisions must often be based on accelerated weathering test as opposed to actual outdoor exposure test results. This situation poses a problem because accelerated weathering tests can yield misleading results,[ii] and they appear to be particularly unreliable for clearcoat/basecoat paint systems.

The apparent decrease in the reliability of accelerated weathering test results for clearcoat/basecoat paint systems relative to monocoat paint systems can be attributed to at least two factors: 1) loss of appropriate weathering performance feedback information, and 2) a dramatic increase in the need for accelerated degradation chemistry to match outdoor exposure degradation chemistry.

The physical behavior of monocoat and clearcoat/basecoat paint systems is fundamentally different during weather exposure. Monocoat paint systems tend to exhibit soft failures, gloss loss and color fade for example, but clearcoat/basecoat paint systems can exhibit hard failures, cracking or delamination, in addition to soft failures. The reason for this difference in behavior resides is pigment location. In monocoat paint systems pigment tends to limit photodegradation to the paint system's surface. During weather exposure, the surface of the paint system degrades and binder erodes away to expose pigment particles that can scatter light. As a result, the rate at which gloss decreases during the early stages of exposure can provide direct insight into long-term weathering performance. In clearcoat/basecoat paint systems, erosion of the paint system surface need not expose hard particles that can scatter light. Therefore, measurements of gloss need provide little insight into long-term weathering performance. In fact, clearcoat/basecoat paint systems can crack or delaminate with little or no warning. It is therefore necessary to weather clearcoat/basecoat paint systems until hard failure is observed in order to correctly define their long-term weathering performance.

The chemical behavior of monocoat and clearcoat/basecoat paint systems is also fundamentally different during weather exposure for the same reason. That is, pigment tends to limit photodegradation to the paint system's surface in monocoat paint systems. Thus, removing the surface of a weathered monocoat paint system by mild abrasion can reveal non-photodegraded material. This behavior could partially explain how accelerated weathering tests that do not accurately reproduce outdoor exposure degradation chemistry could have been used to correctly guide the development of monocoats paint systems that exhibit excellent weathering performance. In clearcoat/basecoat paint systems on the other hand, ultraviolet light absorber (UVA) additive(s) is added to the clearcoat to limit the penetration of UV light like pigments in monocoat paint systems. However, it is now well understood that weather exposure can reduce the concentration of UVA in a paint system's clearcoat and allow UV radiation to penetrate deeper and deeper into the paint system with time. Accordingly, bulk chemical composition changes as well as surface chemical composition changes become possible. When the clearcoat bulk degrades to the point that it can no longer cope with environmental stresses, it cracks. If UV light is allowed to penetrate through the clearcoat to the clearcoat-basecoat interface, degradation of the interface can result in clearcoat peeling. Bulk photodegradation in clearcoat/basecoat paint systems as opposed to surface-limited photodegradation in monocoat paint systems can explain why accelerated exposure degradation chemistry must match outdoor exposure degradation chemistry with great fidelity if an accelerated weathering tests is to produce credible results for clearcoat/basecoat paint systems. Since many paint systems do not undergo the same degradation chemistry during accelerated and outdoor exposure, accelerated test reliability is lower for clearcoat/basecoat paint systems than for monocoat paint systems.

One way to cope with the fundamental differences in the weathering behavior of monocoat and clearcoat/basecoat systems is to add chemical change measurements to the traditional repertoire of weathering performance tests. Research was initiated at Ford Motor Company in the late 1980's to accomplish this task. Early work focused on the development of analytical techniques to characterize the influence of polymer nature, polymer synthesis conditions, crosslinker nature, hindered amine light stabilizer (HALS) nature, and exposure conditions on clearcoat chemical change rates. Analytical techniques were developed for isolated clearcoats but not for clearcoats in complete paint systems. Bauer has reviewed much of this work.[iii] Today, a number of analytical techniques developed for isolated clearcoats have been extended to clearcoats in complete paint systems and in some cases, to every coating layer in complete paint systems. In the future, the feedback information afforded by such analytical techniques will be used to guide the development of an accelerated weathering test that can reproduce the degradation chemistry and mechanical stresses associated with outdoor with sufficient fidelity to restore credibility to accelerated weathering tests for clearcoat/basecoat paint systems.

The present work describes a number of analytical techniques developed to follow chemical changes in parts of paint systems and complete paint systems as a function of exposure. The work concludes with a brief discussion of "reasonable" exposure conditions to obtain credible accelerated weathering test results. Each analytical technique is accompanied by results to illustrate its capability. Because the purpose of this work is to illustrate analysis capability, not test paint systems, the interpretation of results is purposefully held to a minimum.

PHOTOOXIDATION RATE MEASUREMENTS

It is generally agreed that photo-thermal oxidation, photooxidation, is the primary chemistry responsible for the long-term degradation of paint systems during weather exposure outdoors. Accordingly, a number of analytical techniques have been developed to quantify the rate at which paint systems undergo photooxidation-induced chemical changes during exposure.

Transmission Infrared Spectroscopy [iv, v]

Transmission FTIR spectroscopy provides a ready means to follow chemical changes in isolated clearcoat samples as a function of exposure.

In practice, isolated clearcoat samples are prepared by curing resin on infrared-transparent silicon disks. Uncured resin is smeared on a disk with a small glass rod and the resin thickness is adjusted until the strongest peak in the IR spectrum exhibits an absorbance in the 1.5-2.0 range. This usually results in a cured clearcoat film thickness in the 8-12 μm range. There is no need to apply the coating uniformly. In fact, if the resin is applied uniformly, spin-coating for example, interference patterns become a problem and spectra must be recorded at the Brewster angle.[vi] A notched silicon disk is used in combination with a sample holder with a corresponding peg to ensure that spectra are recorded for the same spot as a function of exposure.

Typical transmission FTIR results are illustrated in Figure 1 for samples subjected to SAE J1960 JUN89 weather-ometer® accelerated exposure using borosilicate inner and outer filters at 0.55 W/m^2 light intensity at 340 nm. Two measurements are illustrated, film loss rate and photooxidation product accumulation rate. Both measurements provide generic information about the progress of photooxidation that can be used to compare the photooxidation behavior of clearcoats from different chemical families. No use is made of "fingerprint region" information to identify specific degradation chemistry.

Film loss rates are reported in units of μm of film lost per 5000 hours of exposure. Results are shown for four versions of a series of chemically different clearcoats. The four versions examined are: 1) clearcoat without HALS or UVA additives, 2) clearcoat formulated with the HALS at the concentration intended in practice, 3) clearcoat formulated with the UVA at the concentration intended in practice, and 4) clearcoat with both HALS and UVA additives at the levels intended in practice. Film loss rates were determined by extrapolating plots of integrated -CH intensity versus exposure to 5000 hours of exposure. Integrated -CH intensity was converted to film thickness in μm using calibration charts determined for known film thickness samples. In practice, not all samples actually survive 5000 hours of exposure. As shown in the lower part of Figure 1, film loss rates respond to the presence or absence of additives and the response can be quite different for chemically different clearcoats.

Photooxidation product accumulation rates in units of Δ[(-OH, -NH)/μm] per 5000 hours of exposure are illustrated in the upper part of Figure 1. Δ[(-OH, -NH)/μm] values were obtained by plotting the change in the ratio of integrated -OH, -NH region intensity (3000- 2500 cm^{-1}) to μm of film remaining and extrapolating the curves to 5000 hours of exposure. Integrated -CH absorbance was converted to μm of film as described above. As noted above, not all samples actually survive 5000 hours of exposure. In previous work,[iv] the change in the ratio of integrated -OH, -NH region absorbance to integrated -CH intensity was plotted directly. This treatment operates on the premise that -CH absorbance per μm of film is approximately the same for all clearcoats of interest and would not be appropriate to compare low -CH content materials like partially fluorinated clearcoats with typical clearcoats.

Because Δ[(-OH, -NH)/μm] can increase as the result of hydrolysis as well as photooxidation, Δ[(-OH, -NH)/μm] measurements need not follow photooxidation explicitly. However, the contribution of hydrolysis to the total is usually small in the absence of acid catalysts even under accelerated exposure conditions that over-emphasize hydrolysis. As an illustration, the results shown in the lower part of Figure 2 were obtained for samples exposed behind a 430 nm cut off filter in a proprietary test that employs high intensity light in an exposure protocol similar to SAE J1960 JUN89. As can be seen, the contribution to total Δ[(-OH, -NH)/μm] increase (upper part of Figure 2) is <25% of the total even when the total is small, Clearcoats N and P for example. Thus, for simplicity, Δ[(-OH, -NH)/μm] values are referred to in the present work as a measure of photooxidation product accumulation rate.

As was the case for film loss rates, photooxidation product accumulation rate responds to the presence or absence of additives and substantial differences in response are observed for different clearcoat chemistries, Figure 1 and 2. Interestingly, additive effectiveness does not appear to be distorted by high light intensity. In practice, film erosion and photooxidation product accumulation should be considered together to obtain a clear picture of photooxidation resistance.

The results shown in Figures 1 and 2 demonstrate that transmission FTIR measurements on isolated clearcoats provide a ready means to compare the intrinsic photooxidation resistance of clearcoats from different chemical families and select effective HALS, UVA, and HALS/UVA combinations. However, the results need not be extrapolatable to clearcoat behavior in complete paint systems because clearcoat and basecoat are cured wet-on-wet in complete paint systems. As a result, additives and other lower molecular weight components, crosslinker for example, can diffuse between the two layers to change the additive content and the chemical composition of both layers relative to that of an isolated clearcoat sample.

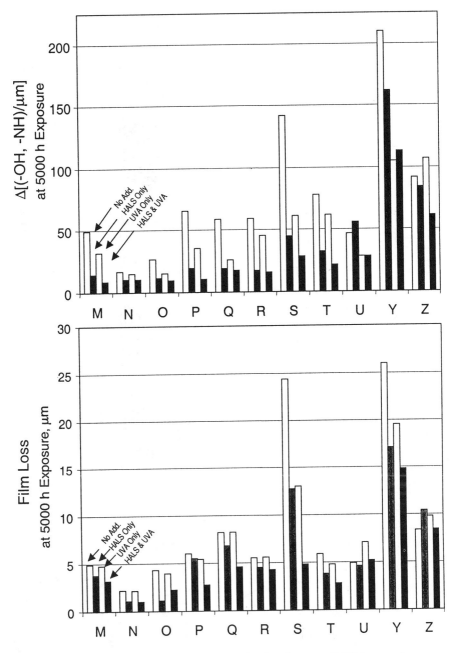

Figure 1. Transmission FTIR Δ [(-OH, -NH)/μm] values per 5000 hours of exposure (upper) and film loss values per 5000 hours of exposure (lower) for chemically different clearcoats: SAE J1960 JUN89 using borosilicate inner and outer filters, and 0.55 W/m² at 340 nm light intensity. Formulation order within each series of four samples is from left to right: 1) no HALS or UVA, 2) HALS only, 3) UVA only, and 4) HALS & UVA.

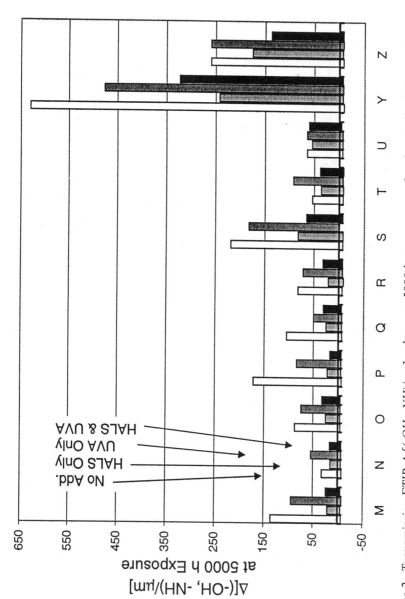

Figure 2. Transmission FTIR Δ [(-OH, -NH)/μm] values per 5000 hours exposure for chemically different clearcoats exposed: a) behind a 430 nm cutoff filter (lower), and b) with no filter (upper). Formulation order within each series of four samples is from left to right: 1) no HALS or UVA, 2) HALS only, and 4) HALS & UVA. Accelerated exposure conditions are proprietary.

Micro-Transmission Infrared Spectroscopy [vii, viii]

As noted above, measurements of any kind on isolated clearcoat samples need not accurately reflect the performance of the same clearcoat in complete paint systems. To illustrate, when an isocyanate-based clearcoat is cured wet-on-wet over a melamine-based basecoat, non-reacted isocyanate is detectable by micro-FTIR spectroscopy more that half way through the basecoat layer. Micro-FTIR spectroscopy provides a means to follow chemical changes as a function of weather exposure across all coating layers in complete paint systems.

In practice, an intact clearcoat/basecoat/primer/e-coat paint system chip is obtained by bending a test panel. De-adhesion usually occurs between the e-coat and the metal substrate unless the paint system is severely degraded. The paint chip is then microtomed edge-on to produce a ~10 µm thick cross-section. The cross-section is mounted in a KBr compression cell and a portion of the gap between the KBr windows is filled with additional KBr. In practice, it can be difficult to reproduce this procedure sample to sample. The difficulty can be overcome by gluing paint system chips from different tests panels together with a reference material prior to microtoming. Microtoming the sample sandwich produces a single cross-section with components from every test panel and a reference material whose spectrum intensity can be used to compare spectrometer response for other test specimens.

Typical results from this type of analysis are shown in Figure 3 for Paint System 1, an acrylic/melamine clearcoat over brown acrylic/melamine basecoat. The sample sandwich consists of paint system chips from 0, 1, 2, 4, and 5-year Florida exposure test panels. The measured intensity ratio (-OH, -NH)/-CH is plotted versus paint system depth in µm across all coating layers because no attempt is being made to compare the degree of photooxidation with the degree of photooxidation in other paint systems. The shaded area between the test specimens indicates the location of the adhesive. The reference material is a thin film of polyacrylonitrile.

The results for the non-weathered test panel indicate that (-OH, -NH)/-CH varies considerably from layer to layer, being highest in the e-coat layer as expected. E-coat cross-linking chemistry involves reaction of amine and epoxy functional groups to produce pendant hydroxyls. It can also be noted that clearcoat thickness varies considerably between allegedly identical test panels.

As illustrated in Figure 3, (-OH, -NH)/-CH ratio first increases in the surface of the clearcoat with exposure and fills in through the clearcoat bulk with continued exposure. After 5 years of exposure, a distinct increase in (-OH, -NH)/-CH is visible at the interface between the clearcoat and the basecoat. The increase indicates that light that can drive photooxidation has penetrated the clearcoat thickness after 4-5 years of exposure. Photooxidation of the clearcoat/basecoat interface is a strong indication that the paint system is at risk of failure by clearcoat peeling.

These results indicate that micro-FTIR spectroscopy measurements on paint

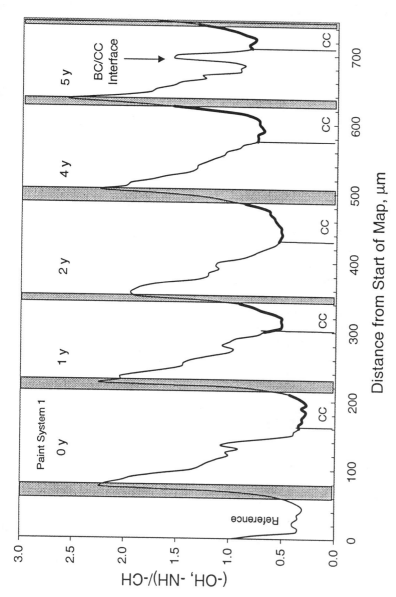

Figure 3. Micro-transmission FTIR map of (-OH, -NH)/-CH ratio for Paint System 1 exposed in Florida for 0, 1, 2, 4, and 5 years. Basecoat, primer and e-coat layers lie to the left of clearcoat in each test panel specimen. Shaded areas indicate adhesive between test panel samples.

system cross sections can be used to follow the progress of photooxidation across all coating layers. However, spatial resolution is limited to ~15-20 μm in the -OH, -NH region and decreases with decreasing wavenumber.

Photoacoustic Infrared Spectroscopy [ix, x]

Photo-acoustic FTIR spectroscopy provides a ready means to examine the chemical composition of a clearcoat's surface in complete paint systems. Analysis depth is determined by modulation frequency. A high modulation frequency leads to shallow depth analysis and vise versa. The results presented in the present work were recorded at a modulation frequency that limits analysis depth to ~8-12 μm in the -OH, -NH region. This depth was selected to reduce spectrum sensitivity to surface contaminants such as wax and dirt and preclude absorbance by underlying basecoat in most paint systems. It should be noted that it is not difficult to envision a paint system where in surface photooxidation products selectively erode away to leave relatively non-degraded material behind. Under these conditions, PAS FTIR would yield a low estimate of the photooxidation rate of the paint system's surface.

In practice, a 1 cm diameter disk is punched from a test panel or actual vehicle component and the PAS FTIR spectrum of the clearcoat surface is recorded. The results illustrated in Figure 4 were obtained for clearcoats in complete paint systems prepared on steel test panels. A 2.5 x 5 cm section was cut from each panel for exposure and sample disks were punched from the same 2.5 x 5 cm section as a function of exposure. The clearcoat series is the same series examined by transmission FTIR spectroscopy on silicon disks. Transmission FTIR and PAS FTIR samples were exposed side-by-side as previously described. Results are shown for three test panels versions: 1) thin clearcoat, ~1 mil (25 μm), over red and silver basecoats, 2) standard thickness clearcoat, ~1.8 mils (45 μm), over red and silver basecoats, and 3) standard thickness UVA-free clearcoat over UVA free red and silver basecoats.

PAS FTIR photooxidation product accumulation rates in units of $\Delta[(-OH, -NH)/\mu m]$ per 5000 hours of exposure were obtained from plots of the change in the ratio of integrated -OH, -NH intensity to integrated -CH intensity as a function of time. $\Delta[(-OH, -NH)/-CH]$ values were multiplied by -CH intensity per μm values determined for free standing films to convert them to $\Delta[(-OH, -NH)/\mu m]$ values. In theory, integrated PAS-FTIR -CH intensity could be used directly for this correction, however, absolute PAS FTIR spectrum intensity can be quite variable over time. Correcting PAS FTIR -CH intensity is designed to allow clearcoats with different -CH content, partially fluorinated clearcoats for example, to be directly compared. The correction is approximate. It assumes that PAS FTIR sampling depth in the -OH, -NH region is constant for all clearcoats and remains constant with exposure. Dividing integrated -OH, -NH absorbance by integrated -CH absorbance serves a different purpose in PAS-FTIR spectrum analysis than transmission FTIR spectrum analysis. In transmission FTIR experiments, division by integrated -CH absorbance compensates integrated -OH, -NH intensity for film loss as the result of weather exposure. In the PAS FTIR experiment, division by integrated -CH intensity compensates for detection efficiency. PAS FTIR spectrum intensity can decrease by as much as 40% as clearcoat surface roughness increases with weather exposure. The use of integrated -CH intensity to correct for signal intensity decrease does appear to be justified. Raman spectroscopy measurements on cross-sections of clearcoats in weathered paint systems indicate no measurable decrease in -CH intensity from the bottom of the clearcoat to its surface as the result of oxidation.[iv]

The results shown in Figure 4 clearly reveal that PAS FTIR spectroscopy can reveal substantial differences in the rates at which clearcoats undergo photooxidative degradation in complete paint systems. Differences in PAS FTIR photooxidation product accumulation rates roughly parallel, but do not match differences in transmission FTIR photooxidation product accumulation rates for the same clearcoats. PAS FTIR analysis reveals no dramatic difference in the photooxidation rate behavior of thin clearcoat and standard thickness clearcoats over red and silver basecoats. And, the absence of UVA does not have a large effect.

Imaging Time-of-Flight Secondary Ion Mass Spectroscopy [v, xi, xii, xiii]

In theory, TOF-SIMS could be used to directly image the ^{16}O added to paint systems by photooxidation. In practice, the small increase in oxygen content, <3% by weight, is difficult to image against the large background of oxygen initially present in typical coatings, >30% by weight. One way to distinguish the oxygen added by photooxidation from that initially present is to expose paint systems in $^{18}O_2$ atmosphere. Under these conditions photooxidation leads exclusively to the formation of ^{18}O-labeled products whose location can be imaged by $^{18}O^-$ ion intensity.

In practice, a 1 cm diameter disk is punched from a test panel and placed in a sealed cell with a quartz exposure window. Vacuum is pulled to remove air and the cell is brought to atmospheric pressure with 25% $^{18}O_2$ in dry nitrogen. After the cell has been exposed, the sample disk is removed and bent to free the paint system from the substrate. The resultant paint system chip is glued between layers of ^{18}O-labeled reference film and 5 mm thick pieces of conductive polypropylene using thin pieces of hot-melt glue. Pressing the ^{18}O-labeled reference/hot-melt glue/paint system chip/hot melt glue/^{18}O-labeled reference sandwich between pieces of preheated conductive polypropylene (<80°C) provides sufficient heat to fuse the assembly. Next, the assembly is microtomed edge-on to produce a smooth surface and TOF-SIMS is used to image $^{18}O^-$ ion intensity across all coating layers with 1-2 μm spatial resolution. The ^{18}O-labeled reference material is an epoxy/acid clearcoat synthesized to contain 1% by weight ^{18}O. The clearcoat is cured on thin Tedlar sheets. The reference material is used to calibrate the $^{18}O^-$-ion signal from the paint system chip and allow

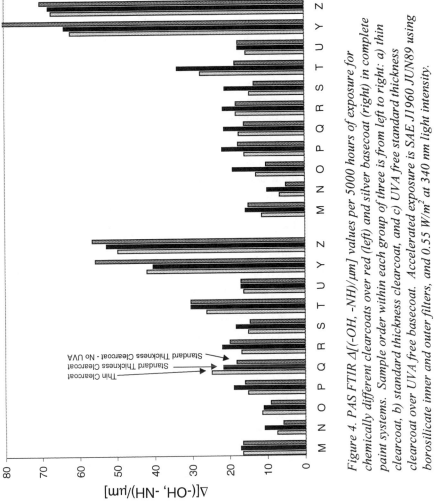

Figure 4. PAS FTIR Δ[(-OH, -NH)/μm] values per 5000 hours of exposure for chemically different clearcoats over red (left) and silver basecoat (right) in complete paint systems. Sample order within each group of three is from left to right: a) thin clearcoat, b) standard thickness clearcoat, and c) UVA free standard thickness clearcoat over UVA free basecoat. Accelerated exposure is SAE J1960 JUN89 using borosilicate inner and outer filters, and 0.55 W/m² at 340 nm light intensity.

$^{18}O^-$-ion signal intensity to be corrected if the sample is not level in the spectrometer. Mounting the ^{18}O-labeled reference/hot-melt glue/paint system chip/hot-melt glue/^{18}O-labeled reference sandwich between pieces of conductive polypropylene is necessary to reduce sample charging during analysis.

The series of $^{18}O^-$-ion images shown in Figure 5 was obtained for an acrylic/melamine clearcoat over a dark gray basecoat paint system, Paint System 2. The test panel had been weathered in Florida for 5 years prior to additional UV exposure. The uppermost image was recorded prior to additional UV exposure in 25% $^{18}O_2$ in nitrogen atmosphere. A low intensity $^{18}O^-$-ion signal from naturally abundant ^{18}O is present. An intense $^{18}O^-$-ion signal from the ^{18}O-labeled reference film is visible on either side of the paint system chip. The middle image shown in Figure 5 was obtained after a sample was exposed facing away from the light during exposure in 25% $^{18}O_2$ in nitrogen atmosphere for 1000 hours. The image indicates that the Paint System 2 undergoes considerable thermal oxidation during accelerated exposure. The temperature within the exposure cell is typically > 65°C. Finally, the lower image in Figure 5 was recorded for a sample facing the light during continuous exposure to borosilicate inner and outer filtered Xenon arc light at 0.45 W/m^2 intensity at 340 nm for 1000 hours. The image clearly reveals the progress of photooxidation across all coating layers. The gradient in $^{18}O^-$-ion signal intensity decreasing from the surface of the clearcoat inward suggests that clearcoat UVA continues to limit the penetration of UV light into the paint system after 5 years of Florida exposure and 1000 hours of accelerated exposure.

A second series of $^{18}O^-$-ion images is illustrated in Figure 6. These images depict the extremes in behavior observed for a series of experiments designed to determine whether long-term weathering performance, >10 years Florida exposure, can be anticipated from the behavior of relatively short-term Florida exposure test panels. The images were recorded for 3-year Florida exposure test panels that were exposed to an additional 1000 hours of accelerated exposure in 25% $^{18}O_2$ in nitrogen atmosphere as described above. The $^{18}O^-$-ion image shown for Paint System 3, acrylic/melamine clearcoat over a charcoal metallic basecoat, upper part of Figure 6, clearly indicates that although this clearcoat is not highly oxidized, the interface between the clearcoat and basecoat is highly oxidized. This behavior suggests that the risk of catastrophic failure by clearcoat peeling is high at long exposure times. Paint System 3 test panels fail by clearcoat peeling after 4-5 years of Florida exposure.

The $^{18}O^-$-ion image shown for Paint System 4, acrylic/melamine clearcoat over a dark red basecoat, in the middle of Figure 6 indicates that its clearcoat as well as the interface between its clearcoat and basecoat is highly oxidized. This behavior suggests that the risk of catastrophic failure by clearcoat peeling or cracking is high. Paint System 4 test panels fail by clearcoat cracking after 4-5 years of Florida exposure.

Finally, the $^{18}O^-$-ion image shown for Paint System 5, acrylic/melamine clearcoat over a charcoal metallic basecoat, in the lower part of Figure 6 indicates that photooxidation is restricted to the paint system's surface. This behavior indicates that the risk of catastrophic failure is low. Paint System 5 test panels continue to perform well after 8 years of Florida exposure.

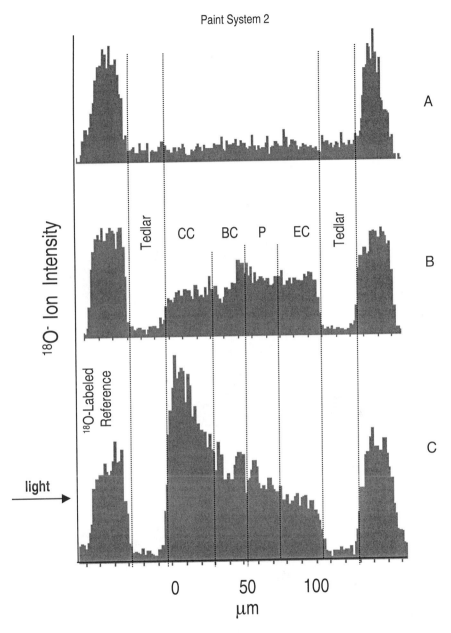

Figure 5. $^{18}O^-$*-ion images obtained for 5-year Florida exposure panel: a) prior to accelerated exposure, b) after 500 hours accelerated exposure in 25%* $^{18}O_2$ *in dry nitrogen atmosphere while facing away from the light, and c) after 500 hours accelerated exposure while facing the light. Accelerated exposure is "Light Only" Xenon arc using borosilicate inner and outer filters, and 0.45 W/m² at 340 nm light intensity.*

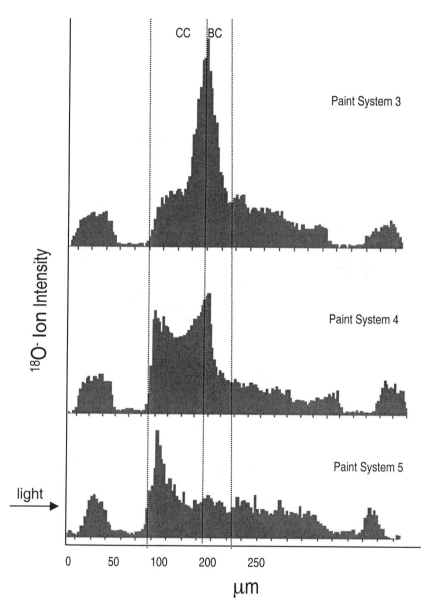

Figure 6. $^{18}O^-$-ion images obtained for 3-year Florida exposure Paint Systems 3, 4, and 5 after 1000 hours exposure in 25% $^{18}O_2$ in dry nitrogen atmosphere to "Light Only" Xenon arc using borosilicate inner and outer filters, and 0.45 W/m² at 340 nm light intensity.

Taken together, results like those described for Paint Systems 3, 4, and 5 above demonstrate that $^{18}O^-$-ion images can be used to identify high-risk paint systems. However, the exposure conditions used raise concern about distorted photodegradation chemistry. Samples are exposed in a sealed chamber at an elevated temperature in the absence of humidity or the means to wash degradation products away. Superimposing artificial exposure on outdoor exposure minimizes this concern, but the concern is not eliminated. Clearly it would be more credible to image the photooxidation products in outdoor exposure test panels directly.

One way to directly image photooxidation products in outdoor exposure test panels using TOF-SIMS would be to selectively react them with an easily detected derivitazing agent. A fluorinated derivatizing agent would be ideal. Unfortunately, our attempts to derivatize hydroxyl substituted photooxidation products with SF_4, trifluoroacetyl chloride, perfluoroacetic anhydride and 3, 3, 3-trifluoropropylsilyl chloride have not produced promising results. Likewise, neutralizing carboxylic acid photooxidation products with fluorinated amines to produce fluorinated salts also yields poor results.

One technique that does show promise is deuterium exchange. This approach is based on the premise that an oxidized polymer should contain more D_2O exchangeable groups than a non-oxidized polymer.

ULTRAVIOLET LIGHT ABSORBER [xiv]

UVA additives are added to the clearcoat in clearcoat/basecoat paint systems to limit clearcoat photooxidation and screen the underlying basecoat from UV light. The ability of a UVA to limit clearcoat photooxidation can be measured by transmission FTIR for isolated clearcoat samples and by PAS FTIR for the clearcoat surface in complete paint systems. Micro-FTIR and $^{18}O^-$-TOF-SIMS can be used to measure UVA effectiveness across all coating layers. Because the risk of paint system failure can be reduced from the onset by selecting long-lived UVAs, a number of techniques have been developed to measure UVA longevity directly.

Transmission Ultraviolet Spectroscopy [xv, xvi]

Transmission UV spectroscopy provides a ready means to determine UVA longevity for isolated clearcoat samples.

In practice, clearcoat resin is smeared on a quartz slide with a glass rod and cured. The cured clearcoat film should be sufficiently thick to produce a starting absorbance of 2-2.5 in the UV region of interest prior to exposure. The samples are exposed and UV absorbance decrease is followed as a function of time. A second series of samples can be prepared to determine initial UV absorbance per µm values. Clearcoat resins are applied as even films on 10 x 30 cm glass plates using a Bird applicator. The samples are cured and the cured films are removed with a razor blade.

Film thickness is measured and combined with UV absorbance measurements to calculate UV absorbance/μm values.

The results shown in Figure 7 were obtained for the same series of clearcoats examined by transmission FTIR and PAS FTIR spectroscopy. Samples containing UVA and UVA with HALS were examined. All samples were exposed side-by-side along with transmission FTIR and PAS-FTIR samples as previously described. Plots of 340 nm region UVA UV absorbance versus exposure time were found to be linear until absorbance decreases below ~0.5. Below ~0.5, absorbance loss rate slows to approach zero asymptotically with continued exposure. Pickett has treated the details of UVA UV absorbance loss kinetics in a companion paper.[xvii] For all practical purposes, the non-linear region of the absorbance decay curves can be ignored to estimate how long clearcoat UVA effectively screens underlying basecoat from UV light. Once total clearcoat UV absorbance drops below 0.5 in the 340 nm region, transmission of 340 nm region light to basecoat is significant, >75%. The linear region of UV absorbance loss curves give the UVA absorbance loss rates used in the present work.

UV absorbance loss rates in the 340 nm region are illustrated in Figure 7. As can be seen, UVA absorbance loss rate and the ability of HALS to reduce UVA absorbance loss rate varies considerably.

UVA absorbance loss rates can be combined with starting absorbance per μm values to estimate how long a given clearcoat thickness can effectively limit UV transmission to basecoat.

The results shown in Figures 7 clearly demonstrate that transmission UV spectroscopy can be used to select superior UVA additives and UVA-HALS combinations. However, as previously noted, measurements of any kind on isolated clearcoat samples need not accurately reflect actual clearcoat behavior in complete paint systems. For example, when clearcoat and basecoat are cured wet-on-wet over plastic substrates, clearcoat UVA can diffuse into the plastic.[xviii]

Solvent Extraction Analysis

One means to determine the UVA content of a clearcoat in a complete paint system is to scrape the clearcoat down to the basecoat and extract the UVA with a solvent. This approach is not valid for covalently bound UVAs or UVAs that become covalently bound during exposure. If there is interest in the nature of the UVAs present, the extract can be analyzed by standard chromatography techniques.[v] If there is no interest in the nature of the UVAs present, only UV absorbance, the UV spectrum of the solvent extract can be recorded directly. The latter case is described below.

In practice, a 2.5 x 5 cm section of test panel is scraped down to basecoat with a razor blade to afford 25-50 mg of clearcoat scrapings. The clearcoat scrapings are added to a tared 25 ml volumetric flask and weighed. 20 ml of CH_2Cl_2 is added to the flask and the flask is allowed to stand overnight. The flask is brought to volume and ~5 ml of the CH_2Cl_2/clearcoat particle suspension is filtered into a 1 cm path length

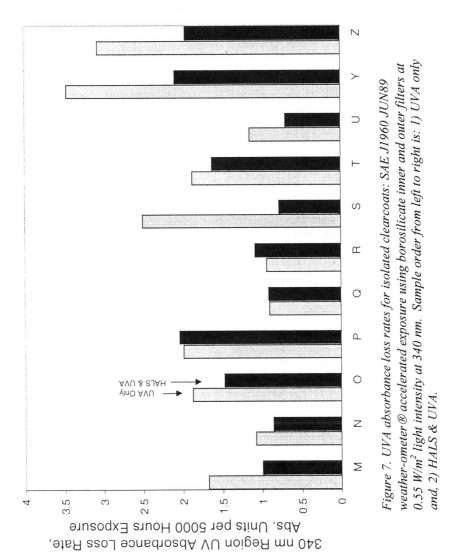

Figure 7. UVA absorbance loss rates for isolated clearcoats: SAE J1960 JUN89 weather-ometer® accelerated exposure using borosilicate inner and outer filters at 0.55 W/m² light intensity at 340 nm. Sample order from left to right is: 1) UVA only and, 2) HALS & UVA.

quartz cuvette and the UV spectrum of the filtrate is recorded. Absorbance/mil can be calculated according to the equation: Abs./mil = [(Abs.$_{meas.}$/(g/ml)(ρ)]/393.3, where Abs.$_{meas.}$ is the absorbance measured for the filtrate in a 1cm path length cuvette, g is grams of clearcoat scrapings, ml is volumetric flask volume, ρ is paint density in g/cc, and 393.3 is mils/cm. Abs./mil values can be converted to total clearcoat absorbance values when clearcoat thickness is known.

Total clearcoat absorbance behavior is illustrated in Figure 8 for three paint systems as a function of Florida exposure. As can be seen, only Paint System 6 limits the transmission of 340 nm region UV light to <1% after 5 years of Florida exposure. The clearcoats in Paint Systems 7 and 8 transmit significant 340 nm region UV light to underlying basecoat after 4-5 years of exposure. Paint System 7 test panels fail by clearcoat peeling after 5 years of Florida exposure. However, Paint System 8 test panels continue to perform well after 5 years of Florida exposure. This behavior suggests that the basecoat in Paint System 8 is less sensitive to UV light than the basecoat in Paint System 7. In the absence of specific information about basecoat sensitivity to UV light, Paint System 6 is the lower risk paint system.

As previously noted, solvent extraction analysis is not appropriate for covalently bound UVA and it can be difficult to determine whether a UVA has become covalently bound as the result of weather exposure. Clearly it would be beneficial to have an analytical technique that could assess clearcoat UV absorbance directly in intact paint systems.

Photoacoustic Ultraviolet Spectroscopy [xix]

PAS UV spectroscopy offers the possibility of directly measuring clearcoat UVA absorbance in intact paint systems. Unlike the PAS FTIR technique, a dispersive spectrometer and high intensity Xenon arc source are used. Spectra are obtained by modulating the output of the monochrometer with a chopper before the light is focused on the sample in the PAS cell. In practice, a 1 cm diameter disk is punched from a test panel and its PAS UV spectrum is obtained. As is the case for PAS FTIR spectroscopy, PAS UV analysis depth is determined by modulation frequency. The results of PASUV measurements where analysis depth was held constant at ~8 μm are illustrated in Figure 9 for Paint System 9. Paint system 9 consists of acrylic/melamine clearcoat over black acrylic/melamine basecoat over primer over e-coat. Test panels were weathered in northern Australia. The results reveal that clearcoat UVA absorbance decreases markedly in the surface region of the clearcoat with exposure, but it has proven to be difficult to translate the PAS UV measurements into quantitative UVA UV absorbance loss rate information for the entire clearcoat. Consequently, this technique has not been pursued to evaluate clearcoat UVA behavior in complete paint systems. It has been used to evaluate process variables effects on UVA UV absorption.[xix]

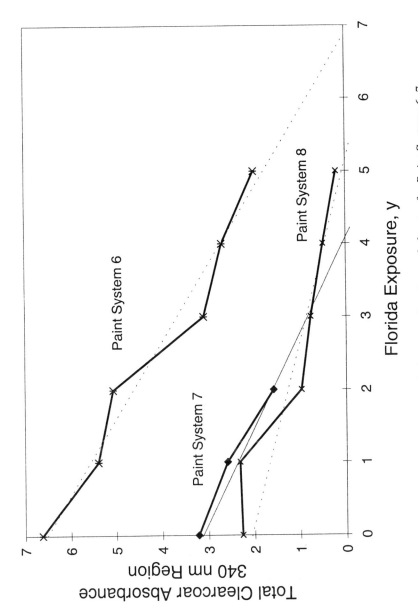

Figure 8. 340 nm region total clearcoat absorbance behavior for Paint Systems 6, 7, and 8 as a function of Florida exposure as determined by solvent extraction analysis.

232

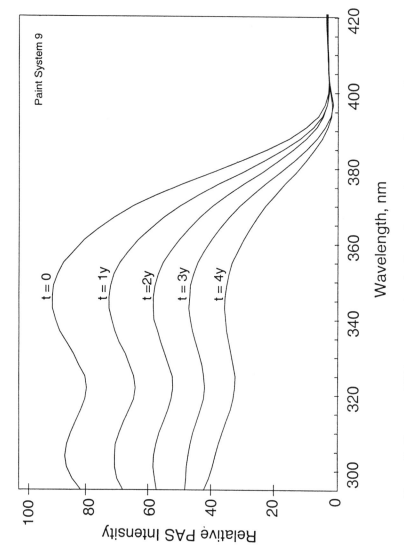

Figure 9. Clearcoat UVA concentration for Paint System 9 exposed in Australia as determined by PAS UV spectroscopy.

Micro-Transmission UV Spectroscopy [xx, xxi]

Micro-transmission UV spectroscopy also offers the possibility of directly measuring clearcoat UVA UV absorbance in intact paint systems. In practice, a 1 cm disk is punched from a test panel and bent to release the paint system. Delamination usually occurs at the e-coat/metal interface unless the paint system is highly degraded. The paint system chip is glued between pieces of ~50 μm thick reference clearcoat containing a known amount of UVA. A high molecular weight, solvent free, polyisocyanate glue is used to ensure that UVA does not diffuse from the reference clearcoat into the paint chip during sample preparation and vice versa. The reference clearcoat/paint system chip/reference clearcoat sandwich is microtomed edge-on to produce a ~10 μm thick cross-section. The cross-section is placed in the micro-UV spectrometer and spectra are recorded with 5 x 10 μm spot size in 5 μm steps from the reference clearcoat across the paint system to the opposite reference clearcoat. Useful UV spectra are usually not recorded for pigmented layers due to light scattering by pigment particles and aluminum flakes.

The UV absorbance of the reference clearcoat is used to determine the actual thickness of the microtomed cross-section and verify that the cross-section is not wedge shaped in the scanning direction. Paint system clearcoat absorbance is divided by actual cross-section thickness to calculate absorbance/μm values. Absorbance/μm is assumed to be constant over the 5μm spot size examined, that is, any gradient in absorbance over this range is ignored.

Micro-UV spectroscopy results are shown in Figure 10 for an acrylic/melamine clearcoat containing a benzotriazole UVA, Paint System 10. A weak gradient in UV spectrum intensity is visible in the non-weathered test specimen. The direction of the gradient suggests that UVA volatilizes during cure. A strong gradient in UV spectrum intensity is observed for the 4-year Florida exposure test specimen. The direction of the gradient clearly reveals that UVA is depleted from the surface of the clearcoat inward during exposure as expected. Total clearcoat absorbance can be determined by summing individual spectral intensities across the layer. Alternatively, analysis spot size can be adjusted to match clearcoat thickness to measure total clearcoat UV absorbance directly after correcting for cross-section thickness. UVA loss rate is determined by the difference in total clearcoat absorbance values for weathered and non-weathered test specimens. If non-weathered test panels are not available, spectrum intensity near the basecoat interface can be used to estimate absorbance for the entire non-weathered layer provided the gradient in spectrum intensity in the weathered test specimen stops before the bottom of the clearcoat is reached.

Micro-UV spectroscopy provides an unambiguous means to determine clearcoat UVA UV absorbance for clearcoats in intact paint systems.

HINDERED AMINE LIGHT STABILIZER

As illustrated in Figures 1 and 2, hindered amine light stabilizer additives can

234

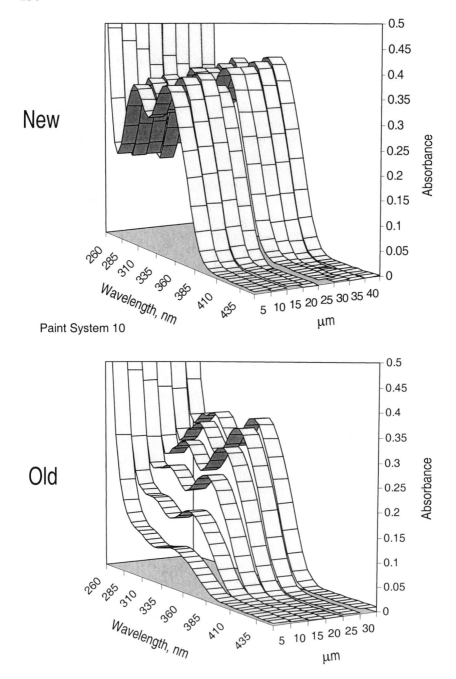

New

Paint System 10

Old

Figure 10. Micro-UV spectra of Paint System 10 clearcoat as a function of depth before exposure (upper) and after 4 years Florida exposure (lower).

play a critical role in determining clearcoat photooxidation resistance. Because it is reasonable to assume that clearcoat photooxidation rate will return to its HALS-free photooxidation behavior when HALS is consumed, HALS longevity is an issue for clearcoats that rely heavily on HALS to achieve photooxidation resistance.

The HALS content of clearcoats is not readily determined by standard solvent extraction techniques because the HALS initially added to a polymer system can be converted to a variety of inhibition cycle products in the process of inhibiting photooxidation. Typical inhibition cycle products include secondary amines, hydroxylamines, amino ethers, and nitroxyl radical derivatives of the specific HALS initially added to the polymer system. The nature of the amino ether derivatives formed depends on the chemical composition of the polymer system being stabilized and can include non-extractable, covalently bound derivatives. An attempt has recently been made to follow ^{15}N-labeled HALS as a function of exposure using ^{15}N CP/MAS Solid-State NMR Spectroscopy.[xxii]

Electron Spin Resonance Spectroscopy [xxiii, xxiv]

One approach to determining HALS longevity in weathered clearcoats is to oxidize the HALS initially added and all of its inhibition cycle oxidation products to a common denominator species, nitroxyl radical. HALS additives are designed to yield a persistent, though reactive, nitroxyl radical in the process of inhibiting photooxidation.

In practice, a 2.5 x 5 cm section of test panel is uniformly scraped down to basecoat with a razor blade to afford 25-50 mg of clearcoat scrapings. The clearcoat scrapings are added to a tared 25 ml volumetric flask and weighed. 20 ml of CH_2Cl_2 is added to the flask and the flask is allowed to stand overnight to ensure complete swelling. The flask is brought to volume and ~0.5 ml of the particle suspension is place in a 3 mm O.D. quartz ESR tube and the ESR spectrum of the suspension is recorded. The double integral area of the ESR spectrum is compared with a calibration chart prepared for known amounts of 4-hydroxyl-2, 2, 6, 6-tetramethylpiperidinyl-N-oxy (TEMPOL) to determine the amount of nitroxyl radical present in the sample. The amount of nitroxyl radical detected reveals the "steady-state" concentration of nitroxyl radical present in the clearcoat prior to peracid oxidation. Next, a small amount of p-nitroperbenzoic acid, < 30 mole excess relative to parent HALS concentration, is added to the ESR tube. The tube is stirred with a small glass rod and the ESR spectrum of the sample is repeatedly recorded over a half hour period. In most cases, nitroxyl radical signal intensity increases rapidly during the first 5 minutes of oxidation and then slowly decays. The maximum in nitroxyl radical signal intensity is used to calculate nitroxyl radical concentration. When the total amount of clearcoat available for analysis is small, the entire procedure can be carried out directly in the ESR sample tube.

The analysis described above yields two measures of HALS content: 1) the steady-state concentration of nitroxyl radical present in the clearcoat prior to oxidation, and 2) the total amount of parent HALS and all of its inhibition cycle products that can be oxidized to nitroxyl radical. This latter quantity is referred to as "Active HALS." In theory, plots of steady-state nitroxyl radical concentration versus exposure time could reveal HALS longevity because steady-state nitroxyl radical concentration is linked to the amount of HALS present in the system. However, in practice, the link need not be direct. That is, the presence of nitroxyl radical strongly infers that HALS is present to inhibit photooxidation, but the absolute concentration of nitroxyl radical present need not reflect the total amount of HALS remaining in the system. "Active HALS" analysis appears to provide a direct measure of the amount of HALS remaining to inhibit photooxidation to the extent that all HALS based species capable of inhibiting photooxidation are readily oxidized to nitroxyl radical by p-nitroperbenzoic acid. It should be noted that a number of parent HALS are not readily oxidized to nitroxyl radical by peracids and do inhibit photooxidation. Most HALS inhibition cycle products appear to be readily oxidized to nitroxyl radical by peracids.

The series of "Active HALS" analysis results shown in Figure 11 were obtained for clearcoats in Paint Systems 11 and 12 as a function of outdoor exposure. The clearcoat in Paint System 11 is an acrylic/melamine formulated with Tinuvin 292, an effective tertiary amine HALS that is not readily oxidized to nitroxyl radical by p-nitroperbenzoic acid. The clearcoat in Paint System 12 is an acrylic/urethane formulated with Tinuvin 123, an effective amino ether HALS that is readily oxidized to nitroxyl radical by p-nitroperbenzoic acid. The results shown for Paint System 11 clearly indicate that the amount of HALS available to inhibit photooxidation decreases to a low level after 5 years of northern Australia exposure while exposure in Belgium and Florida produces smaller decreases. It can be noted that although Tinuvin 292 is not readily oxidized to nitroxyl radical by peracids, a substantial fraction of the Tinuvin 292 initially added to the clearcoat can be oxidized to nitroxyl radical after only 3 months of exposure. The results shown for the clearcoat in Paint System 12 indicate that HALS will be available to inhibit photooxidation at very long exposure times. However, bulk clearcoat HALS analysis results need not provide sufficient information to correctly assess HALS performance. Experiments wherein clearcoat was scraped in step-wise fashion from the surface of the clearcoat inward have revealed steep gradients in "Active HALS" concentration increasing from the surface of the clearcoat inward.[xxv] Here, bulk analysis could not reveal that the surface region of the clearcoat is HALS deficient.

PHOTOOXIDATION, UVA, AND HALS MAPS

In-plane microtomy [v, xiii, xxvi, xxvii, xxviii] can be used to produce 5 µm thick slices of all coating layers in complete paint systems on steel test panels. When the individual paint system slices are analyzed by FTIR, UV, and ESR spectroscopy, the results can be used to construct 5 μm resolution maps of the disposition of photooxidation, UVA, and HALS across all coating layers in complete paint systems.

In practice, a 2.5 cm thick polypropylene block is mounted in an in-plane microtome and microtomed flat. Next, a 2.5 x 5 cm sample is sawed from a steel test panel and adhered to a flat polypropylene block using double-sided tape. And finally, the paint system on the test panel is progressively sliced into 5-10 μm thick slices. Slice thickness determines spatial resolution.

Photooxidation, UVA, and HALS analysis results are illustrated in Figures 12-14 for powder clearcoat over white waterborne basecoat over primer over e-coat, Paint System 13. A single 3-year Florida exposure test panel was available for analysis. A section of test panel that was shielded from sunlight by exposure rack clamps, but not shielded from heat and humidity, was analyzed for comparison with a section of test panel exposed to sunlight, heat, and humidity. The clearcoat on both test panel sections proved to be too soft to accurately microtome consistent thickness slices. That is, if the clearcoat slices are assumed to be 5 μm thick, then the sum of the slices accounts for approximately half of the total clearcoat actually present to emply that some of the slices were thicker than 5 μm. Nevertheless, the analysis results still provide a clear picture of paint system weathering performance. Basecoat, primer, and e-coat layers were easily microtomed.

As shown in Figure 12, transmission FTIR spectra indicate that the surface of the clearcoat is moderately oxidized compared to the surface of the section of test panel shielded from sunlight. There is also some evidence that light penetrates to the clearcoat/basecoat interface, basecoat slice #1. Transmission UV absorbance measurements, Figure 13, indicate that sufficient UVA remains in the clearcoat to screen underlying basecoat from UV light in the 340 nm region for many years of additional exposure. And finally, HALS analysis results, Figure 14, indicate that sufficient HALS remains in the clearcoat to inhibit photooxidation for many years of additional exposure. When these results are combined with the results of fracture energy measurements [xxix, xxx] designed to assess the mechanical repercussions of photooxidative degradation, it can be concluded that the risk of catastrophic paint system failure at long exposure times, >10 years in service, is minimal. The fracture energy value determined for Paint System 13 is contrasted with the fracture energy of other paint systems in Figure 15. Nichols et al describe fracture energy measurements in detail in a companion paper.[xxxi]

ACCELERATED WEATHERING EXPOSURE CONDITIONS

While the analytical techniques described above can reliably follow the chemical composition and mechanical property changes that occur in paint systems as the result

238

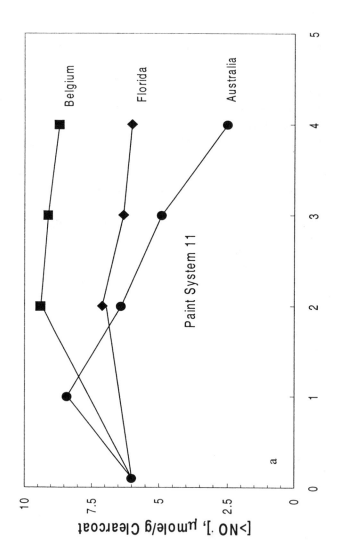

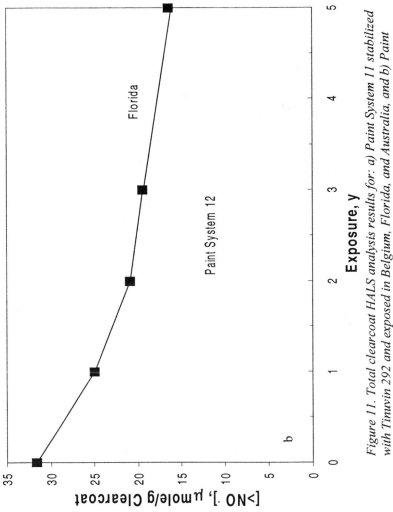

Figure 11. Total clearcoat HALS analysis results for: a) Paint System 11 stabilized with Tinuvin 292 and exposed in Belgium, Florida, and Australia, and b) Paint System 12 stabilized with Tinuvin 123 and exposed in Florida..

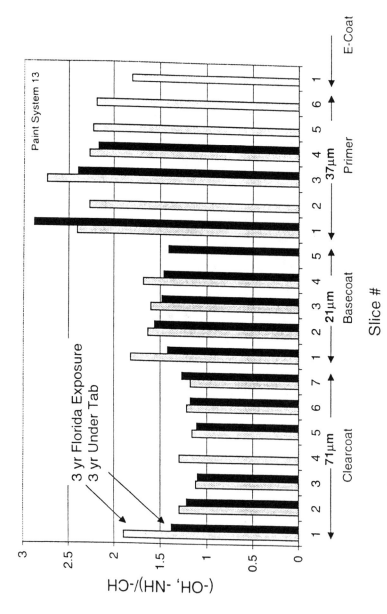

Figure 12. Transmission FTIR (-OH, -NH)/-CH ratio for ~5 μm in-plane slices of Paint System 13: a) section under tab during 3 years Florida exposure, and b) section in the open during 3 years Florida exposure.

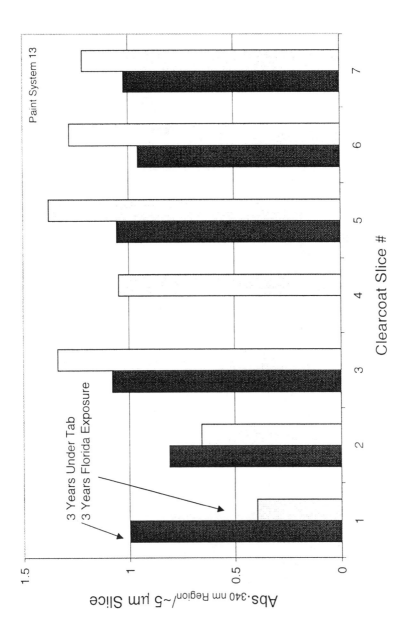

Figure 13. 340 nm region UV absorbance for ~5 μm in-plane slices of Paint System 13 clearcoat: a) section under tab during 3 years Florida exposure and b) section in the open during 3 years Florida exposure.

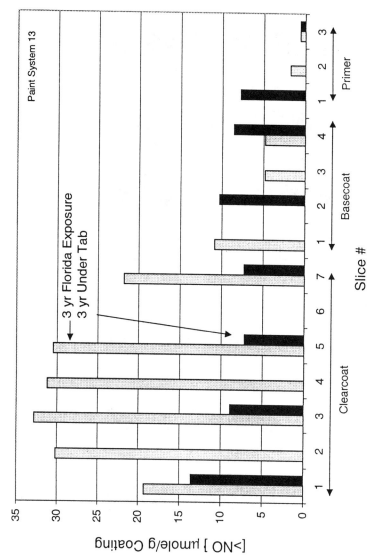

Figure 14. Peracid oxidation HALS analysis values for ~5 μm in-plane slices of clearcoat in Paint System 13: a) section under tab during 3 years Florida exposure and , b) section in the open during 3 years Florida exposure.

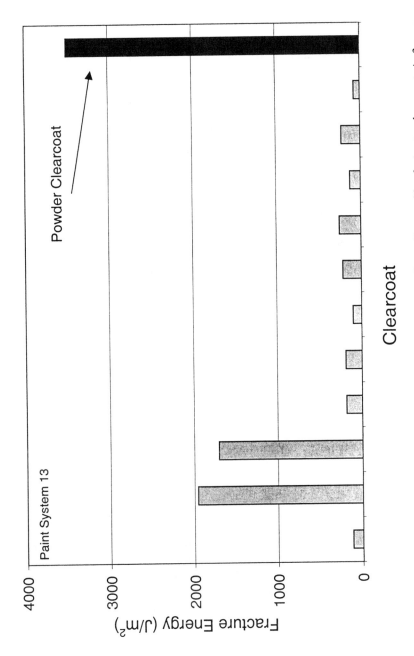

Figure 15. Clearcoat fracture energy for 3-year Florida exposure Paint System 13 relative to clearcoats in 2-year Florida exposure paint systems.

of exposure, the credibility of all results is directly linked to the credibility of the test specimens being assayed. As of this writing, outdoor exposure test specimens are assumed to be credible although even this assumption is open for discussion.[xxxii, xxxiii] On the other hand, the credibility of samples produced by accelerated exposure poses an issue because as noted in the introduction, because it is not always practical to wait for the results of actual outdoor exposure tests. We have long maintained that in order for accelerated exposure test specimens to be credible, exposure conditions must lead to the same kind and balance of chemical changes that occur during actual outdoor exposure.[ix, xxxiv] This contention is reinforced by the two examples that follow.

The results shown in Figures 1, 2, and 4 for polyester/urethane Clearcoats Y and Z clearly indicate that both clearcoats rapidly photooxidize during borosilicate inner and outer filtered Xenon arc exposure and should therefore exhibit inferior long-term weathering performance. Both clearcoats are formulated with a phthalate extender. In actuality, both clearcoats exhibit superior long-term weathering performance both during Florida exposure and in service. An explanation for the false Xenon arc exposure prediction may lie in the overlap of the UV spectra of the clearcoats with the output of the light source used.

The UV spectra of additive-free Clearcoats Y and T are shown in Figure 16 superimposed on the wavelength distribution curves of Miami noon sunlight and borosilicate inner and outer filtered Xenon arc light. Clearcoat T, an acrylic/melamine is included because its accelerated test behavior matches its Florida exposure behavior. As can be seen, Clearcoat Y's UV spectrum overlaps the output of the lamp considerably more than sunlight while the overlap of Clearcoat T's UV spectrum with sunlight and the lamp output is minimal. This suggests that the accelerated test light source could drive more photooxidation chemistry that cannot be driven by sunlight in Clearcoat Y than Clearcoat T. And indeed, PAS FTIR spectra of Florida and borosilicate inner and outer filtered Xenon arc exposed samples of Clearcoat Y do not match one another while the PAS FTIR spectra of Clearcoat T exposed under the same two conditions do match. This behavior strongly suggests that the false Xenon arc exposure results obtained for Clearcoats Y and Z can be directly attributed to the shorter than natural UV light component that remains in borosilicate inner and outer filtered Xenon arc light. This particular case is straightforward because the UV absorbance of a major component, phthalate extender, approaches the wavelength cut off of sunlight, ~295 nm. In most cases, photooxidation is driven by trace chromophores whose UV spectrum is not easily recorded.

As a second example, when Clearcoat N over waterborne basecoat is subjected to Xenon arc weather-ometer® accelerated exposure according to SAE J1960 JUN89 using borosilicate inner and outer filters, and 0.55 W/m^2 light intensity at 340 nm, it photooxidizes rapidly without UVA and the addition of UVA slows its photooxidation dramatically, Figure 17. This behavior suggests that Clearcoat N is very dependent on

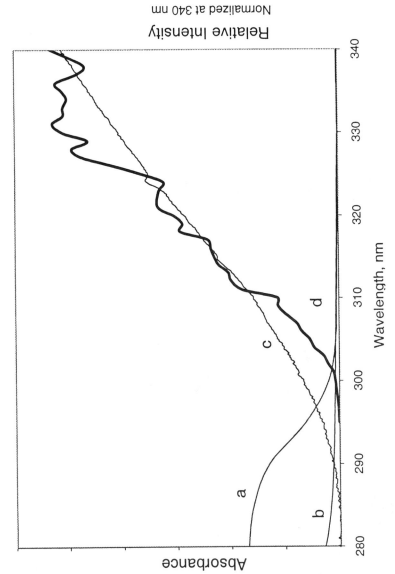

Figure 16. UV spectra of additive free Clearcoats Y (a) and T (b) superimposed on wavelength distribution of inner and outer borosilicate filtered Xenon arc light (c) and sunlight (d).

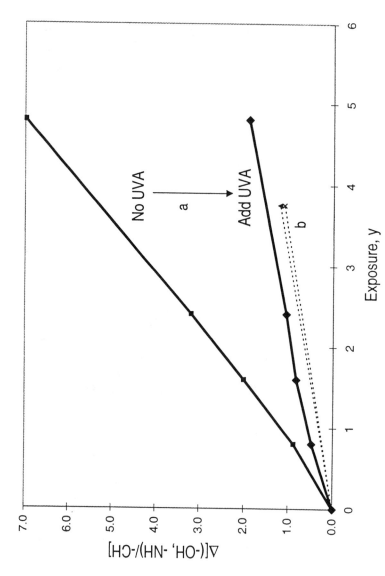

Figure 17. PAS FTIR Δ[(-OH, -NH)/-CH] values for Clearcoat N over waterborne basecoat with and without clearcoat UVA: a) subjected to SAE J1960 JUN89 using borosilicate inner and outer filters at 0.55 W/m² light intensity at 340 nm, and b) EMMAQUA exposure.

UVA to gain photooxidation resistance, a conclusion that is at odds with Clearcoat N's behavior when tested as an isolated clearcoat, Figures 1 and 2, and when Clearcoat N is tested over solvent borne basecoat, Figure 4. These latter results indicate that Clearcoat N is intrinsically photooxidation resistant and not dependent on UVA to gain photooxidation resistance.

An entirely different behavior is observed when Clearcoat N over waterborne basecoat is subjected to EMMAQUA exposure. Photooxidation is slow and the addition of UVA has little or no effect, Figure 17. Because EMMAQUA exposure employs reflected sunlight, this behavior suggests that the filtered Xenon light used excites a chromophore present when Clearcoat N is applied over waterborne basecoat and absent when Clearcoat N is applied over solvent borne basecoat. In essence, the UVA protects the chromophore from light that is not present in sunlight and therefore the photooxidation rate behavior observed during artificial exposure need have little bearing on the paint system's actual weathering performance. The actual weathering performance of Clearcoat N over the waterborne basecoat used is not known.

These examples strongly suggest that much greater attention will have to be paid to the wavelength distribution match between artificial light sources and sunlight if test specimen behavior is to accurately reflect actual long-term weathering performance in accelerated weathering tests.

CONCLUSIONS

The present work describes a number of analytical techniques designed to follow chemical composition changes for isolated clearcoat samples, clearcoats in complete paint systems, and all coating layers in complete paint systems as a function of exposure. It is argued that chemical composition change rate measurements can be combined with fracture energy measurements of tolerance for chemical composition change to dramatically reduce the risk of introducing new paint systems that will exhibit hard failures in service.

When accelerated exposure is used to prepare test specimens, result credibility relies on how closely accelerated exposure induced chemical composition changes match outdoor exposure chemical composition changes.

References

[i] *Surface Coatings Science & Technology*; Paul, S., Ed., 2nd Edition; John Wiley & Sons: New York,
NY, 1996; pp. 653-669.
[ii] Chess, J. A.; Cocuzzi, D. A.; Pilcher, G. R.; Van de Streek, G. N. In *Service Life Prediction of Organic*

248

Coatings: A Systems Approach; Bauer, D. R.; Martin, J. W., Eds.; ACS Symposium Series 722, Oxford University Press: 1999, pp. 130-148.

[iii] Bauer, D. R. *J. Coat. Technol.*, **1994**, 66, 57.

[iv] Gerlock, J. L.; Smith, C. A.; Cooper, V. A.; Dusbiber, T. G.; Weber, W. H; *Polym. Degrad. Stab.*, **1998**, 62 (2), 225-234.

[v] Adamsons, K.; Litty, L.; Lloyd, K.; Stika, K.; Swartzfager, D.; Walls, D.; Wood, B. In *Service Life Prediction of Organic Coatings: A Systems Approach;* Bauer, D. R.; Martin J. W. Eds.; ACS Symposium Series 722, Oxford University Press: 1999, pp. 257-287.

[vi] Born, M.; Wolf, E. *Principles of Optics*, 4th edition, Pergamon Press: N. Y., 1970, p. 43.

[vii] Israeli, Y.; Lacoste, J.; Lemairre, J.; Singh, R. P; Sivaram S. *J. Polym. Sci. Polym. Chem.* **1994**, 32, 485.

[viii] Lemiare J.; Siampiringue, N., In *Service Life Prediction of Organic Coatings: A Systems Approach;* Bauer, D. R.; Martin, J. W. Eds. , ACS Symposium Series 722, Oxford University Press: 1999 pp. 246-256.

[ix] Bauer, D. R.; Paputa Peck; M. C.; Carter, R. O. III *J. Coat. Technol.*, **1987**, 89, 103-109.

[x] Dittmar, R. M.; Palmer, R. A.; Carter, R. O. III *Appl. Spectrosc. Rev.*, **1994**, 29(2), 171.

[xi] Gerlock, J. L.; Prater, T. J.; Kaberline, S. L.; deVries, J. E. *Polym. Degrad. Stab.*, **1995**, 47(3), 405-411.

[xii] Gerlock, J. L.; Prater, T. J.; Kaberline, S. L.; Dupuie, J. L.; Blais, E. J.; Rardon. D. E. *Polym. Degrad. Stab.*, **1999**, 65, 37-45.

[xiii] Hadersdorfer, A.; Jurgetz, A.; Lamers, P.; Kohler, H. H. *Farbe and Lack*, in press.

[xiv] Valet, A. In *Light Stabilizers for Paints;* Zorll, U. Ed.: Vincentz Verlag: Hannover, Germany, 1997.

[xv] Pickett, J. E.; Moore, J. E. In *Polymer Durability: Degradation, Stabilization, and Lifetime Prediction;* Clough, R. L.; Billingham, N. C., Gillen, K. T. ACS Advances in Chemistry Series 249, Washington, DC: American Chemical Society: 1995 pp 287-301.

[xvi] Gerlock, J. L.; Smith, C. A.; Nunez, E. M.; Cooper, V. A.; Liscombe, P.; Cummings, D. R.; Dusbiber, T. G.; *ibid.*, pp 335-347.

[xvii] Pickett, J. E. in companion paper.

[xviii] Tokash, R. D. *Proceedings of the 3rd Annual ESD Advanced Coatings Conference*, **1993**, 173-183.

[xix] Carter, R. O. III *Opt. Eng.*, **1997**, 36 (2), 326-331.

[xx] Frank, H. P.; Lehner, H. *J. Polym. Sci., Part C*, **1970**, No. 31, 193-203.

[xxi] Dudler V.; Muinos, D. *Polymer Preprints*, ACS Div. of Polym. Chem., **1993**, 34 (2), 221-222.

[xxii] Rokosz, M. J.; Gerlock, J. L.; Kucherov A. V.; Belfield, K. D.; Fryer N. L.; Moad, G. *Polym. Degrad. Stab.*, **2000**, 70, 81-88.

[xxiii] Commereuc, S.; Schiers, V.; Verney, J.; Lacoste, *J. Appl. Polym. Sci.*, **1998**, 69, 1107-1114.

[xxiv] Kucherov, V. A.; Gerlock, J. L.; Matheson, R. R. Jr. *Polym. Degrad. Stab.*, **2000**, 6, 1-9.

[xxv] Gerlock, J. L.; Kucherov, A. V.; Smith, C. A. *Polym. Deg. Stab.*, in press.

[xxvi] Bohnke, H.; Avar, L.; Hess, E., *J. Coat. Tech.*, **1966**, 63, 53.

[xxvii] Haacke, G.; Andrawes, F. F.; Campbell, B. H. *J. Coat. Tech.*, **1966**, 68, 57.

[xxviii] Gerlock J. L.; Kucherov A. V.; Nichols, M. E. *J. Coat. Tech.*, in press.

[xxix] Nichols, M. E.; Darr, C. A.; Smith, C. A.; Thouless, M. D.; Fischer, E. R. *Polym. Degrad. Stab.*, **1998**, 60, 291-299.

[xxx] Nichols, M. E; Gerlock, J. L.; Smith, C.A. *Prog. Org. Coat.*, **1999**, 35, 153-159.

[xxxi] Nichols, M. E.; Tardiff, J. L. in companion paper.

[xxxii] Martin, J. W in companion paper.

[xxxiii] Bauer, D. R. *Polym. Deg. Stab.* **2000,** 69, 297-306.

[xxxiv] Bauer, D. R. *Polym. Deg. Stab.* **2000,** 69, 307-316.

Chapter 12

Permanence of UV Absorbers in Plastics and Coatings

James E. Pickett

Corporate Research and Development, General Electric Company,
1 Research Circle, Schenectady, NY 12301

ABSTRACT

Ultraviolet absorbers (UVA's) undergo photo-degradation with quantum yields on the order of 10^{-6} even in glassy, unreactive matrices such as poly(methyl methacrylate) (PMMA). The kinetics of UVA loss in a film or coating can be described by the equation: $A = \log_{10}[(1 - T_0)10^{(A_0 - kt)} + 1]$ where A is the absorbance at time t, T_0 is the initial transmission, A_0 is the initial absorbance, and k is a zero order rate constant. The rate constant can be found as the slope of $\log_{10}(10^A - 1)$ plotted vs. exposure. The derivation and implications of these equations and a general review of UVA degradation chemistry are discussed, including the effects of the matrix material, concentration, and hindered amine light stabilizers. The rate of UV absorber loss is of relatively minor consequence when the absorber is used as a bulk additive in a polymer, but is of critical importance in the service lifetime of coatings. The kinetics of photochemical UV absorber loss and use of loss rates in coating lifetime predictions are discussed in detail.

[†] Adapted from a paper presented at the 7th Annual ESD Advanced Coatings Technology Conference in Detroit, MI Sept. 28-29, 1998. Used with permission from ESD, The Engineering Society.

The photostability of UV absorbers (UVA's) recently has been recognized as key to the lifetime of coatings [1,2,3,4]. The types of molecules typically used as UVA's are shown in Figure 1. These compounds are extraordinarily photostable considering that they are organic molecules. However, the pioneers of UV stabilizers recognized that they are not truly permanent. In 1961, Hirt, Searle, and Schmitt wrote:

> The protective absorbers are not everlasting; they do photodecompose, but at a much slower rate than the materials which they are designed to protect. The photo-decomposition was found to be dependent on a number of factors including the substrate in which the absorber is dispersed and the wavelength of irradiation [5].

Unfortunately, little subsequent work was published to flesh out and substantiate this insight, and the photostability issues of UVA's disappeared from the common wisdom. Questions about coating failure mechanisms arose in the late 1980's and early 1990's that prompted reexamination of UVA lifetimes. We and other groups have been investigating the photostability of UVA's to better understand how UVA's degrade, the kinetics of the process, and how this knowledge allows better prediction of coating lifetimes [6].

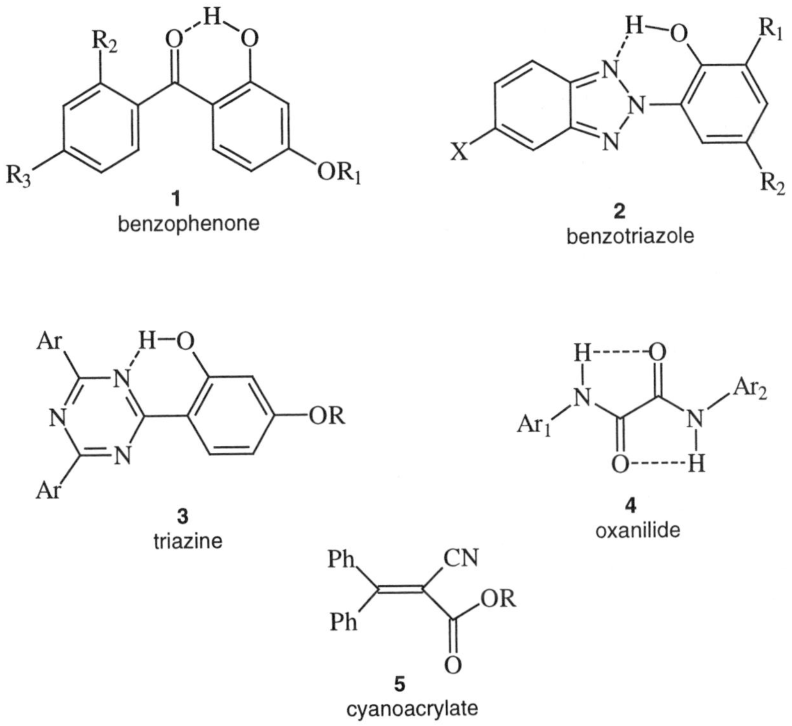

Figure 1. Structures of commonly-used commercial UVA's.

Photophysics of UV Absorbers

UVA's are added to coatings and plastics to absorb ultraviolet radiation in the wavelength range of 290 to 400 nm in order to protect materials from weathering. Usually the absorbers have little absorption at wavelengths >380 nm to avoid imparting a yellow color to the products. A key feature of absorbers 1-4 shown in Figure 1 is the presence of a strong intramolecular hydrogen bond between an O-H or N-H group and an oxygen or nitrogen 4 or 5 atoms away. This hydrogen bond allows the energy absorbed by the molecule to be dissipated harmlessly as heat as shown in Figures 2 and 3. The ground state of the UVA is excited to its first singlet state upon absorption of a photon but rapidly undergoes Excited State Intramolecular Proton Transfer (ESIPT) to form the excited state of a tautomer (S'_1) which has a smaller energy gap with its ground state (S'_0) than the "normal form" tautomer. The small energy gap between S'_1 and S'_0 makes radiationless decay more facile, and the excited state energy rapidly is lost as vibrational energy to the matrix. The proton then is transferred back to its original position in a very rapid keto-enol tautomerization step.

Anything that disrupts the ESIPT process will result in a longer lifetime for the S_1 excited state and a greater chance of the molecule undergoing irreversible photochemistry. Highly polar matrices can lead to more *inter*molecular hydrogen bonding and poorer photostability. In addition, media that are basic could result in deprotonation of a small equilibrium concentration of UVA molecules. Even if the concentration were very low, this small population would be highly susceptible to photolysis and cause rapid loss of absorber.

The photochemistry of UVA's is difficult to follow because the rate of degradation is very slow, and many of the primary photoproducts are much less photostable than the starting molecule. Therefore, the concentration of primary products is very small, and only highly degraded fragments usually are recovered. Benzophenones have yielded benzoic acid [5,7] and benzotriazoles have yielded benzotriazole [8,9,10] as shown in Figure 4. The phenolic moieties that would result from photolysis reactions such as these would be very unstable and undergo rapid photo-oxidation. Studies of the photolysis of both benzophenones and benzotriazoles generally show that free radicals alone have little effect on the stability. However, the combination of UV radiation, radicals, and oxygen results in rapid degradation, although the actual mechanisms involved remain unclear [8,10,11]. It should be noted that these photoproducts have very little absorbance in the wavelength range of 290 to 400 nm and cannot function as UVA's. Thus, the absorbance is observed to disappear cleanly as the UVA photodegrades.

The photophysics and photochemistry of the cyanoacrylate class (Structure 5) is less well understood. The excited state probably involves charge separation and weakening of the double bond as shown in Figure 5. The weakened double bond would allow enhanced vibration and rapid energy dissipation. The double bond is also susceptible to radical or nucleophilic addition that would result in loss of the chromophore.

1
1'
2
2'
3
3'
4
4'
hν
-Δ

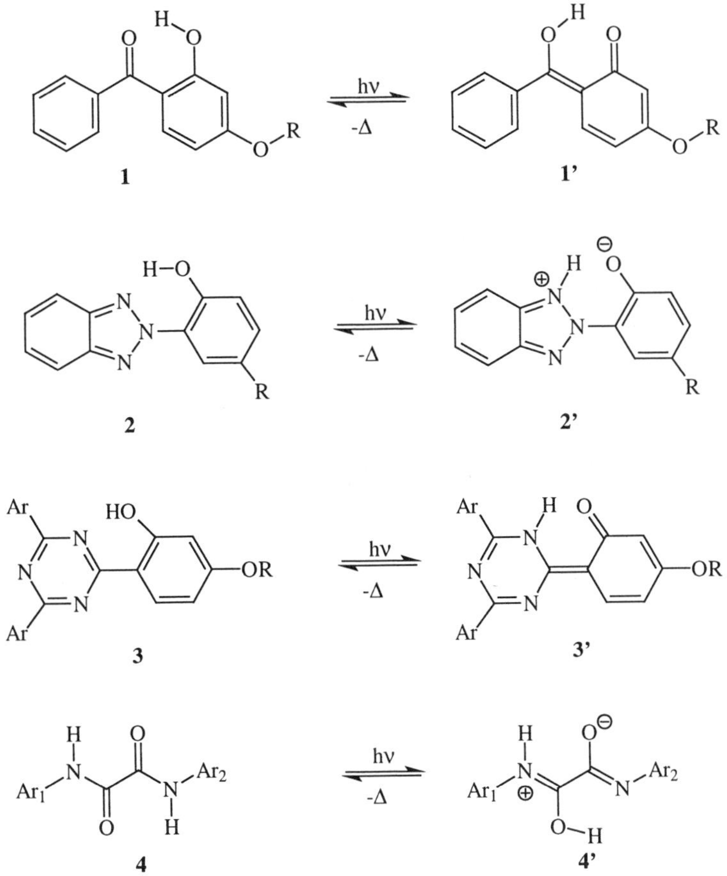

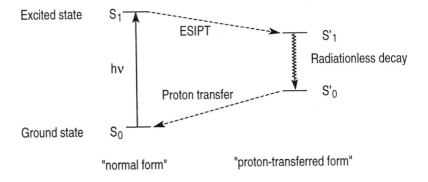

Figure 2. Proton-transfer process in protic UVA's.

Excited state
Ground state
S1
S0
hν
ESIPT
S'1
Radiationless decay
S'0
Proton transfer
"normal form"
"proton-transferred form"

Figure 3. Schematic of energy dissipation mechanism for protic UVA's.

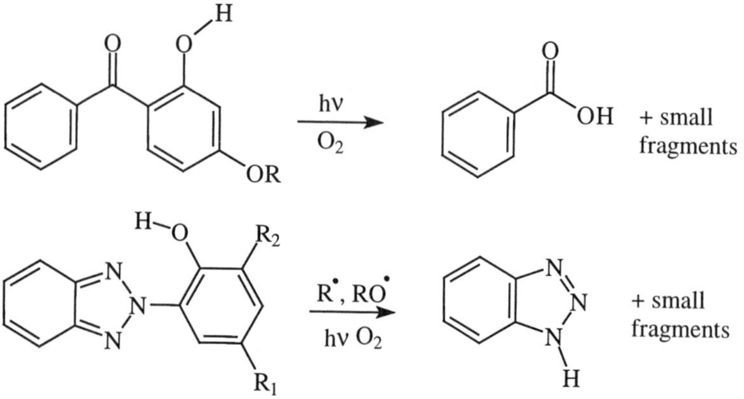

Figure 4. *Photochemistry of benzophenone and benzotriazole UVA's.*

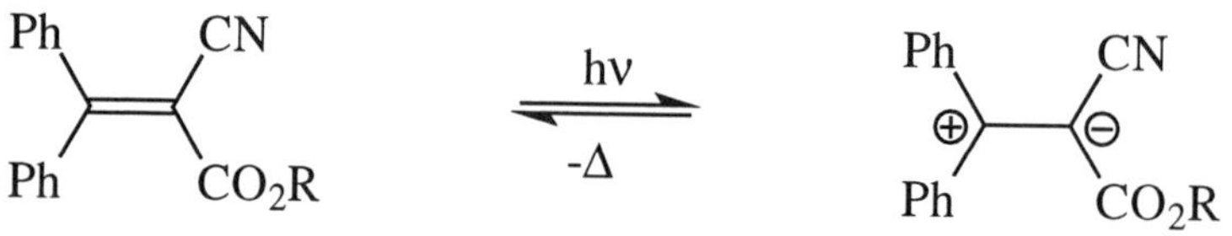

Figure 5. *Possible energy dissipation process for cyanoacrylate UVA's.*

UVA's, of course, have very low quantum yields for decomposition. A number of studies of various absorbers in various media show quantum yields on the order of 10^{-5} to 10^{-7} with the lowest values found in very non-polar solvents [10,12,13,14,15]. UVA's in most polymer matrices have quantum yields on the order of 10^{-6}. Very polar solvents can increase the quantum yield to 10^{-4}. Two studies [1,13] have shown pronounced wavelength dependence for both benzotriazoles and benzophenones with light of 300 nm having quantum yields 3-20x higher than light of 350 nm.

Kinetics of UV Absorber Photodegradation

The kinetics of UVA loss have been described in detail in several papers [1,3,16]. In this analysis, we will consider the UV absorber in a film or coating and the effect of UVA loss on the total absorbance or transmission of the film. The rate of absorbance loss can described by Equation 1 where k is a rate constant and T is the fraction of light transmitted by the film ($T = I/I_o$). Note that the rate is proportional to the fraction of light absorbed ($1-T$) and not the absorbance ($A = -log[T]$). The rate constant incorporates the quantum yield for degradation and the incident light flux integrated over all wavelengths. Substituting the relation $T = 10^{-A}$ and integrating gives absorbance as a function of time as shown by Equations 2 or 3 where A_o and T_o are the initial absorbance and transmission, k is a rate constant, and t is time (or exposure). The rate constant, k, can be determined by the method of Iyengar and Schellenberg [17] in which Equation 3 is rearranged to give Equation 4. Thus, plotting $log(10^A-1)$ as a function of time or exposure should give a straight line with slope k and an intercept related to the initial absorbance. This is demonstrated in Figures 6 and 7. Figure 6 shows absorbance loss data for a PMMA film containing 2% of a benzophenone UVA, Cyasorb® 531, plotted as a function of exposure in a xenon arc Weather-ometer®. Curvature is evident in the data, and a line calculated from Equation 3 using a rate constant of 0.18 A/1000 kJm^{-2} fits the data nicely. Figure 7 shows the same data plotted according to Equation 4. The data show an excellent fit to this linear relationship.

$$dA/dt = -k(1-T) \qquad \text{(Eqn. 1)}$$

$$A_t = \log_{10}\left[\left(1 - T_0\right)10^{\left(A_0 - kt\right)} + 1\right] \qquad \text{(Eqn. 2)}$$

$$A_t = \log_{10}\left(10^{\left(A_0 - kt\right)} - 10^{-kt} + 1\right) \qquad \text{(Eqn. 3)}$$

$$\log(10^{A_t} - 1) = -kt + \log(1 - T_0) + A_0 \qquad \text{(Eqn. 4)}$$

In a highly absorbing coating or film, the UVA acts as an inner filter so that the molecules near the surface absorb more photons and degrade more rapidly than those that are deeper into the film. Consequences of this are shown in Figures 8 and 9. When the absorbance of a coating is >1, essentially all of the incident UV photons are absorbed and the loss of absorbance appears linear with time; that is, it shows *zero order* kinetics. The rate of loss is dependent on only the light exposure. This type of loss is typical of a coating as a whole. When the absorbance is <0.1, then there is little consumption of light through the thickness of the film, and the rate of loss is dependent on the amount of UVA present. The loss appears linear on a log scale; that is, it shows *first order* kinetics. This type of loss is typical of the surface of a coating or a stabilized plastic.

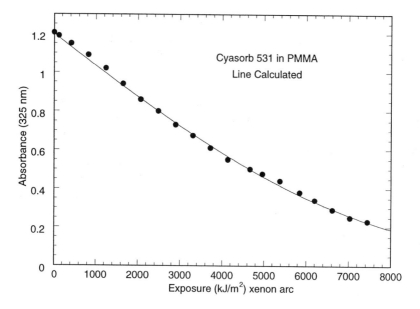

Figure 6. *Absorbance of a PMMA film containing 2% of 2-hydroxy-4-octyloxybenzophenone (Cyasorb 531) upon exposure to borosilicate-filtered xenon arc weathering. The smooth line is calculated from Equation 3 using a rate constant of 0.18 A/1000 kJ/m².*

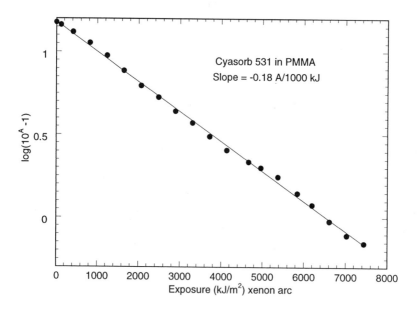

Figure 7. *Data from Figure 6 plotted according to Equation 4.*

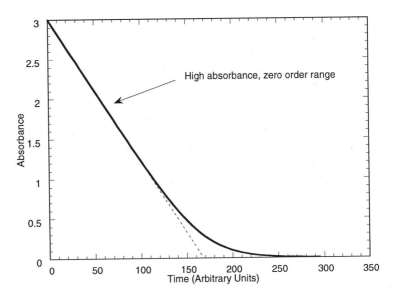

Figure 8. Calculated UVA loss for a highly absorbing coating plotted as zero order kinetics. The absorbance in the range $A > 1$ can be described by $A_t = A_o - kt$.

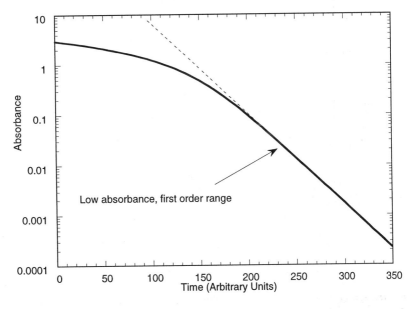

Figure 9. Calculated UVA loss for a highly absorbing coating plotted first order kinetics. The absorbance in the range $A < 0.1$ can be described by the equation $log(A_t) = log(A_o) - kt$.

We have observed both kinds of kinetic behavior when measuring the loss of UVA from a silicone hardcoat containing a benzophenone-type UVA during outdoor weathering [1]. Figure 10 shows the loss of absorbance from the entire coating applied to polymethyl methacrylate (PMMA) and exposed outdoors. The sample exposed in Mt. Vernon, IN showed approximately linear loss of absorbance, as predicted. A series of samples tested in Florida of the same coating applied to polycarbonate were tested by attenuated total reflectance IR which analyzes only the very surface of the sample; most of the information comes from the top few tenths of a micron where the absorbance due to the UVA is < 0.1. The loss could be seen to obey first order kinetics as shown in Figure 11.

Descriptions of UVA lifetime in a material can be ambiguous [16]. The concept of a half-life is applicable only to the region of absorbance <0.1 where first order kinetics apply. In addition, the common practice of plotting the fraction of remaining absorbance as a function of time on a linear scale gives very misleading results since the slope is entirely dependent on the initial absorbance, and that information is lost in such plots. The rate constant as defined in Equation 4 does give an unambiguous value for a particular UVA in a particular matrix independent of initial absorbance and concentration.

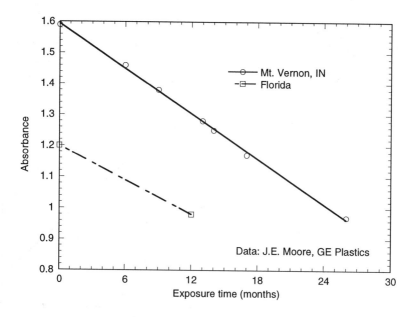

Figure 10. Loss of absorbance in a silicone hardcoat upon outdoor exposure (from Ref. 1).

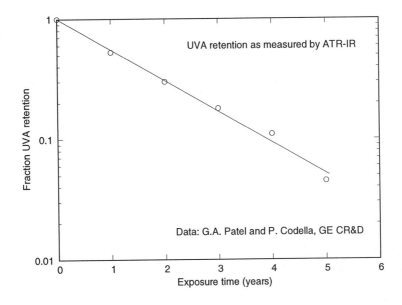

Figure 11. *Loss of UV absorber from the top 0.3 μm of a silicone hardcoat upon Florida weathering (from Ref. 1).*

Factors Affecting UVA Degradation Rates

The most important factor affecting the lifetime of most UVA's is the matrix in which it resides [7,18]. A quickly degrading matrix leads to rapid decomposition of the UVA in it probably due to free radical attack. Polar matrices are often detrimental due to disruption of the critical internal hydrogen bond. The magnitudes of these effects are not very predictable as shown in Table 1. The rankings of UVA stability can reverse in different coatings.

Table 1.Rates of UVA loss in various matrices upon borosilicate-filtered xenon arc exposure.

Absorber	Class	Rate (A/1000 kJm^{-2})		
		PMMA	Silicone hardcoat*	UV-cured acrylic
Cyasorb® 531	benzophenone	0.18	0.10	0.42
Cyasorb® 5411	benzotriazole	0.11	0.32	0.28
Uvinul® 3039	cyanoacrylate	0.14	0.31	0.42
Sanduvor® VSU	oxanilide	0.14	0.31	N/A
Cyasorb® 1164	triazine	0.085	0.81	0.22

* Absorber with a trialkoxysilyl group in place of an alkyl group, but otherwise the same chromophore.

Table 1 shows that there was some differentiation among the various classes of UVA in the relatively polar, but unreactive PMMA matrix with the triazine degrading most slowly and the benzophenone degrading most quickly. The silicone hardcoat consisted of condensed methyl trimethoxysilane and colloidal silica and was slightly basic. Some residual hydroxyl groups are present as well. In this coating, the ordering of benzophenone-type and triazine-type UVA is reversed. The UV-cured acrylic was based on hexanediol diacrylate, and degraded fairly rapidly under these conditions. The order of stability was approximately the same as for the PMMA, but all the loss rates were much higher. These results underscore the importance of testing each candidate UVA in the particular matrix in which it is to be used.

We have found no dependence on concentration, at least in PMMA films. The only effect of hindered amine light stabilizers (HALS) was that if the HALS stabilized the matrix, as for a UV-cured acrylic coating for example, then the stability of the UVA was improved as well. Similar results have been reported by Decker [4,15,19] and by Gerlock [2]. We have also examined the effects of structure on UVA lifetime and found that the effects of alkyl substituents on the benzophenone and benzotriazole classes generally are subtle and are probably highly dependent on the matrix [16]. The triazines, however, can show dramatic substituent effects. Cyasorb[®] 1164 (Structure 3, Ar = 2,4-dimethylphenyl) degrades somewhat more slowly than benzotriazoles in most matrices we have tested. However, Tinuvin[®] 1577, which lacks the methyl groups, has a nearly undetectable rate of loss in a PMMA matrix. The alkyl groups are reported to increase the lifetime of the first excited state [20,21] which leads to faster degradation. Tinuvin[®] 1577 appears to be the most stable commercially available UVA.

Effect of UVA Loss on Material Lifetime

UVA's can be lost by diffusion or by photochemical decomposition. Diffusion can play a major role in plastics or coatings in which the test or use temperature is greater than the glass transition temperature (T_g). In this case diffusion is fast, and most UVA's can be rapidly removed from the surface [22,23,24]. This can be a problem for low T_g materials such as polyolefins, plasticized PVC, and UV-cured coatings. However, the rates of diffusion decrease dramatically at T_g and can become vanishingly small when materials are well below T_g [25,26]. UVA loss by diffusion during use is inconsequential for most thermally cured coatings and engineering thermoplastics.

Photochemical UVA loss occurs in the surface of plastics. The effect on plastics lifetime is difficult to predict, but probably is not of great consequence for current generation materials. This is because the lifetime of high performance UVA's in the surface layer is on the order of two years or so, and most engineering plastics undergo surface erosion within that time span thus continually "renewing" the surface. However, as materials are developed with greater inherent stability, more stable UVA's may have to be developed to protect them. UVA's are of much less importance in polyolefins where HALS are the primary stabilizers.

UVA stability can be critical to coating lifetimes as recently described by Bauer [3, 27]. An unpigmented coating protects the substrate by absorbing the incident ultraviolet radiation. Some always leaks through to reach the surface of the substrate, but at high absorbance the transmission is small. An example of a coating with an initial absorbance of 3.0 (transmission of 0.001) is shown in Figure 12. As the UVA in the coating degrades upon exposure, the transmission remains low until the absorbance reaches 1 or so whereupon it increases rapidly. However, the important factor is not the absorbance or transmission at any particular time, but rather the integral of the transmission which is the total UV dose that reaches the surface. When the transmitted dose reaches some critical value, the substrate will be degraded sufficiently to cause coating failure. The light dose, D, as a function of time or exposure can be calculated by converting Equation 2 or 3 to transmission and integrating to give Equations 5 and 6. Note that the transmitted dose, and hence the theoretical lifetime of a coating, is determined only by the initial absorbance (or transmission) and the rate of UVA loss. The remaining variable is the dose, D_{fail}, that is required to reach the substrate to cause coating failure (by delamination, for example). This can be determined by making a coating containing no UVA at all and measuring the time or exposure at which the coating fails. Equation 6 can be solved for time as shown in Equation 7. If the initial transmission of the coating, T_0, the rate of UVA loss, k, and the transmitted dose required to cause coating failure, D_{fail}, are all known, then Equation 7 gives a prediction for the maximum possible lifetime of the coating. Earlier failure can always occur if other mechanisms such as hydrolysis are operable or if other failure modes such as cracking or hazing are important.

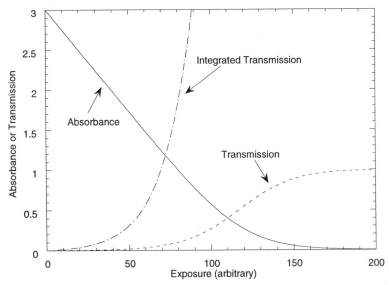

Figure 12. Absorbance, transmission, and integrated transmission for a hypothetical coating.

$$D = t + \frac{1}{k}\left(A_t - A_o\right) \qquad \text{(Eqn. 5)}$$

$$D = t + \frac{1}{k}\log_{10}\left[T_0 + \left(1 - T_0\right)10^{-kt}\right] \qquad \text{(Eqn. 6)}$$

$$t_{fail} = \frac{1}{k}\log_{10}\left[\frac{10^{kD_{fail}} + T_0 - 1}{T_0}\right] \qquad \text{(Eqn. 7)}$$

A plot of transmitted doses for a hypothetical coating and substrate is shown in Figure 13, and a summary of calculated fail times is shown in Table 2. The effects of changing the stability of the substrate, rate of UVA loss, and the initial absorbance of the coating are examined. The D_{fail} values, which would be determined by testing a coating with no UVA, are chosen as one and two years of outdoor exposure for this example. The UVA loss rates are typical for commercial absorbers in relatively unreactive matrices. One sees that especially for the high absorbance coatings, very little light is transmitted until late in the exposure resulting in rapid and catastrophic failure. This failure would be unanticipated if it were not known that the UV absorbance of the coating was decreasing during exposure.

Fail times calculated using Equation 7 for hypothetical coating systems are shown in Table 2. Interesting effects on the calculated failure times can be seen. If a coating with no UV absorber fails in 1 year, then Entries 2, 3, and 4 show that initially blocking 90%, 99%, and 99.9% of the light increases the lifetime to 3, 5.5, and 8 years respectively. If the UV absorber were truly permanent, a lifetime of 1000 years would be expected for the coating in Entry 4. Comparing Entries 3 and 6 or 4 and 7 shows the effect of a more stable UV absorber. Decreasing the loss rate by a factor of two increases the lifetime by roughly a factor of 1.6 for this example. This factor will be different for coatings with other initial absorbances and loss rates. The bottom half of the table shows the effect of increasing D_{fail} from one year to two years, that is, doubling the stability of the substrate. The lifetime of the coated system is not doubled, but depending on the other variables, the lifetimes are increased by only 15 to 40% in this hypothetical case. Vastly different numbers are obtained if different rates of UV absorber loss are chosen. Equation 7 obviously is not linear in any variable, so intuition is of little value in estimating the magnitude of improvements. However, this model generally shows that the lifetime of the coated system is determined primarily by the lifetime of the coating itself and UV absorber in it. The stability of the substrate is an important, but secondary, concern.

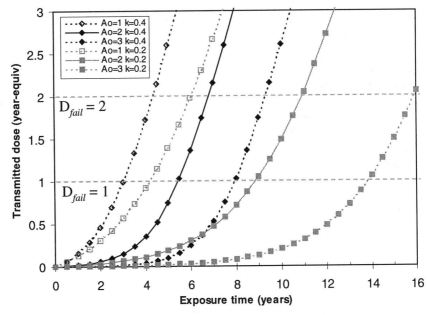

Figure 13. *Transmitted light doses for hypothetical coating systems with various initial absorbances and UVA degradation rates calculated using Equation 6.*

Table 2. Calculated failure times for the coating systems shown in Figure 13.

Entry	D_{fail} (years)	k (A/year)	A_o	T_o	t_{fail} (years)
1	1	-	0	1	1.0
2	1	0.4	1	0.1	3.0
3	1	0.4	2	0.01	5.5
4	1	0.4	3	0.001	8.0
5	1	0.2	1	0.1	4.2
6	1	0.2	2	0.01	8.9
7	1	0.2	3	0.001	13.8
8	2	-	0	1	2.0
9	2	0.4	1	0.1	4.3
10	2	0.4	2	0.01	6.8
11	2	0.4	3	0.001	9.3
12	2	0.2	1	0.1	6.0
13	2	0.2	2	0.01	10.9
14	2	0.2	3	0.001	15.9

Summary and Conclusions

UV absorbers can undergo photochemical degradation when their normal means of energy dissipation through excited state intramolecular proton transfer (ESIPT) is disrupted. For many UVA's the quantum yield of degradation is on the order of 10^{-6}, however, the rates of degradation are highly dependent on the matrix and must be determined experimentally for each system. For most UV absorbers, one can expect a loss of at least 0.3 absorbance units per year of Florida exposure. The loss can be as high as 1 absorbance unit per year in a highly degrading coating.

The rates of UVA loss are best expressed rate constants which can be obtained by plotting the absorbance of a coating as $\log(10^A - 1)$ as a function of exposure. If the system obeys the kinetics described above, then the plot should be a straight line with the slope being the rate constant. This rate constant plus the initial absorbance of a coating can be used to estimate the theoretical lifetime of a coating. Other factors such as cracking, hazing, or hydrolysis may cause failure long before the theoretical lifetime is reached.

Acknowledgments

I thank my colleagues at GE for many helpful discussions, especially Arnold Factor, Greg Gillette, and Randall Carter. I particularly thank Mr. James Moore for sharing pioneering work on the loss of UVA's from coatings and contributing valuable data and insight.

References

1. J.E. Pickett, *Macromol. Symp.* **1997**, *115*, 127-141.
2. J.L. Gerlock, C.A. Smith, E.M. Nunez, V.A. Cooper, P. Liscombe, D.R. Cummings, T.C. Dusibiber. In: R.L Clough, N.C. Billingham, K.T Gillen, ed. *Polymer Durability: Degradation, Stabilization, and Lifetime Prediction.* ACS Advances in Chemistry Series 249. Washington DC: American Chemical Society, 1995. pp 335-348.
3. D.R. Bauer, *J. Coatings Tech.* **1997**, *69*, 85-95.
4. C. Decker, K. Moussa, T. Bendaikha, *J. Polym. Sci. A Polym. Chem.* **1991**, *29*, 739-74.
5. R.C. Hirt, N.Z. Searle, R.G. Schmitt, *SPE Transactions* **1961**, 26-30.
6. For a comprehensive review see J.E. Pickett, In: S.H. Hamid, ed. *Handbook of Polymer Weathering Second Edition,*. Marcel Dekker, 1999. *in press.*
7. J.E. Pickett, J.E. Moore, *Polym. Degrad. Stabil.* **1993**, *42*, 231-244.
8. J.L. Gerlock, W. Tang, M.A. Dearth, T.J. Korniski, *Polym. Degrad. Stabil.* **1995**, *48*, 121-130.
9. M.A. Dearth, T.J. Korniski, J.L. Gerlock, *Polym. Degrad. Stabil.* **1995**, *48*, 111-120.

10. J. Sedlar, J. Petruj, J. Pac," *1ˢᵗ International Symposium on Photochemical Processes in Polymer Chemistry*, IUPAC, Louvain, Belgium, June 1972.

11. K.B. Chakraborty, G. Scott, *Europ. Polym. J.* **1979**, *15*, 35-40.

12. J. Petruj, J. Sedlar, M. Parizek, J. Pac, *J. Photochem.* **1973/74**, 2, 393-400.

13. J. Catalan, J.C. Del Valle, F. Fabero, N.A. Garcia, *Photochem. Photobiol.* **1995**, *61*, 118-123.

14. G. Rytz, R. Hilfiker, *Eighteenth Annual International Conference on Advances in the Stabilization and Degradation of Polymers*, Luzern (Switzerland), June 19-21, 1996.

15. C. Decker, S. Biry, K. Zahouily, *Polym. Degrad. Stabil.*, **1995**, *49*, 111-119.

16. J.E. Pickett, J.E. Moore, *Angew. Makromol. Chem.*, **1995**, *232*, 229-238.

17. R. Iyengar, B. Schellenberg, *Polym. Degrad. Stab.*, **1998**, *61*, 151.

18. J.E. Pickett, J.E. Moore, In: R.L. Clough, N.C. Billingham, K.T. Gillen, eds., *Polymer Durability: Degradation, Stabilization, and Lifetime Prediction*. ACS Advances in Chemistry Series 249. Washington DC: American Chemical Society, 1995. pp 287-301

19. C. Decker, K. Zahouily, *Polym. Mat. Sci. Eng.* **1993**, *68*, 70-71.

20. J. Keck, H.E.A. Kramer, H. Port, T. Hirsch, P. Pischer, G. Rytz, *J. Phys. Chem.* **1996**, *100*, 14468-14475.

21. J. Keck, G.J. Stüber, H.E.A. Kramer, *Angew. Makromol. Chem.* **1997**, *252*, 119-138.

22. J. Luston, In: G. Scott, ed. *Developments in Polymer Stabilization-2*. London: Applied Science Publishers, 1980. pp 185-240.

23. N.C. Billingham, P.C. Calvert, In: G. Scott, ed. *Developments in Polymer Stabilization-3*. London: Applied Science Publishers, 1980. pp 139-190.

24. N.C. Billingham, In: J. Pospisil, P.P. Klemchuk, eds. *Oxidation Inhibition in Organic Materials. Volume I*. Boca Raton: CRC Press, 1990. pp 249-297.

25. M. Johnson, R.G. Hauserman, *J. Appl. Polym. Sci.* **1977**, *21*, 3457-3463.

26. D.R. Olson, K.K. Webb, *Macromol.* **1990**, *23*, 3762.

27 D.R. Bauer, In: D.R. Bauer, J.W. Martin, eds., *Service Life Prediction of Organic Coatings*, ACS Symposium Series 722, American Chemical Society, 1999. pp 378-395.

Chapter 13

Mechanical Properties of Painted TPO Plastic Resulting from Morphological Interphase Variations

Rose A. Ryntz

Visteon Corporation, 401 Southfield Road, Dearborn, MI 48121

The physical behavior of painted and/or unpainted thermoplastics subjected to applied stress in field applications varies dramatically depending upon material selection and processing parameters. Of particular concern when trying to relate chemical structure of the paint/plastic to end-use properties is the "interphase" miscibility within the plastic alloy. In this paper, the "interphase" management of elastomer, as dispersed within a poly(propylene) matrix, is discussed and related to resultant surface damageability, e.g., scratch (of unpainted specimens) and friction induced paint damage (a compressive shear loading event) of painted specimens. The role of polymer processing, in particular injection molding shear velocity, as well as the paint process conditions utilized, on the "interphase" between the elastomer/poly(propylene) matrix, will be discussed. It was determined that by controlling molecular weight, molecular weight distribution, crystallinity, and melt viscosities of the elastomer/poly(propylene) matrix, surface damageability of the fabricated part could be controlled.

In injection molded plastics, in particular where semi-crystalline polymers are involved, residual stresses can occur in the top few microns of the plastic depending upon injection molding conditions and polymer composition utilized [1]. It was reported by Schonhorn and others [1-7] that heterogeneous nucleation and crystallization of polymer melts against high-energy surfaces, e.g., metals, metal oxides, and alkali halide crystals, results in marked changes in the surface region morphology. More specifically, it was claimed that transcrystallinity [8,9] (transcrystallinity is reported to consist of elongated spherulites originating at the polymer surface and propagating for several microns normal to the surface) could be induced in polyethylene and polypropylene by crystallizing the respective melts in intimate contact with specially prepared aluminum and copper. Formation of a transcrystalline layer is favored by conditions that induce a high density of nuclei at the surface. A close arrangement of growth centers causes the spherulites to grow in a columnar fashion with little lateral development [10].

Microstructures of semi-crystalline polymers vary from spherulitic to lamellar and single crystals, in size of structural units, depending upon [11]:

- molding temperature
- low patterns
- aging and heat treatment
- nucleating agents

The distribution of spherulite sizes varies as a function of depth. In injection molded polypropylene, for example, not only does the crystal size vary but also the crystal type. Hexagonal spherulites (less perfect flat, concave boundaries) were shown to be more prominent at the surface. They were characterized by greater susceptibility to etching in SEM analysis. Monoclinic spherulites, a more crystalline, ordered arrangement, were found to occur in the bulk morphology in conical shapes. In thermoplastic olefin (TPO, a blend of polypropylene and ethylene propylene diene rubber (EPDM)) that had been injection molded, Bonnerup [12] found that not only was the polypropylene surface crystallinity affected but that the EPDM phase separated in the crystallization process and was present in very low concentrations at the surface.

Bikkerman [13] found that two solids in contact couldn't fail exactly at the interface between them. Hence, if failure occurs at or near an interface at a relatively small-applied stress a *weak boundary layer* is assumed to have been present. Weak boundary layers are believed to arise if:

- there is an incompatibility between two polymers so that they remain separate;
- the surface roughness amplitude between two similar solids is 3000 angstroms or less; and
- a similar T_g for both polymers exists so that annealing eliminates shrinkage and thermal stresses. Fracture toughness

and crack propagation are related to weak boundary layer management.

Injection molding is one of the most widely utilized methods of polymer processing. This process includes injecting molten polymer into a cold mold followed by packing under high pressure and subsequent cooling to solidification. The filling and cooling stages have an important effect on the rheological properties since the viscoelastic nature of the polymer results in development of shear and normal stresses and large elastic deformation during filling with subsequent incomplete relaxation during the cooling stage. The resultant residual stresses, which determine the orientation in the final molded part, are dependent upon the thermal, rheological, and relaxation properties of the polymeric material as well as the processing conditions [14].

Molecular orientation in injection molding has been modeled by many researchers [15, 16] to explain the complex molecular orientation distribution observed. Most models incorporate flow and heat transfer mechanisms coupled with molecular theories. The orientation in the surface skin is related to steady elongational flow in the advancing front, whereas the orientation in the core is related to the shear flow, behind the front, between two solidifying layers [15]. Coupled with the elongational and shear-induced orientations, a molecular relaxation process takes place that is determined by the rate of heat transfer. Internal stresses that develop within the injection-molded part are the result of thermal, flow, and pressure histories[17]. The melt temperature of the polymer was found to cause two maxima in residual stress (R.S.) [18]. The second one reverses from compressive to tensile. In general, most changes occur in the surface regions, while R.S. decreases with increasing melt temperature, as is the case in zones far away from the gate [19]. It was found that R.S. are compressive in the surface layers and tend to decrease upon increase in mold temperature and distance from the gate.

Residual stress can be measured by optical birefringence. Birefringence has a number of causes. The polarizability of chemical bonds change when they are stressed, giving rise to the photo-elastic response upon which measurements of residual stress may be measured. On the other hand, chemical bonds are directional and in a non-isotropic material the presence of favored bond orientations will produce birefringence, so that in an injection molded article containing frozen in molecular orientation this may provide a much larger contribution than the residual stresses [20]. The birefringence effect should be higher closest to the mold wall, dependent of course upon the heat diffusivity of the polymer and the molding conditions.

Injection molding conditions, e.g., melt temperature, aging and heat treatment, surface geometry, etc., have been shown to be significant in determining the final cohesive strength of a TPO substrate [21]. The crystallinity and phase separation that occurs in a TPO sample once molded can be controlled to some degree based on the molding history. The degree of cohesive strength in a TPO sample, however, seems to be a direct result of the molecular weight, molecular weight distribution, and miscibility of the blends of elastomer and PP that are chosen.

In this paper, an attempt is made to relate structure of the TPO blend chosen to surface damageability caused by scratching, and friction induced compressive shearing ("gouging"). The role of interphase management, e.g., control of miscibility between alloying agents, appears to be the major factor affecting the ability of the plastic part to resist surface damage caused by external forces.

Data and Results

The TPOs evaluated in this study were made by physically dry-blending the components (Table 1) at 20 wt% elastomer in 80 wt% PP (where filler was utilized it was done so at 10 wt%) and melt extruding the blend through a Werner-Pfleiderer twin screw extruder using a general compounding profile and a single strand die. The extruder was preheated to the following barrel conditions: 210°C, 220°C, 220°C, 220°C, 220°C, 225°C. The screw was run at a constant speed of 115 rpm. Poly(propylene) and elastomer were obtained from Exxon Chemical, Bayport, TX or Dow-DuPont Elastomers, Midland, MI. Melt flow rates (MFR) of each component were measured according to ASTM D1238-96, 230/2.16. Melting points and percent crystallinities were determined on a DuPont Model TA Modulated Differential Scanning Calorimeter. Molecular weights and molecular weight distributions were supplied by Exxon Chemical and DuPont Dow Elastomers. Talc (Cimpact) was obtained from Luzenac Minerals, Denver, CO. Wolastonites were obtained from Nyco Minerals, Inc., Charlottesville, VA. The supplier performed all particle size measurements.

Each blend was molded into plaques on a Cincinnati Milacron 110 ton, 5 ounce injection molding machine at injection velocities of 1.27 and 7.62 cm/sec. Tensile testing on unpainted plaques was performed on an Instron Model 6025 tensile tester equipped with a 454.5 kg load cell and Instron series IX computer controlled software. Work of fracture data was calculated from the total area under the stress-strain curve. Sample specimens utilized in the work of fracture calculations were 5 x 17.8 cm bars modified with varying ligament lengths inscribed through use of double-sided notches obtained with a razor blade. Essential work values were calculated according to the method described by Wu [22].

Flexural modulus (ASTM D790-96) and Izod impact (ASTM D256-93, method A) data were measured on unpainted samples.

Plastic moduli (H_{plas}) and degree of plastic deformation (plasticity) (W_r/W_t) were measured on unpainted specimens with a Fisherscope H100 with a load of 1000 mN utilizing a Vickers indentor.

Optical microscopy analysis of 30 micron thick unpainted specimens, obtained by cryogenically microtoming the specimens (direction of cut was parallel to the flow direction in the plaque) using Histoprep media, was measured on a Leica microscope equipped with cross-polarizing filters. The samples were mounted in Canada balsam between two microscope slides prior to analysis. Depth of boundary layers within the specimens was measured utilizing Optimus optical imaging software.

Scanning electron microscopy (SEM) was performed on selected samples on cryogenically fractured surfaces of TPO under a magnification of 1500x.

Scratch testing was performed according to the Ford Laboratory Test Method (FLTM) BN 108-13. In this FLTM, the panel is placed on a movable platen onto which is placed a beam containing a scratch pin. The beam is 250 mm long and is equipped with a scratch pin that consists of a highly polished steel ball (1 mm +/- 0.1 mm in diameter). The beam is loaded with a weight ranged of 7.0 N. The beam is driven by compressed air to draw the pin across the surface of the plaque to generate a scratch. Sliding velocity was maintained at approximately 100 mm/sec and all tests were performed at 25°C. The samples were conditioned at 25°C for 24 hours prior to scratching. Scratched samples were analyzed with an interferometer at intervals of 1 hour and 24 hours after scratching. Scratches produced ranged from 1 to 3 μm in depth depending upon the plaque evaluated. Scratch deformations are reported as depth of deformation (in microns) as referenced to the unscratched surface.

Panels were painted through use of an adhesion promoter (7.5 μm dry film thickness of a chlorinated polyolefin primer), white basecoat (37.5 μm dry film thickness, a one-component (1K) melamine crosslinked basecoat or a two-component (2K) urethane basecoat, and a 2K urethane clearcoat (50 μm dry film thickness), which were applied wet-on-wet and subsequently baked for 30 minutes at 121°C ambient.

Friction induced paint damage resistance ("gouge") was measured on either a Ford proprietary friction induced paint damage apparatus (*STATRAM*) or a commercially available FIPD device (*SLIDO*) utilizing a 2^3 design of experiments (DOE), varying parameters such as sliding velocity, temperature, acceleration, and compressive (vertical) or traction (horizontal) force [23, 24]. Results are presented as either the area of gouge damage or as the percentage of panels that failed in the painted TPO FIPD testing obtained from the design of experiments (3 replicates under each experimentally designed run were measured). Table 1 lists the physical properties of the resins utilized to make the TPO blends in this study. As can be seen in the Table, the elastomer types were chosen to reflect varying solubility characteristics with the PP as well as varying crystallinity and melt flow rates. The following studies will look at the influence of melt flow ratio of elastomer/PP in the TPO samples as well as the effect of elastomer crystallinity on physical properties attained.

Table 1
Properties of TPO Components

Polymer*	Density (g/cm³)	MFR (dg/min)	DSC Melt (°C)	% Crystallinity	Mn	Mw/Mn
1042 PP	0.9049	1.9	162.2	46	67070	3.51
JSR07P	0.8589	<1	20.7	3	27098	4.29
4033 EB	0.8837	1.3	61.9	14.8	50475	1.96
3125EB	0.9124	1.2	108.3	46	45562	1.96
3022 EB	0.9057	9	100	39	230409	2.63
8150 EO	0.868	1	56	16	76000	2
8180 EO	0.863	1	50	11	76000	2

*where PP=base poly(propylene); EP=ethylene-propylene copolymer;
EB=ethylene-butylene copolymer; EO=ethylene-octene copolymer

Table 2 lists the physical properties of the unfilled TPO blends formulated. The "gouge" resistance is presented as a percentage of samples that exhibited cohesive delamination when tested under 2 loading events (136.6 kg and 272.7 kg) at a temperature of 68.3°C and a sliding velocity of 0.32 cm/sec. Figure 1 depicts the type of cohesive ripping ("cavitation") attained as a result of compressive shear loading placed on the 1K/2K white basecoat/2K clearcoat black TPO panel. The type of failure is very indicative of a "stick-slip" phenomenon typically seen in friction induced failures, therefore the failure is referred to as a "friction-induced paint dâmage" scenario (herein referred to as "gouge"). The gouge depth can be on the order of tens of microns to hundreds of microns into the TPO substrate.

272

Figure 2 depicts the gouge damage evidenced in the series of TPO panels described in Table 2. The panels were all painted with the same 1K/2K basecoat/clearcoat system and subjected to the same compressive shear loading conditions (272.7 kg, 68.3°C and a sliding velocity of 0.32 cm/sec). As can be viewed in the Figure, as the crystallinity of the elastomer (Exact 3125, ethylene-butene based) increases and the melt viscosity of the elastomer approaches the melt viscosity of the poly(propylene), described as the melt flow rate (MFR) ratio, the ensuing gouge resistance increases. It is postulated that MFR ratio will affect the dispersion size of elastomer in the matrix, as described by Silvis [25], due to the viscoelastic breakup of particles under shear conditions. A MFR ratio of 1 should allow optimum dispersion.

Table 2
Properties of Unfilled TPO Plaques

TPO Type	Injection Rate (cm/sec)	Gouge* (% Fail)	Hp (N/mm²)	Tensile Strength (Pa x 10⁻²)	WBL Depth (microns)
1042 PP	1.27	0	112	67.9	51
	7.62	0	120	66.9	8
4033 EB	1.27	12.5	54.7	48.6	54.7
	7.62	12.5	76.2	45.7	20
07P EP	1.27	62.5	44.9	44.2	44.9
	7.62	87.5	62	38.4	24
8150 EO	1.27	25	44.4	40.1	44.4
	7.62	25	52.5	40.3	52.5
8180 EO	1.27	25	43.7	40.3	43.7
	7.62	25	53.8	39.8	53.8
3022 EB	1.27	6	56.6	45.4	56.6
	7.62	6	59.8	43.5	59.8
3125 EB	1.27	0	81.5	52.5	81.5
	7.62	0	80.3	51.4	80.3

*1K/2K basecoat/clearcoat (BC/CC)

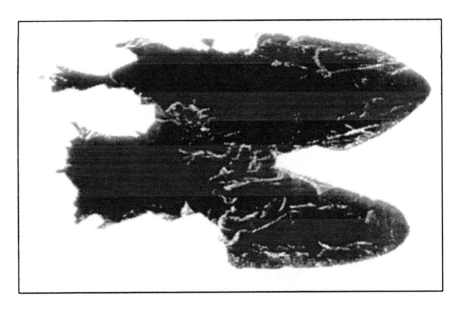

Figure 1
Friction Induced Paint Damage

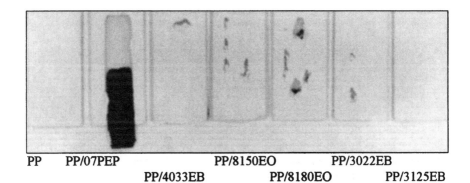

PP   PP/07PEP   PP/8150EO   PP/3022EB
         PP/4033EB   PP/8180EO   PP/3125EB

Figure 2
Gouge Resistance of Painted Polymer Blends

Figure 3 graphically depicts the attained gouge resistance of 1K/2K painted plaques from the variety of TPO blends tested as a function of tensile strength. In this Figure, it can be seen that as the tensile strength increases so too does the gouge resistance.

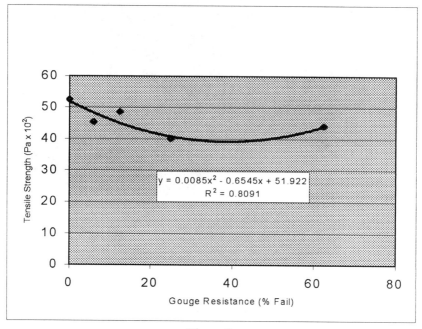

Figure 3
Gouge Resistance as a Function of Tensile Strength of TPO

Figure 4 displays the tensile strength as a function of the boundary layer depth achieved in each blend at an injection velocity of 1.27 cm/sec. The results here concur with those reported by Fujiyama [26] in that the tensile strength of an injection molded specimen increases linearly with boundary layer thickness obtained within the molded specimen.

Tensile strength, however, is not the only factor contributing to gouge resistance. As shown in Figure 5, increased surface hardness of the unpainted plastic also contributes to increased gouge resistance. This result is not unexpected, either, in that stress applied to a harder surface can be more easily

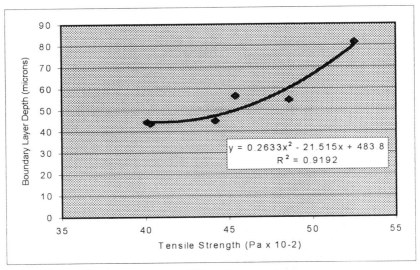

Figure 4
Tensile Strength as Affected by Boundary Layer Depth

dissipated across the top surface interface. If the applied stress does not reach a depth to which the boundary layer interface is "weakened", then the resultant ripping of this weakened interface will not be achieved.

Figure 6 displays a TEM micrograph of the interphase region of the 1042 poly(propylene)/3125 ethylene-butene elastomer. In this micrograph the dispersion of elastomer is clearly evidenced by the "darker" osmium tetroxide stained region. The area adjacent to the "dark" elastomer region and the "lighter" poly(propylene) region is of particular interest. Shaded region of intermediate staining can be seen in which the boundary between the elastomer and the poly(propylene) are not succinct. In these areas, regions of interlamellar entanglement can be seen. It is the interlamellar entanglements that are believed to be responsible for the increased cohesive integrity.

In Table 3, the physical properties of a variety of filled TPO samples are shown. The fillers evaluated include wolastonites of varying filler size and filler surface pretreatment. The Nyad G Special and Nyad 400HAR are surface pretreated with silane sizing agents to affect miscibility in the resultant blend. The Nyglos wolastonites are not surface pretreated. Cimpact is a commercial grade of talc available from Luzenac. All fillers were dry blended in the extruder with the melt blend at a 10 % by rate ratio.

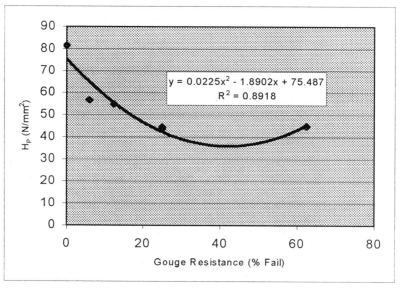

Figure 5
Gouge Resistance as a Function of TPO Hardness

Representative interphase Region of "interlamellar" entanglement

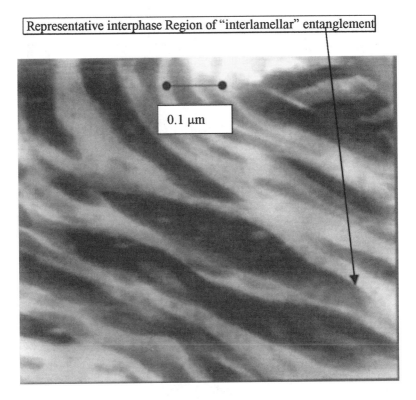

0.1 µm

Figure 6
Interphase Development within Elastomer/PP Matrix

Table 3
Properties of Filled TPO Plaques

TPO Blend	Filler Type	Injection Rate (cm/s)	Particle Size (micron)	Hplas (N/m m²)	Plasticity	WBL (micron)	Gouge *(mm² failure)	Scratch Depth (micron)
1042/ 3125	none	1.27	none	101	0.48	45	25.8	15.9
		7.62		115	0.49	82	86.4	6.34
	Nyglos 5	1.27	4.5	104	0.49	102	24.2	22.3
		7.62		110	0.5	78	52.5	20.62
	Nyglos 8	1.27	8	106	0.49	102	31.2	27.62
		7.62		119	0.5	65	59.4	30.64
	Nyglos 4	1.27	3.5	104	0.51	106	14.9	22.56
		7.62		113	0.5	65	46.4	17.1
	Nyglos 12	1.27	12	105	0.49	115	25.6	30.64
		7.62		117	0.49	61	78.9	28
	Nyad G Special	1.27	40	104	0.51	123	39.7	25.62
		7.62		113	0.49	78	91.1	26.8
	Nyad 400HAR	1.27	15	109	0.48	135	26.5	27.64
		7.62		114	0.5	78	70.5	22.82
	Cimpact 699	1.27	1.2	106	0.49	151	7.6	22.04
		7.62		116	0.5	115	64.1	12.38

*2K/2K white BC/CC where mm² is area of failure cohesively within the substrate

As shown in Figure 7, decreasing filler size can increase gouge resistance of samples molded at the lower injection velocity (1.27 cm/sec). These findings cannot be accounted for based on surface hardness nor boundary layer depth alone, as was the case for unfilled samples.

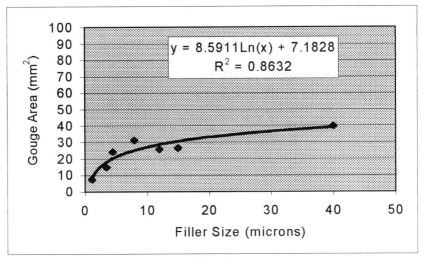

Figure 7
Gouge Resistance as a Function of Filler Size

It appears, as shown in Figure 8, that fillers may act as a locus for possible fracture within the sample, thereby lowering gouge resistance. Figure 8 shows that in the filled Nyglos 12 TPO material that regardless of injection velocity at which the sample was fabricated, the filler acts as a locus for fracture and cavitation. It is interesting to note, however, that injection velocity plays a huge role in the type of fracture that occurs. As the injection velocity is increased (from 1.27 cm/sec to 7.62 cm/sec) the amount of "fibrillar cavitation" increases as a result of compressive shear force imposed. The cavitation can be seen in the micrograph representative of the TPO sample fabricated at 7.62 cm/sec injection velocity. Small areas of white fibrillated polymer can be seen in various areas within the sample upon fracture, whereas in the micrograph of the sample fabricated at the lower injection velocity the same pattern is not seen.

Scratch resistance was also monitored on the unpainted filled TPO samples described in Table 3. Figure 9 displays the imparted scratch as analyzed by an interferometer. The micrograph depicts the scratch as viewed in three-dimensional topographic interference. As can be seen in the depth of scratch analysis, as material is removed from the scratch "ditch" it is plowed into "shoulders" on either side of the scratch "ditch". The depth of the scratch, as listed for the variety of filled TPO samples in Table 3, is taken as the area of the "ditch" only as would be deduced from the unperturbed surface profile into the scratch "ditch".

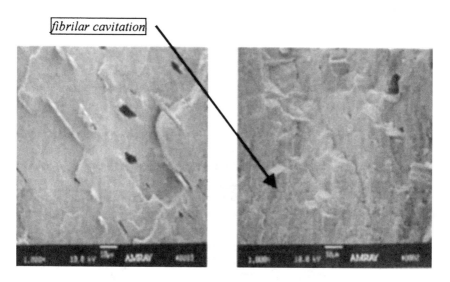

Figure 8
SEM Fracture of Filled TPO as a Function of Injection Velocity

Figure 9
Scratch Analysis of Unpainted Plastic

Surface hardness appears to not only be an important attribute of plastics as related to gouge resistance, but also is shown to have an effect on scratch resistance. As shown in Figure 10, as the surface hardness increases within the unpainted plastic the scratch resistance decreases. This may be the result of increased fracture behavior within the sample as it is stressed by the scratching event. Filled samples exhibited worse scratch resistance than their unfilled counterparts. Analogous results were reported by Chu [27] in filled poly(propylene) plastics subjected to scratching phenomenon. In his work, Chu attributed the lessened scratch resistance of filled poly(propylene) materials to increased plastic deformation, with potential fracture in the filler/poly(propylene) interface regions. Deeper, fractured scratches were related to higher "whitening" behavior, as a result of greater fracture and filler-poly(propylene) disbondment.

Summary

Stresses applied to filled or unfilled semi-crystalline thermoplastic alloys, in the form of compressive shearing events, e.g., scratched, gouges, etc., often result in cohesive debondment within the weak interface of the system. In cases where the plastic is painted, friction induced damage results in cohesive "ripping" of the top tens to hundreds of microns of the plastic surface. In unfilled specimens that are unpainted, the applied stress, in the form of a

282

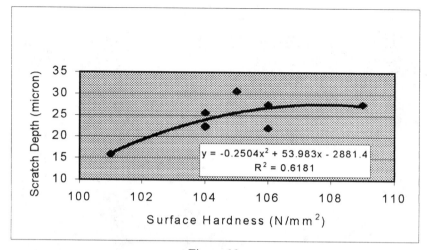

$$y = -0.2504x^2 + 53.983x - 2881.4$$
$$R^2 = 0.6181$$

Figure 10
Scratch Resistance as a Function of Surface Hardness

scratch, results in plastic deformation of the surface and potential fracture within the top few microns of the surface. In both cases, it is believed that the resultant cohesive failure is a result on improperly controlled "interphases" that develop within the plastic during the extrusion and injection molded fabrication of the plastic sample.

As shown in Figure 11, the "interphase" region is defined as the area adjacent to the elastomer dispersion within the poly(propylene) matrix of a thermoplastic olefin material. The interphase can be controlled through judicious choice of poly(propylene) and elastomer, where crystallinity, molecular weight, molecular weight distribution, and viscosity ratio of the elastomer to poly(propylene), are controlled. Injection velocity used to fabricate the molded samples, however, is also shown to influence the resultant interphase strength, due to the shear stresses imparted on the molten matrix during the mold filling process.

References

1. Allen, K.W. *Analytical Proceedings* **1992**, *29*, 389.
2. Ryntz, R.A.., *Progress in Organic Coatings* **1996**, *27*, 241.
3. Schonhorn, H. *Macromolecules*, **1968**, *1*(2), 145.
4. Kwei, T.K., Schonhorn, H.L., and Frisch, H.L. *J.Appli. Phys.*, **1967**, *38* (6), 2512.

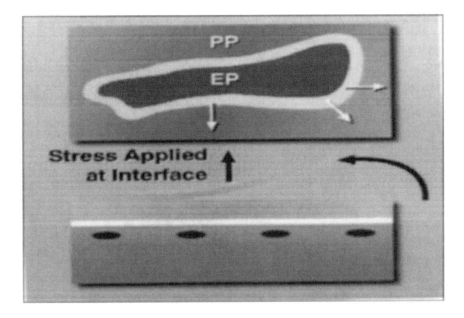

Figure 11
Cavitation in Elastomer/PP Alloy under Applied Load

284

5. Schonhorn, H. *J. Polymer Phys.B*, **1967**, *5*, 919.
6. Schonhorn, H. and Bryan, F.W. *J. Polym. Sci.,A-2*, **1968**, *6*, 231.
7. Schonhorn, H. *J. Polym. Sci. B*, **1964**, *2*, 465.
8. Sharples, A. Introduction to Polymer Crystallization, St. Martin's Press, New York, **1966**, p. 21.
9. Kumaraswamy, G; Verma, R.K.; and Kornfield, J. *Review of Scientific Instruments*, **1999**, *70* (4), 2097.
10. Fitchmun, D.R. *J. Polym. Sci. Part B*, **1969**, *7* (4), 301.
11. Bartosiewicz, L. and Mencik, Z. Internal Ford Communication **1974**.
12. Bonnerup, C. and Gatenholm, P. J. *Polymer Sci.Part B*, **1993**, *31*, 1487.
13. Bikkerman, J. J. The Science of Adhesive Joints 2nd Edition, Academic Press, New York, Chapter 7, **1968**.
14. Isayev, A. I. and Hieber, C. A. *Rheol. Acta*, **1980**, *19* , 168.
15. Tadmor, Z. *J. Applied Polym. Sci.*, **1974**, *18*, 1753.
16. Flaman, A. A. M. *Polymer Eng. and Sci.*, **1993**, *33* (4), 193.
17. Hastenberg, C. H. V., Wildervanck, P. C., and Leenen, A. J. H. *Polymer Eng. and Sci,*, **1992**, *32* (7), 506.
18. Siegmann, A., Buchman, A., and Kenig, S. *Polymer Sci. and Eng.*, **1982**, *22* (9), 560.
19. White, J. R. *Polymer Testing* , **1984**, *4*, 165.
20. Ryntz, R. A., Ramamurthy, A. C., and Mihora, D. J. *J. Coating Technology*, **1994**, *67* (840), 35.
21. Ryntz, R.A., *J. Vinyl & Additive Technology*, **1997**, *3* (4), 1.
22. Wu, J. and Mai, Y.-W., *Polymer Engineering and Science*, **1996**, *36* (18), 2275.
23. Ryntz, R.A., Everson, M., Pollano, G., *Progress in Organic Coatings*, **1997**, *31* , 281.
24. Ryntz, R.A. and Mihora, D.J., *SAE Technical Transaction Paper*, **1999**, #99-M-136.
25. Silvis, C., *Proceedings of TPOs in Automotive 1995 International Conference,* Executive Conference Management, Novi, MI **1995**.
26. Fujiyama, M. and Kimura, S. *Obunsha Ronbunshu, Eng. Ed,.* **1975**, *4* (10), 777.
27. Chu, J. *ANTEC '99*, Society of Plastic Engineers Annual Technical Congress Proceedings, **1999**, p. 2943.

Chapter 14

Determining the Particle Size Distributions of Titanium Dioxide Using Sedimentation Field-Flow Fractionation and Photon Correlation Spectroscopy

S. Kim Ratanathanawongs Williams[1], Belinda Butler-Veytia[2], and Hookeun Lee[1]

Departments of [1]Chemistry and Geochemistry and [2]Chemical Engineering and Petroleum Refining, Colorado School of Mines, Golden, CO 80401

The particle size distributions (PSD) of titanium dioxide (TiO$_2$) have been measured using sedimentation field-flow fractionation (SdFFF) and photon correlation spectroscopy (PCS). Two types of TiO$_2$ with different surface coatings were analyzed. Unlike molecular-size analyte that are easily and homogeneously dissolved in solution, particulate samples must be suspended in solution. The sample preparation step should cause no changes, e.g., aggregation, to the original state of the sample. A procedure has been developed to reproducibly prepare TiO$_2$ suspensions of well-dispersed particles. Both SdFFF and PCS show the presence of two distinct size populations at approximately 200 and 100 nanometers.

Introduction

Particle sizing techniques can generally be categorized as nonfractionation or fractionation methods (1). The former includes microscopy, electrozone sensing, and light scattering. These techniques tend to be rapid but of low resolution since measurements are made of the entire particle mixture. The fractionation techniques include disc centrifugation, hydrodynamic chromatography, and field-flow fractionation. High resolution is achieved since the different particle populations are separated and then sized. This work involves a comparison of two techniques: sedimentation field-flow fractionation (SdFFF) and light scattering, particularly, photon correlation spectroscopy (PCS). This combination of techniques has previously been applied to emulsions (2,3).

Sedimentation FFF

Sedimentation field-flow fractionation (SdFFF) is a chromatography-like elution technique that separates colloids and particles on the basis of differences in effective mass (2-9). The SdFFF channel is wrapped inside a centrifuge basket as shown on the left side of Figure 1 and the assembly is spun at controlled speeds. Special rotating seals allow a stream of carrier liquid to be pumped through the channel. Under conditions of laminar flow, the carrier liquid has a parabolic flow profile with the highest velocity streamlines at the center of the channel and decreasing velocities towards the walls. A small pulse of sample particles is injected and transported into the channel by the flow of carrier liquid. However, the displacement velocity of different particles through the channel is unequal because of their different diffusion coefficients and interactions with the centrifugal field. The particles that interact more strongly with the field and/or possess slower diffusion coefficients will form equilibrium layers that are located in the slow flow regions near the accumulation wall. These sample components (e.g., band x in Figure 1) will thus be transported out of the channel more slowly than those further from the wall (band y). Particles whose densities are greater than that of the carrier liquid will move towards the outside channel wall whereas particles less dense than the carrier fluid will float towards the inner wall. Whether particles float or sink, separation is governed by the same principles. This separation of submicron particles is referred to as the normal mode and the theory is discussed in the following section. For particles larger than 1 μm, the steric mode separation mechanism prevails (8). The elution order is opposite to that of normal mode. The diameter at which the transition between normal and steric mode occurs is conditions dependent and ranges between 0.2 μm and 3 μm. Consequently, it is important to initially view the sample using a microscope to confirm the size extremes and the selected FFF experimental parameters.

Photon Correlation Spectroscopy

Photon correlation spectroscopy (PCS), also known as dynamic light scattering, quasi-elastic light scattering, autocorrelation spectroscopy, or intensity fluctuation spectroscopy, is a sizing method that calculates diffusion coefficients based on the fluctuations in light scattering intensity of a sample suspension (10-18). These fluctuations can be described by a time-dependent function, also referred to as the autocorrelation function. In the PCS system (Figure 2), the sample scatters light, which is collected by a detector at a 90-degree angle to the incident laser beam. The light intensity is correlated and adapted according to the characteristics (i.e. refractive property, viscosity of the carrier solution) of the sample.

Photon correlation spectroscopy is a rapid method that is applied to an ensemble of particles. Consequently, the resolution is low and at least a two-fold difference in size is required in order for PCS to register the presence of distinctly different populations. This method has also been reported to yield inaccurate particles sizes for mixtures of monodispersed suspensions spanning a large size range (14). The resolution and polydispersity limits in PCS vary somewhat depending on the intensity of the laser used.

Photon correlation spectroscopy and SdFFF each have different strengths: high speed for PCS and high resolution for SdFFF. Thus the choice of techniques depends upon the complexity of the sample and the information sought (19).

Theory and Working Equations

Sedimentation FFF (normal mode)

The retention equation in FFF relates the experimental retention time t_r to the void time t^0 and the retention parameter λ (2).

$$\frac{t_r}{t^0} = \frac{1}{6\lambda\left[coth(1/2\lambda) - 2\lambda\right]} \tag{1}$$

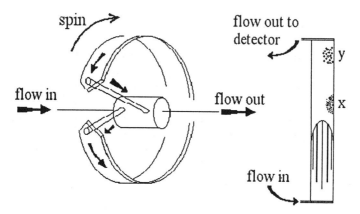

Figure 1. Illustration of curved ribbon-like channel used in sedimentation FFF and magnification of a short section of this channel.

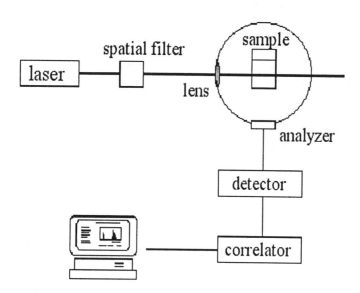

Figure 2. Schematic diagram of the PCS apparatus.

For well retained particles ($t_r > 5$ t^0), λ is small and the following approximation is valid.

$$t_r \cong \frac{t^0}{6\lambda} \tag{2}$$

The parameter λ in SdFFF is related to particle mass m or effective spherical diameter d by

$$\lambda = \frac{kT}{wG(|\Delta\rho|/\rho_p)m} = \frac{6kT}{\pi wG|\Delta\rho|d^3} \tag{3}$$

where k is the Boltzmann constant, T is the absolute temperature, G is the centrifugal acceleration, w is the channel thickness, ρ_p is the particle density, and $\Delta\rho$ is the difference in density between the particle and carrier liquid.

Equations 2 and 3 can be combined to obtain

$$t_r \cong \frac{t^0 wG(|\Delta\rho|/\rho_p)}{6kT} m = \frac{\pi t^0 wG|\Delta\rho|}{36kT} d^3 \tag{4}$$

Equation 4 shows that the normal mode retention in SdFFF is dependent on the particle size and the density difference between the sample particles and the carrier liquid.

The high resolution achievable in normal mode is due to the third power dependence $t_r \propto d^3$. This sensitivity also increases the accuracy of particle size measurements, but it can also yield lengthy analysis times for polydisperse materials. For example, a 10-fold size difference between the smallest and largest particles in the sample will result in a 10^3 difference in retention times. Larger particles tend to be excessively retained and their concentration in the outlet stream is low due to their elution from the system over extended time intervals. These problems can be solved by using field strength programming (20,21). A high initial field strength is required to resolve smaller particles, while a lowering of field strength reduces the excessive retention and dilution of larger components. In a field programming run, the field strength S is held constant at an initial level S_0 for a time-lag period t_1 and decreased according to a power function of elapsed time t. The power decay function is

$$S = S_0 \left(\frac{t_1 - t_a}{t - t_a} \right)^8 \tag{5}$$

where t_a is a constant.

Photon Correlation Spectroscopy

The time dependent correlation function of the light scattered by particles in a liquid suspension is defined as (18)

$$\langle I(0)I(\tau) \rangle = \lim_{T \to \infty} \frac{1}{T} \int_0^T I(t)I(t-\tau)dt \tag{6}$$

where I is light intensity, T is the total experiment duration, t is time at which measurements begin, and τ is the correlation delay time. Since scattered light is composed of photons, the measured intensity corresponds to the number of photons per unit time. An equation analogous to Equation 6 can be written for the photon autocorrelation function

$$C(\tau) = \langle n(0)n(\tau) \rangle = \lim_{N \to \infty} \frac{1}{N} \sum_{j=1}^{N} n_j n_{j-m} \qquad m = 1,2,3...M' \tag{7}$$

where $n(t)$ represents the number of photons measured at time t, N is the total number of samples, and M' is the number of points at which an autocorrelation function is measured. Equation 7 can be subsequently used as a basis for deriving the following photon autocorrelation function for rigid, monodisperse, and compact particles (18)

$$C(\tau) = (n)^2 \left[1 + b \exp(-2Dq^2\tau) \right] \tag{8}$$

where b and Dq^2 are experimental constants. The term D represents the diffusion coefficient whereas q is the amplitude wave vector for a particular fluctuation and measurement angle and is readily calculated using the Bragg equation (18). Equation 8 assumes that a large number of particles are present in the observed scattering volume and therefore the results can be approximated by a Gaussian distribution. The constants b and Dq^2 are determined by fitting the measured autocorrelation function to a single exponent. The D value is then used to calculate the hydrodynamic diameter d via the Stokes-Einstein equation

$$D = \frac{kT}{3\pi\eta d}$$

where k is the Boltzmann constant, T is the absolute temperature, and η is the viscosity of the medium in which particles are suspended.

The analysis of mono- and polydisperse samples by PCS is notably more difficult for the latter. Since the scattering intensity is not measured for single particles, size distribution information is obtained by deconvoluting the measured autocorrelation function into its single exponential components. This can be a complicated process and requires at least a two fold difference in particle size in order for two distinct size populations to be reported.

Experimental

FFF Carrier Liquid

The carrier liquid was deionized distilled water containing 0.1% (v/v) FL-70 (Fisher Scientific, Fair Lawn, NJ) and 0.02% (w/v) sodium azide (Sigma, St. Louis, MO). The pH of the carrier liquid was 9.85. Sodium azide (NaN_3) is a mutagenic chemical and should be handled with care.

Samples and Sample Preparation

Two TiO_2 samples, designated as A and B, were analyzed. Sample A is used in plastics and has a small amount of phosphate as the first treatment and a 1.7 mass % alumina as the second treatment. Sample B is a pigment marketed for architectural paints and has approximately 1.5 % mass silica as the first treatment and 3 % mass alumina as the second treatment (22). Rutile titanium dioxide has a refractive index of 2.73 and a density of 4.2 g/cm^3 (23,24). Silica and alumina have densities of 2.18 and 3.9 g/cm^3, respectively (24,25). Using these nominal densities and compositions, the densities of samples A and B were calculated as 4.20 and 4.16 g/cm^3, respectively.

A 0.1% (w/v) TiO_2 suspension was prepared by weighing approximately 0.01 g of sample into a glass vial. Ten milliliters of the carrier liquid (0.1% FL-70 and 0.02% sodium azide) were added. The suspension was sonicated for 3 minutes at an output control of 60W and 40% duty cycle. Aliquots of this 'stock' suspension were used to prepare the samples analyzed by SdFFF and PCS. For SdFFF, the stock suspension was further diluted with carrier liquid. The final sample suspension contained 0.01% TiO_2, 0.1% FL-70, and 0.02% NaN_3. The TiO_2 suspensions used in PCS experiments were made by transferring 150 microliters of the 0.1% stock suspension and 3 mL of distilled deionized water to a cuvet. The resulting suspension was 0.005% (w/v) TiO_2 and 0.005%FL-70 and 0.001% NaN_3. The suspensions were sonicated prior to SdFFF analysis. The PCS suspensions were analyzed immediately after preparation to prevent particle settling in the cuvets.

Instrumentation

Separation takes place in the rectangular channel with triangular ends shown in Figure 1. The major walls of the channel are made from highly polished Hastelloy C (a nickel rich steel). The channel had a void volume V^0 of 3.41 mL, rectangular length L of 87.1 cm, breadth b of 2.1 cm, and thickness w of 0.0186 cm. The rotor radius representing the distance between the channel and axis of rotation in the system is 15.1 cm. The SdFFF system is similar to the S101 instrument produced by FFFractionation, LLC (Salt Lake City, Utah). The system includes an HPLC pump (Model 410, Kontron Electrolab, London, UK), a high flowrate flushing pump (Model QD-1, Fluid Metering, Oyster Bay, NY)), a Spectra 100 UV detector (Spectra-Physics, San Jose, CA) set at 254 nm, and a PC-compatible computer that controls the centrifuge speed and collects data. Photon correlation spectroscopy (PCS) was done using a Brookhaven Instrument ZetaPlus equipped with a 15 mW He-Ne laser (632.8 nm). Sonication was performed using a Model W-225R Sonicator Cell Disruptor (Heat Systems-Ultrasonics, Inc., Plainview, NY) with a microtip.

SdFFF Analyses

The experimental procedure in SdFFF commenced with the injection of a sample plug into the channel. The channel flow is then turned off and the channel is spun at a desired speed for a set period of time. In this stopflow stage, the sample components remain in the vicinity of the channel inlet where they are relaxed to their equilibrium positions above the accumulation wall. After relaxation, the channel flow is resumed and the various sample species are swept through and out of the channel at different times. After each analysis, the channel was flushed with the carrier solution using a high flowrate pump set at ~10 mL/min. The flowrate during the separation and elution process were measured prior to each analysis.

PCS Analyses

The instrument was turned on and allowed to warm up for at least 10 min prior to use to ensure constant chamber temperature. Data acquisition times ranged between 5-20 minutes. The sample suspensions yielded particle counts within the recommended range of 50-300 kilocounts per second (19).

Results and Discussion

Sample preparation is a critical step in obtaining reproducible and accurate size measurements of colloidal and particulate samples. The dispersibility of the TiO_2 samples was examined in three solutions: deionized water, Tamol 731 + sodium hexametaphosphate $((NaPO_3)_6)$, and FL-70 + NaN_3. Tamol, an anionic surfactant produced by Rohm and Haas, is commonly used in paint formulations (26,27) and $(NaPO_3)_6$ is a recommended dispersant for alumina and titanium dioxide (24). The combination of FL-70 (a complex mixture of the surfactant Tergitol, sodium carbonate, and other chemicals) and NaN_3 (a bactericide) has proven to be successful in numerous FFF applications. The latter two solutions readily dispersed the coated TiO_2 samples. In addition, particles that settled to the bottom of the sample vial over time were easily redispersed with sonication.

The sonication times were varied and it was found that 60 watts at a 40% duty cycle for 3-5 min provided well-dispersed samples. Aggregation was induced when the suspension was sonicated longer than 10 min.

The long-term stability (at least 1 month) of the TiO_2 sample suspended in 0.1% FL-70 and 0.02% NaN_3 resulted in its use as a 'stock' solution. The PCS sample was prepared by sonicating an aliquot of this stock solution and diluting 20x with deionized water. The resulting suspension contained 0.005% TiO_2, 0.005% FL-70, and 0.001% NaN_3 and was prone to aggregation when the particles settled. Thus, samples were prepared immediately prior to PCS analysis. The PCS size distributions of the two TiO_2 samples are shown in Figure 3. Each figure is a compilation of five analyses with 3 replicates per analysis. Bimodal distributions are evident in both types of TiO_2. Sample A is composed of 37 nm and 191 nm particles while sample B diameters are 47 nm and 192 nm. Experiments were done with different data acquisition times to verify that the sedimentation velocities of these dense particles did not affect the measurement of diffusion coefficient and size. No distinctive trend was observed.

The main criteria used in selecting the FFF carrier liquid are good dispersing ability and high sample recovery. The Tamol-$(NaPO_3)_6$ solution was a good TiO_2 dispersant but a poor SdFFF carrier liquid. Sample adsorption onto the channel wall resulted in complete sample loss. The most effective carrier liquid for SdFFF analysis was the 0.1%FL-70 and 0.02%NaN_3 solution. Hence, the FL-70 and NaN_3 containing solutions were used in both sample preparation and separation.

The SdFFF fractograms shown in Figure 4 were obtained using field programming. A high initial rpm was used to resolve the early eluting components. The rpm was then decreased (using the power function described in Equation 5 to hasten the elution of highly retained sample components.

The fractograms can be converted to size distribution curves using Equations 4 and 5. This conversion requires a prior knowledge of the particle densities. In this work, the average density of each TiO_2 sample was determined from the nominal particle compositions. The calculated densities of TiO_2 samples A and B based on manufacturer's nominal particle compositions and assuming spherical particles are 4.20 and 4.16 g/cm^3, respectively. The SdFFF particle size distribution curves obtained using these densities are shown in Figure 5. As with the PCS results, both TiO_2 samples have very similar size distributions.

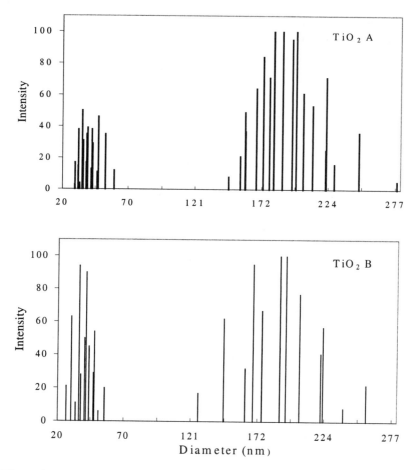

Figure 3. PCS results of two TiO₂ samples. The carrier liquid was 0.005% Fl-70 and 0.001% sodium azide and the data acquisition times were between 5 and 20 minutes.

295

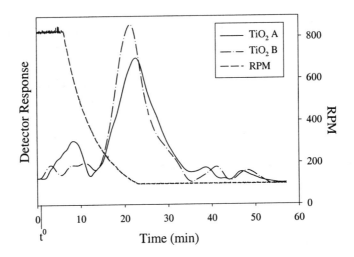

Figure 4. Superimposed SdFFF fractograms of two TiO₂ samples. The carrier liquid was 0.1% Fl-70 and 0.02% sodium azide and the channel flow rate was 4.1 mL/min. The field strength was programmed using the following power program parameters: initial rpm = 828, hold rpm = 100, t₁ =6.2 min, tₐ =-49.6 min, p = 8.

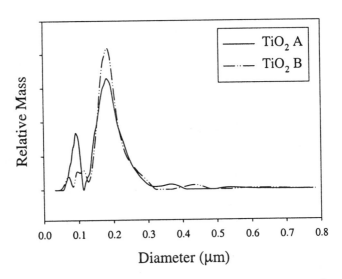

Figure 5. Superimposed SdFFF particle size distribution curves of two TiO₂ samples using Δρ values calculated from nominal compositions (sample A: Δρ=3.20 g/mL., sample B: Δρ=3.16 g/mL).

A second method used to determine the average densities of the coated TiO_2 samples involved introducing the diameter measured by PCS and the retention time measured by SdFFF (peak maxima in Figure 4) into the programmed field strength form of Equation 4. The resulting density values of TiO_2 A and B are 3.80 and 3.43 g/cm^3, respectively. Since $d \propto (\Delta\rho)^{1/3}$, the difference between the two sets of densities results in only a 4% difference in diameter. Caldwell et al. (2,3) have used a more detailed approach in their studies of emulsions. They collected fractions at the outlet of the SdFFF channel and measured the diameter of each fraction using PCS. These authors found that while two different emulsions yielded similar SdFFF fractograms, i.e., had similar distributions of particle mass, PCS showed different diameters for fractions collected at the same retention times. This means that the particles may be of the same mass and elute at the same t_r, but they possess different d and $\Delta\rho$ values. This approach is applicable to TiO_2 and could yield important information on the uniformity of the coating on the particles.

Table 1. Diameters of Bimodal TiO_2 Samples

	$TiO_2 A$	$TiO_2 B$
PCS	37, 191	47, 192
SdFFF	86, 180	88, 181

The PCS and SdFFF diameters of the major population are comparable as shown in Table 1. Both techniques show the presence of small particles. However, PCS and SdFFF are not in agreement with the diameter of the small particles. The former shows peaks in the vicinity of 40 nm and the latter at 90 nm. It is possible that both minor populations exist. Since the typical target diameter for TiO_2 is ~220 nm, the SdFFF conditions used in this work were optimized for analyzing pure TiO_2 particles (density of 4.28 g/cm^3) larger than 80 nm. In other words, a 40 nm particle would not be retained. New optimum conditions need to be calculated and employed. (With current instrumentation, the smallest retainable TiO_2 particle is 27 nm.) This illustrates a potential pitfall for SdFFF, that is, a distinct and unexpected population of small particles may not be observed if the separation conditions were set up for a different range of diameters. On the other hand, an inherent weakness of PCS is that at least a two-fold difference in size is needed for two distinct populations to be registered. Hence, the 90 nm and 191 nm populations may be too close in size to be distinguished by PCS. Another potential source of error arises from the PCS software automatically fitting all data to a Gaussian curve. Studies have also shown that PCS results for polydisperse samples can be erroneous (14).

The main advantages of SdFFF are its high resolution and the possibility of collecting fractions eluting from the channel outlet for further analysis. The main advantages of PCS are its rapidness and ease of use. Consequently, PCS is more suitable for quality control work but SdFFF is required when investigating potential explanations to anomalous results.

Acknowledgements

We thank the National Institute of Standards and Technology for financial support and Dr. Bruce Honeyman for use of the PCS instrument. We also thank Dr. P. Stephen Williams for use of FFF optimization and retention time calculation software.

References

1. *Modern Methods of Particle Size Analysis*; Barth, H. Ed.; Wiley: New York, 1984.
2. Caldwell, K.D.; Li, J.M.; Li, J.T. *Proc. Polym. Mater. Sci. Eng.*, **1993**, *69*, 404.
3. Caldwell, K.D.; Li, J. *J. Coll. Interf. Sci.*, **1989**, *132*, 256.
4. Giddings, J.C. *Science* **1993**, *260*, 1456.
5. Giddings, J.C. *Anal. Chem.* **1995**, *67*, 592A.
6. Sharma, R.V.; Edwards, R.T.; Beckett, R. *Water Res.* **1998**, *32*, 1497.
7. Giddings, J.C.; Ho, J. *Langmuir*, **1995**, *11*, 2399.
8. Giddings, J.C.; Williams, P.S. *Am. Lab.* **1993**, *95*, 88.
9. Kirkland, J.J; Yau, W.W.; Doerner, W.A., *Anal. Chem.*, **1980**, *52*, 1944.
10. *Dynamic Light Scattering: Application of Photon Correlation Spectroscopy*; Pecora, R., Ed.; Plenum Press: New York, 1985.
11. *Dynamic Light Scattering: The Method and Some Applications*, Brown, W., Ed.; Clarendon Press: Oxford, 1993.
12. Finsy, R. *Adv. Coll. Interface Sci.* **1994**, *52*, 79.
13. Phillies, G.D.J. *Anal. Chem.* **1990**, *62*, 1049A.
14. Filella, M.; Zhang, J.; Newman, M.E.; Buffle, J. *Coll. Surf. A: Physicochem. Eng. Aspects* **1997**, *120*, 27.
15. Ovod, Vladimir I.; Mackowski, D. W.; Finsy, R. *Langmuir* **1998**, *14*, 2610.
16. Willemse, A. ; Merkus, H.; Scarlett, B. *J. Coll. Interf. Sci.* **1998**, *204*, 247.
17. Chu, B.;Gulari, Es.; Gulari, Er. *Phys Scr.*, **1979**, *19*, 476.
18. Weiner, B.B., "Particle Sizing Using Photon Correlation Spectroscopy", in *Modern Methods of Particle Size Analysis*; Barth, H. Ed.; Wiley: New York, 1984.
19. Williams, S.K.R.; Lee, H.; Turner, M.M. *J. Magn. Magn. Materials* **1998**, *194*, 248.
20. Williams, P.S.; Giddings, J.C. *Anal. Chem.* **1987**, *59*, 2038.
21. Williams, P.S.; Giddings, J.C. *Anal. Chem.* **1994**, *66*, 4215.
22. McKnight, M., Building Materials Division, National Institute of Standards and Technology: Gaithersburg, MD, personal communication

23. *Surface Coatings, Science, and Technology*: Paul, S., Ed.; John Wiley & Sons: New York, 1996, Ch 3.
24. Bernhardt, C. *Adv. Coll. Interface Sci.* **1998**, *29*, 79.
25. *Handbook of Chemistry and Physics*; CRC Press, Inc: Boca Raton, 1998.
26. Flick, E.W. *Handbook of Paint Raw Materials* 2nd ed.; Noyes Publication: New Jersey, 1989.
27. Flick, E.W. *Industrial Surfactants*; Noyes Publications: New Jersey, 1988.

Chapter 15

Characterization of Physical and Chemical Deterioration of Polymeric Bridge Coatings

Y. C. Jean[1], H. Cao[1], R. Zhang[1], J. P. Yuan[1], H. M. Chen[1], P. Mallon[1],
Y. Huang[1], T. C. Sandreczki[1], J. R. Richardson[2], J. J. Calcara[2], and Q. Peng[2]

[1]Department of Chemistry, University of Missouri at Kansas City,
Kansas City, MO 64110
[2]Department of Civil Engineering, University of Missouri at Columbia,
Kansas City, MO 64110

The environmental durabilities of several commercial bridge coatings are being investigated by exposing samples to accelerated UV irradiation. Primary microscopic techniques include positron annihilation spectroscopy (PAS), which detects and characterizes nanometer-scale physical holes/defects, and electron spin resonance (ESR) spectroscopy, which detects broken chemical bonds. For the PAS tests, significant decreases of sub-nanometer defect parameters are observed as a function of exposure time and of depth from the surface to the bulk. This is interpreted as a result of a loss of free-volume and holes fraction and an increase in cross-link density of the polymers during the degradation process. For the ESR tests, direct free radical signals are observed as a function of irradiation time and chemical environments. A high sensitivity of PAS and ESR tests to the early stage of degradation is observed.

Introduction

Service life is relatively short for bridge coatings because bridges are subjected to ice, wind, de-icing salts, sunlight, water, industrial plant exhaust, and automotive chemicals[1]. Furthermore, bridges are subjected to higher frequency dynamic vibration due to traffic loads than non-bridge structures. This repeated dynamic stressing of the steel elements and surface coatings is likely an additional factor contributing to coating degradation. Protective coatings for steel and structural systems typically used for transportation often consist of a multilayered structure with three or four elements: a topcoat (finish), an intermediate coat, a primer, and in some cases a surface sealer for the substrate, which is either metal or concrete [2-3]. Coatings are multifunctional materials, which contain organic or inorganic binder, metals, oxides, and other minerals.

The purpose of this research is to provide the fundamental knowledge, which will lead to long-term durability of protective coating systems for structural materials. This research aims directly at the areas of (1) identification of degradation mechanisms of polymers in coatings and subsequent development of appropriate models for performance predictions, and (2) investigation of synthesis /structure /property /performance relationships for coating systems.

Durability is a primary concern for protective coating systems [4]. There is an incomplete understanding of the origins of poor durability in most coating systems. Existing methods of assessment of durability and degradation are chiefly macroscopic approaches, measuring mechanical properties: adhesion, hardness, pulling strength, etc. Most knowledge of coating degradation and failure is based on these evaluations of performance. However, the microscopic origins, mechanisms, and progress of degradation are not yet ascertained for coating systems. Only recently have spectroscopic and physical methodologies begun to be used to investigate the underlying course of coating degradation [5].

There are two fundamental processes involved in coating degradation, chemical and physical changes, although they are interrelated in most cases [6]. For example, photo-degradation is a chemical process well known to degrade paints [6]. Polymers in paint absorb UV radiation, which leads to bond breakage yielding free radicals. These radicals initiate chain reactions in polymers, which eventually lead to the degradation and failure of polymer paints. This chemical process is a complicated one which depends on the composition/formulation of the coating. The process becomes more complicated in the presence of other environmental factors, such as oxygen, moisture, pollutants, temperature changes, and long-term dynamic stress loadings.

A series of microscopic physical defects occurs simultaneously with, and is caused by, chemical reactions. The defects involved with chemical degradation initiate at a very small scale, on the order of 0.1 nm, as bonds are broken and atoms are displaced from polymer structures. Essential information about these atomic- and molecular-level defect properties is currently unavailable because of the extremely small scale and very brief duration of the phenomena involved.

Defect sizes and distributions as a function of distance from the top-most surface layer down to the bulk of the coating system are being investigated by using an innovative physical method, positron annihilation spectroscopy (PAS). Radical formation leading to degradation of topcoat/finish coat systems is being investigated using electron spin resonance (ESR) spectroscopy. In this way, the origins of degradation are investigated in terms of chemical defects at the atomic level.

In this paper, we report the recent results from a series of PAS and ESR studies on commercial coatings which are subjected to the accelerated UV irradiation, i.e. xenon-lamp light, and QUV lamp.

Experimental

Fifteen coating samples as shown in Table 1 were prepared, of which fourteen are topcoats and one (sample 3) is midcoat. Those coatings are classified into four categories according to chemical composition: polyurethane, acrylic, epoxy and metallic-aluminum-containing paint. The detailed description of these commercial coating materials can be found in the product data sheets available at companies' websites.

Table 1. Coating samples studied

No.	Coating Type	Coating Name	Vender	Description
1	Polyurethane	MC-Ferrox A	Wasser, Seattle, WA	moisture-cured aliphatic polyurethane
2		MC-Luster	Wasser	moisture-cured aliphatic polyurethane
3		541-D-101	Valspar, Baltimore, MD	Moisture-cured urethane intermediate coat
4		Acrolon 218 HS	Sherwin-Williams, Cleveland, OH	polyester modified acrylic polyurethane
5		Poly-lon 1900	Sherwin-Williams	polyester-aliphatic polyurethane
6		Carboline 133 HB	Carboline, St. Louis, MO	aliphatic polyurethane
7		Carbothane 134 HG	Carboline	acrylic aliphatic polyurethane
8		Devthane 359	ICI Devoe, Louisville, KY	acrylic aliphatic polyurethane
9	Acrylic	Carboline 3359	Carboline	Waterborne acrylic
10		Devflex 4218	ICI Devoe	Waterborne acrylic
11		Devflex 4206	ICI Devoe	Waterborne acrylic
12	Epoxy	Epolon II	Sherwin-Williams	catalyzed polyamide epoxy
13		Macropoxy 646	Sherwin-Williams	high solid polyamide epoxy
14		Devran 224 HS	ICI Devoe	catalyzed polyamide epoxy
15	Aluminum Coating Paint	Silver-Brite	Sherwin-Williams	metallic aluminum in petroleum resin

The coatings were thoroughly mixed and then applied to aluminum sheets using a pressurized spray gun (Binks 95, Binks Manufacturing Company, Franklin Park, IL) connected with a nitrogen gas tank regulated at 60 psi. The thicknesses of the coatings were determined to be ~ 20 μm using profilometry.

Two types of artificial light sources were applied in the accelerated UV irradiation of the coating samples: a QUV chamber with UVB-313 fluorescent lamps, and a xenon arc-lamp light source (SLM Instrument, Inc., Urbana, IL; with a luminescence 32,000 W/m^2 over the wavelength range 250 nm to 1,200 nm, corresponding to 1,900 W/m^2 from 250 to 350 nm).

The technique of Doppler broadened energy spectra (DBES) of positron annihilation coupled with a slow positron beam was employed to measure the first three coating samples. The DBES experiments were performed at Brookhaven National Laboratory and at the University of Missouri-Kansas City[7]. The energy resolution was 1.5 keV at 0.511 MeV (corresponding to the positron 2γ annihilation peak). The total counts for each DBES spectrum was 0.5 million with the counting rate of 4000 cps. The obtained DBES data are characterized by S-parameter, as a measure of the momentum broadening. The S-parameter is defined as the ratio of the central area to the total counts after the background is properly subtracted. It provides a qualitative measurement of sub-nanometer defects, such as free volumes and holes, of polymers in coatings. The change of S-parameter, $-\Delta S = S_t - S_0$, where S_t and S_0 are the value after and before irradiation, respectively, gives information about the change of physical defects due to weathering.

The positron annihilation lifetime (PAL) experiments were performed at the intense slow-positron facility in the Electrotechnical Laboratory (ETL) in Japan [8]. The PAL data were fitted into four lifetimes: τ_1 (~0.125 ns, p-Ps); τ_2(~0.4 ns, positron), τ_3 (1-3 ns, o-Ps in coatings) and τ_4 (>10 ns, o-Ps in vacuum). Detailed description of DBES and PAL can be found in our previous paper s [9-11].

ESR experiments were performed on the topcoats to detect the free radicals involved in the photochemical reactions. The spectra were recorded using a Bruker ER-200-D X-band ESR spectrometer. Spectral scans were 15 mT, and were averaged from 3 to 10 scans at a scan rate of 0.30 to 0.75 mT/s. Modulation amplitude was typically 0.5 mT.

Results and Discussion

PAS:

Bridge coatings are multi-layered systems and each layer is a complex mixture of polymers, solvents, pigments, stabilizers, binding agents, and other additives used to achieve desired properties. In our previous paper [9], a three-layer depth profile of the S-parameter in aircraft topcoat MIL-C-85285B was observed: a surface skin polymer layer, an intermediate layer and the bulk. Figures 1-3 show the results of S defect parameter in three different coatings both un-irradiated (virgin) and one hour xenon lamp light irradiated as a function of positron incident energy.

The variation of S-parameter vs. positron incident energy is: a sharp increase near the surface (<20 nm), then a decreases. This variation is a general feature for polymeric coating systems. This has been interpreted in terms of a multi-composition [9]. This multi-composition feature in the topcoat can be interpreted as being due to a concentration gradient of the pigment from the surface to the bulk. This feature is also observed in the slow positron lifetime results discussed later. Below 20 nm from the surface, there exists a polymer skin layer. The small value of S at the surface is due to the back-diffusion of slow positrons implanted and not detected by the solid state detector. This is a general phenomenon for neat polymers. The decrease of the S parameter inside 20 nm is a result of positron annihilation with paint, which contains pigments. For example, the 3d electrons from Ti in TiO_2 has a higher electron momentum which will lead to a smaller value of the S parameter than that in polymers, which contain lower momentum O, H, and N atoms.

It is interesting to observe a smaller value of the S-parameter in Xe-irradiated samples than in virgin samples. A decrease of the S parameter due to UV irradiation is observed.

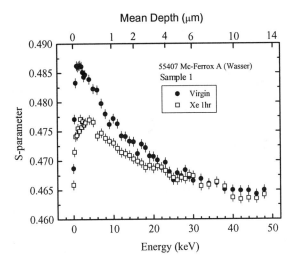

Figure 1. S-defect parameter vs positron incident energy in un-irradiated (virgin) and irradiated coating sample 1.

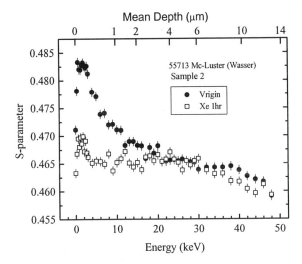

Figure 2. S-defect parameter vs positron incident energy in un-irradiated (virgin) and irradiated coating sample 2.

This observation is also consistent with our existing findings for aircraft coatings under different UV wavelengths [9-11]. The magnitude of the S-parameter reduction, $-\Delta S = S_t - S_0$, is calculated and $-\Delta S$ vs. the depth from the surface is plotted in Figure 4 for different samples. The decrease of $-\Delta S$ with depth (d) indicates an attenuation of UV intensity as the light enters the samples.

304

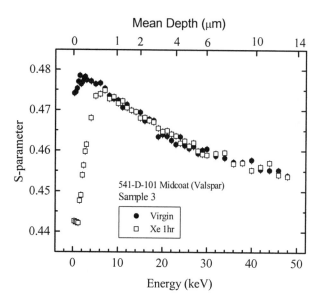

Figure 3. S-defect parameter vs positron incident energy in un-irradiated (virgin) and irradiated coating sample 3.

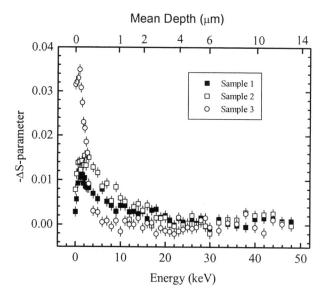

Figure 4. The reduction of S parameter (–ΔS) due to 1 hr of Xe light irradiation vs. positron incident energy in three coating samples 1,2, and 3.

The absorption of UV light can follow an exponential function according to Beer's law: $-\Delta S_t = -\Delta S_{max}(1-e^{-\varepsilon d})$, where $-\Delta S_{max}$ is maximum change inside the thin polymer layer (>20 nm). However we found that the variation of $-\Delta S$ vs d is more rapid than single exponential can describe for coating systems. For a good fit to the data, two exponential functions are required for Xe-irradiated data. Figure 5 shows such a two-exponential fit with $-\Delta S_{max}$=0.053, and 0.065 with ε=5.7 and 0.46 μm^{-1}, respectively.

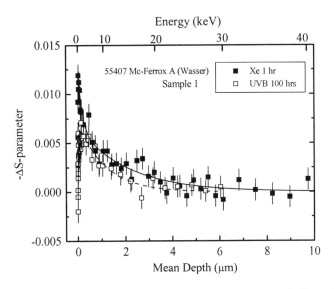

Figure 5. The reduction of S parameter ($-\Delta S$) due to 1 hr of Xe light and 100 hrs of UVB (313 nm) irradiation's vs. mean depth from the surface in sample 1 coating. The solid line is fitted with two exponential functions for Xe-irradiated data. The dashed line is fitted with single exponential for UVB-irradiated data

The need of two exponentials is due to the spread of UV wavelengths from a Xe-lamp light source. In general for a monochromic light source, $-\Delta S$ vs d can be fitted by a simple exponential function according to Beer's law. For example, in Figure 5, we also present $-\Delta S$ vs the mean depth for the sample subjected to a QUVB source (which peaks at 313 nm with a spread of about 40 nm). In the case of 313 nm UV light, we found $-\Delta S_{max}$=0.071 and ε=0.85 μm^{-1}. The ε value agrees with the extinction coefficient measured by using UV spectroscopy. The current results support our original idea that the S parameter is a measure of defects induced by UV irradiation. From the magnitude of $-\Delta S$ shown in Fig. 4 for these three different paints, we rank the quality of durability against UV irradiation as sample 1 > sample 2> sample 3.

From these observations, we obtain the following results for the PAS method: (1) S is a new physical parameter to test the durability of coatings in terms of sub-nanometer defects induced by UV irradiation. (2) The positron technique provides a defect profile from the surface to the bulk at a depth precision about 10% of the mean depth. (3) The sensitivity for detecting deterioration is a few minutes for the accelerated Xe-light source, 10 hrs for the 313 nm UVB source, and 50 hrs for the 340 nm UVA source. (4) The detection can be applied to both transient and permanent defects by performing *in situ* or *ex situ* experiments, respectively.

Slow positron annihilation lifetime (PAL) experiments were also performed using the same series of samples to obtain the free-volume and hole sizes. In this method, the measured ortho-Ps (o-Ps) lifetime, τ_3 has a direct correlation with the free-volume and hole size which is typically a few tenths of a nm [12]. The o-Ps lifetime (τ_3) and its intensity (I_3) were resolved from the data using a multi-exponential analysis. From these results, we then calculated the hole radius R of free volumes according to a correlation equation[10] as shown in Figures 6-9. In virgin coatings, o-Ps lifetime ($\tau_3 \sim 2$ ns) shows different variations as a function of depth: it can be larger or smaller near the surface as plotted in the left plots of Fig. 7-9. However Xe-light irradiation significantly reduces the defect size as seen from the consistently smaller values compared with the virgin data at the same energy. This is interpreted as smaller holes being formed after UV irradiation. Similarly, a large change in o-Ps intensity as a function of irradiation time is observed. I_3 systematically decreases with Xe light irradiation. Since o-Ps intensity has been suggested to be correlated with free-volume fraction in polymers [12], the decrease of o-Ps intensity can be interpreted as a reduction of free-volume and hole fraction induced by photodegradation of the coating material. The decrease of I_3, which is a measure of o-Ps formation, is consistent with the decrease of the S-parameter, which is a measure of p-Ps formation. A possible explanation for the decrease of the free-volume fraction is that there is an increase in crosslink density during the degradation process. In our previous paper [12], we observed a direct correlation between the free radical intensity and the S-parameter. The transient free radicals recombine and crosslink to form a terminal structure with a higher density. The highly crosslinked structure restricts the free motion of the polymer chains and therefore reduces the free-volume fraction.

Figures 9 compares the depth profile of $-\Delta S$ and $-\Delta I_3$ of the topcoat after 1 hrs of Xe-light irradiation for the first sample. The consistency of these two parameters show that both DBES and PAL coupled with slow positron technique can provide qualitative information on the sub-nanometer defect profile of coating materials. Figure 10 compares $-\Delta f_v$ (the loss of free-volume hole fraction) for the three different samples. The magnitudes of degradation for those three paints indicated by $-\Delta f_v$ values are consistent with $-\Delta S$ in Figure 4.

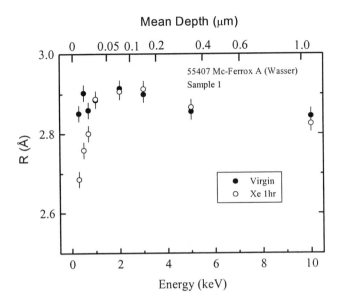

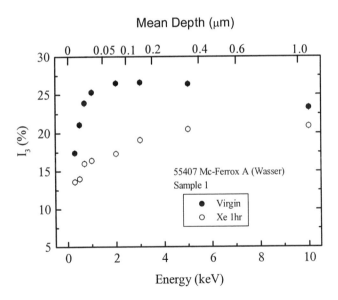

Figure 6. The free-volume hole radius (left) and its intensity (right) vs positron incident energy in coating sample 1.

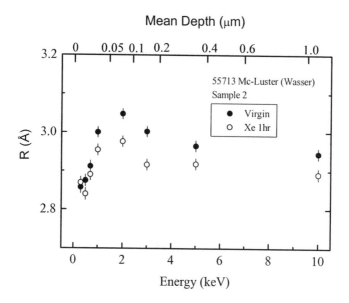

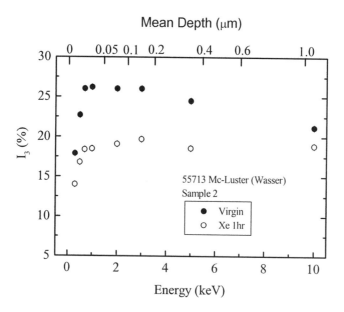

Figure 7. The free-volume hole radius (left) and o-Ps intensity (right) vs positron incident energy in sample 2 coating

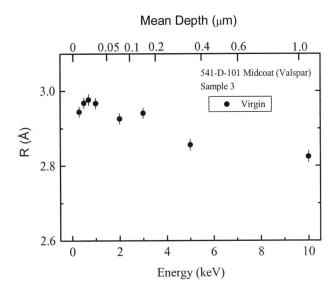

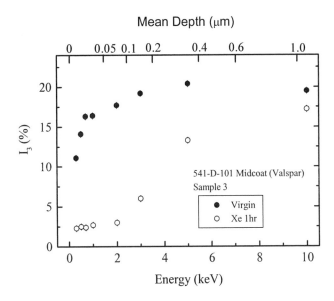

Figure 8. The free-volume hole radius (left) and o-Ps intensity (right) vs positron incident energy in coating sample 3. The R of irradiated sample was found to be no difference from the virgin values.

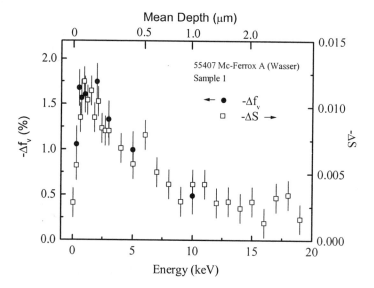

Figure 9. Comparison of the reduction in the defect parameter (-ΔS) and the loss of free-volume hole fraction (-Δf$_v$) vs positron incident energy in coating sample 1.

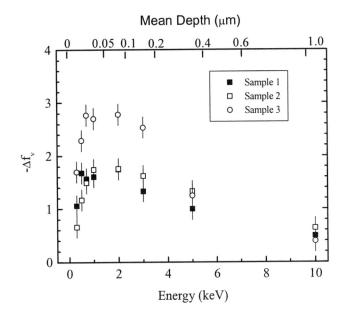

Figure 10. Comparison of the loss of free-volume hole fraction (-Δf_v) vs positron incident energy in three coating samples 1,2, and 3

ESR:

Samples for use in the ESR studies were applied to aluminum sheets. They were then exposed to xenon-lamp irradiation under two different sets of conditions. In the first case, fourteen commercial topcoat materials were irradiated for 30 minutes under vacuum at -196°C. They were then examined by ESR to detect the presence of free radicals, indicating photo-induced bond cleavage. Currently, we attempted to correlate the change of ESR spectral amplitudes, Δ, with susceptibility-to-bond cleavage, which is an important process in photodegradation. However, the data presented here have not yet been correlated with changes in coating performance (e.g. loss of gross). These experiments are currently in progress. The results are summarized in the bar chart shown in Figure 11. The seven different polyurethane samples identified in the chart had significantly different numbers of free radicals following irradiation,
with spectral amplitudes varying by as much as a factor of three. The Poly-lon 1900 sample appeared to be the most susceptible to bond cleavage, in that it had the highest number of free radicals. The Carboline 133 HB had the lowest number of radicals. The three epoxy samples displayed an even wider range of susceptibilities to bond cleavage, with spectral amplitudes varying by as much as a factor of four. The Epolon II sample had the highest number of free radicals, and the Devran 224 HS had the least. The three different acrylics displayed a narrower range of susceptibilities, with

312

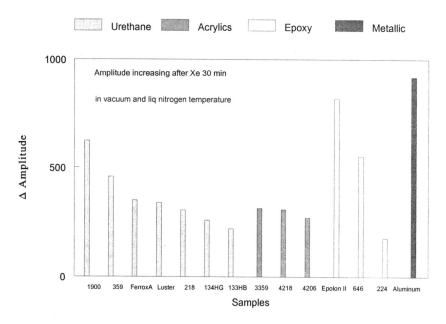

Urethane   Acrylics   Epoxy   Metallic

Figure 11. Amplitude of ESR signals after 30 min of Xe-light irradiation on 14 commercial samples (Table 1) in vacuum at 77 K.

the spectral amplitudes varying by less than ten percent among the samples. All the acrylics had a low-to-moderate susceptibility to bond cleavage compared to the polyurethane and epoxy samples. Finally, the metallic aluminum-containing sample, Silver Brite, showed the highest susceptibility to free radical formation of all the samples, with a spectral amplitude approximately five times greater than that of the most stable sample of the fourteen, Devran 224 HS. The above data was acquired in order to quickly determine the relative stabilities of commercial paint formulations to short-wavelength light. The results in Figure 11 indicate that knowledge of a topcoat's chemical class (e.g., polyurethane or epoxy) is not enough to predict its photo-stability, and that is possible to formulate highly stable polyurethanes and epoxies. The performance of the acrylics was nearly as good at those of the best epoxy and polyurethane.

Seven of the above topcoats were down-selected for further investigation. These include two each of the polyurethanes (Poly-lon 1900 and Carboline 133HB), epoxies (Epolon II and Macropoxy 646), and acrylics (Devflex 4218 and Devflex 4206), plus the one aluminum-containing topcoat (Silver Brite). This down-selected set of samples was exposed to xenon-lamp irradiation under a second set of conditions, viz., up to 30 minutes at room temperature under 1 atmosphere of O_2. From a set of related

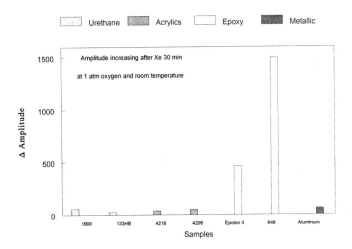

Figure 12. Amplitude of ESR signals after 30 min of Xe-light irradiation on 7 commercial samples (Table 1) at room temperature and under 1 atm oxygen.

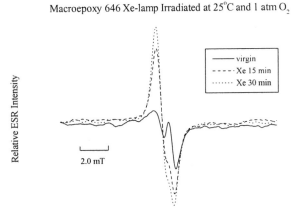

Figure 13. ESR spectra from sample 1 coating following different times of Xe-irradiation.

experiments on polyurethane clearcoats, it is known that the temperature alone (up to 150 °C) does not result in measurable radical production. Under these conditions, photo-generated free radicals exist in a mobile environment, and therefore freely react with their surroundings. The result is that the radicals observed are only those net radicals that remain after a sequence of reactions with neighboring chains. These are often secondary (or later) radicals formed from the reaction of the initial photo-generated primary radicals. A reason for performing this experiment, is that the added molecular mobility and the presence of oxygen enable the process of photo-oxidative degradation to occur. This process can include branching chain reactions that result in extensive bond cleavage per photon absorbed. The results of this investigation are shown in Figure 12. There are at least two main features to be noted. First, most of the samples have a very similar net number of radicals following exposure to the photo-oxidative environment. This is in marked contrast to the widely varying numbers of radicals observed under vacuum at -196°C (Figure 13). Second, one of the epoxy samples (Macropoxy 646) produced an unusually high number of observable radicals under these exposure conditions, with the net number of radicals for this sample being nearly three times higher than the average for this group of samples. ESR spectra from the Macropoxy 646 sample are shown in Figure 13, where the large increase in signal intensity (due to broken chemical bonds) with exposure time is evident. Without examining a second set of these same samples following xenon irradiation at room temperature in vacuum it is not possible to comment on the role of O_2 in determining the lifetimes of the radicals in the coatings. For example, two alkyl-peroxy radicals can combine to form a stable non-radical product, plus molecular oxygen. This decay pathway is only available for radicals produced in a photo-oxidative environment, and not in an oxygen-free environment. This second set of experiments is in preparation.

A major aspect of this project involves the correlation of photo-sensitivity of the different coatings with their molecular structures. To accomplish this, topcoat resins are being analyzed using high performance liquid chromatography (HPLC) along with nuclear magnetic resonance (NMR) and Fourier-transform infra-red (FTIR) spectroscopies. Chromatography allows separation of commercial formulations into their separate chemical components, and the spectroscopic techniques identify the molecular structures of the components. In a related study, we have determined that the polyester polyol component of a particular polyurethane clearcoat is the component susceptible to photo-induced bond cleavage, and that an aliphatic radical is a major component of the ESR spectrum.

Conclusion

In this paper, we have presented a series of results on the photodegradation of polymer-based coatings induced by Xe-light irradiation as studied by slow positron annihilation and electron spin resonance methods. The S-parameter and o-Ps intensity from the positron method are found to decrease with exposure time. Current results show that the slow positron technique is a sensitive tool for detecting coating degradation in a time much earlier than for any existing testing methods. The electron spin resonance results show promise for identifying the specific chemical bonds or locations which are responsible for the deterioration of coatings. Combining these two atomic and molecular testing methods will be pursued next in a systematic way for both commercial products and model compounds. The next major research activity is to perform engineering tests and to correlate change in macroscopic engineering properties with observed nano-scale changes detected by PAS and ESR.

Acknowledgement

This research is supported by National Science Foundation (CMS-9812717) and by the Air Force Office of Scientific Research under Contract Numbers F49620-97-1-0162 and F49620-98-0309. We also appreciate for collaborations with Drs. R. Susuki, T. Ohdaira, and B. Nielsen. Fruitful discussions with Profs. K.L. Cheng and E.W. Hellmuth are acknowledged.

References

1. Federal Highway Administration, "Recording and Coding Guide for the Structural Inventory and Appraisal of the Nation's Bridges," U.S. Department of Transportation, Federal Highway Administration, Report No. FHWA-PD-96-001, 1995.
2. Weismantel, G.E. *Paint Handbook*, Ed. Charlesworth, G.B. and Weismantel, G.E., Chapter 18, New York: McGraw-Hill, 1981.
3. D. Satus, *Coating Technology Handbook*, New York: M. Dekker,1991.
4. *Polymer Durability: Degradation, Stabilization, and Lifetime Prediction*; Clough, R.E.;Billingham, N.C.;Gillen, K.T. Eds.; *Adv.Chem.Ser. #249*, Washington, D.C.: Amer.Chem.Soc., 1996.
5. For example, see: *Polymer Spectroscopy*, Ed. Fawcett, A.H. New York: Wiley & Sons Sci., 1996.
6. Rabek, J.R. *Polymer Degradation and Stabilization: Photodegradation of Polymer, Physical Characterization and Applications*, New York: Springer Pub., 1996.
7. Zhang, R.; Cao, H.; Chen, H.M.; Mallon, P.; Sandreczki, T.C.; Richardson, J. R.; Jean, Y.C.; Nielsen, B.; Suzuki, R.; Ohdaira,T. J. Rad. Chem. Phys. **2000**, 58, 639-644.
8. Suzuki R.; Kobayashi Y.; Mikado T.; Ohgaki H.; Chiwaki M.; Yamazaki T.; Tomimasu T., Mater. Sci. Forum, **1992**, 105-110, 1993-1996.
9. Cao H.; Zhang R.; Sundar C. S.; Yuan J.-P.; He Y.; Sandreczki T.C.; Jean Y.C.; Nielsen B. Macromolecules, **1998**, 31, 6627-6635.
10. Cao H.; He Y.; Zhang R.; Yuan J.-P.; Sandreczki T. C.; Jean Y.C.; Nielsen B. J. Polym. Sci. B: Polym. Phys. Ed., **1999**, 37, 1289-1305.
11. Cao H.; Yuan J.-P.; Zhang R; Huang C.-M.; He Y.; Sandreczki T. C.; Jean Y. C.; Nielsen B.; Suzuki R.; Ohdaira T. Macromolecules, **1999**, 32, 5925-5933.
12. Jean Y.C., In Positron Spectroscopy of Solids, pp.563-580, Dupasquier, A., Mills, Jr., A.P. Eds. IOS Press, Amsterdam, 1995.

Chapter 16

Fundamentals of the Measurement of Corrosion Protection and the Prediction of Its Lifetime in Organic Coatings

Gordon Bierwagen[1], Junping Li[1], Lingyun He[1], and Dennis Tallman[2]

Departments of [1]Polymers and Coatings and [2]Chemistry, North Dakota State University, Fargo, ND 58105

I. Introduction

In this paper, we will consider, first, the fundamentals of the measurement of corrosion protection of organic coatings. This first portion will emphasize electrochemical characterization methods for the corrosion protective properties of organic coatings, especially electrochemical methods, and the information about the coating such characterization gives us. In the second portion of the paper, we will consider the use of two of these methods, electrochemical noise methods (ENM) and electrochemical impedance spectroscopy (EIS) to consider film thickness effects on corrosion. In the third section, we will show how one can pred corrosion protection lifetime of organic coatings from ENM and EIS data obtained in immersion and Prohesion exposure.

II. Basics of Measuring Corrosion Protection Properties of Organic Coatings

A. Introduction

The corrosion protection properties of organic coatings are some of the most important properties of the entire coatings system. In a standard (top coat + primer + metal pretreatment) coatings system, the topcoat provides water resistance and barrier properties, the primer provides corrosion protection when the coating system is damaged as well as adhesion to the metal substrate, and the metal pretreatment provides some damage protection as well as a surface to which the primer adheres strongly. Each layer in this complex three-layer structure must perform properly in order to achieve optimal corrosion protection. Therefore, the testing of corrosion protection is a very important component of the entire battery of tests performed on this coatings system.

B. Exposure Testing – Qualitative Analysis

Much of the testing of the corrosion protective properties of organic coatings is done by natural outdoor exposure at sites of high corrosivity or accelerated exposure in an exposure cabinet. The former is performed at sites such as Laque, NC, the DOD test site on Key West, FL, or sites of high industrial corrosivity such as Chicago, Cleveland, or Pittsburgh. Coated test panels, which can be scribed or unscribed, are exposed for controlled periods of time, and qualitatively examined for signs of failure such as blister density, visual signs or corrosion, creep of visible corrosion from a scribe mark on the coating. One difficulty of outdoor panel exposure is the time required to achieve interpretable results. Another difficulty of such field-testing is that only qualitative information that is generally available from these exposures (see below). Image analysis of the panels helps to provide improved ranking accuracy, and the use if *in situ* electrodes during the exposure provides some quantitative electrochemical information during the exposure.[1]

In exposure cabinet testing, scribed or unscribed panels are put in a cabinet which either continuously exposes the panels to a spray of electrolyte solution at elevated temperature, or exposes them in a cyclic manner to electrolyte spray, then to dry periods, both at elevated temperature.[2] If the standard salt spray test, run according to ASTM B117, is used, the electrolyte solution is 5% NaCl, and the temperature is 95°F. In cyclic exposures, the ProHesion™ test protocol is the most commonly used, but others have been used. For all of these exposure tests, the evaluation is usually a visual, qualitative ranking based on blister count in a fixed area of the exposed face of the panel, percent area of exposed panel area corroded, and the distance the corrosion has moved from a scribe mark on the panel face. A picture of panels in a Q-Fog™ Test Cabinet[1] is given in Figure 1.

The cyclic test methods have proved superior to continuous exposure as in the ASTM B117 protocol.[3] Valuation of the results of these methods are very non-numerical in nature and have not always yielded rankings the agree with actual field exposure. Recently, image analysis of the panels has been utilized to perform the evaluation,[4] and this has improved the accuracy and reproducibility of the ratings. Still, exposure evaluation with these evaluation methods has not given an lifetime predictive capability and the regular salt spray testing has given many incorrect results on non-chromated, environmentally benign systems.

---

[1] Q-Panel Corporation, Cleveland Ohio

318

Figure 1. Coated Metal Panels in Exposure in Q-Fog ™ Test Cabinet

C. Electrochemical Testing

Because of the problems inherent in exposure testing methods, especially their subjective nature, quantitative electrochemical testing has become quite widely used.[5] This increased use is also due to the fact that the instrumentation for this testing has recently become easy to use and less expensive than it has been in the past. The major electrochemical test methods used for examining coatings systems and metal pretreatments are electrochemical impedance spectroscopy (EIS), electrical noise methods (ENM), and potentio-dynamic scanning.[5,6]

1. EIS Methods

EIS measures the complex impedance of a coated metal panel immersed in a test electrolyte as a function of the signal frequency.[7,8] Advantages of EIS are its relative ease of use, reproducibility, and the fact that one can compare physical models for coated metal systems with experimental results via equivalent circuit modeling. The electrochemical cell used for this type of test is pictured in Figure 2.

The format most commonly used for data representation is Bode plot format, where the modulus, $|Z|$, or the phase angle, φ, of the polar representation of the complex impedance is plotted vs. log measurement frequency. An example of this for $|Z|$ vs. frequency is given in Figure 3.

This figure shows several features of EIS results from organic coatings. The first is that high performance coatings show mainly capacitive performance, a linear negative slope in $|Z|$ vs f, through to low frequencies (f < 0.01 to 0.0001 Hz), and high $|Z|$ values at low frequencies. The phase angle is $-90°$ at high frequencies, consistent with capacitive behavior shown with the modulus, and it increases towards $0°$ at low

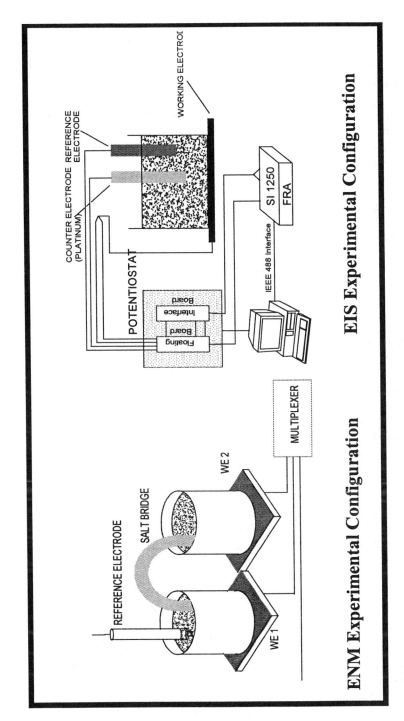

Figure 2. Test Cell Configurations for Electrochemical Measurements

frequencies, showing the system is acting more like a pure resistor at low frequencies. For poor or degrading coatings, the purely capacitive behavior of |Z| ceases at higher frequencies, and the system begins acting like a resistor, |Z| independent of f, at these frequencies. The phase angle shifts from – 90° at frequencies such as 10 – 100 Hz. Further details of the interpretation of EIS data are given by Scully, et al.,[9] Murray[10], and de Wit, et al.[11,12] In general, EIS has become one of the must used electrochemical measurements used for coatings characterization, but the data generated by this method must be treated with great care, and multiple samples used to insure the measurement contains no artifacts. EIS data is especially sensitive to imperfections in samples and film thickness.

2. ENM Studies

ENM is the other electrochemical method besides EIS that can used to characterize the corrosion protective properties of coatings. It is a method that has been used since 1986 for examining organic coatings and their performance.[13] The experimental method consists of the monitoring the spontaneous fluctuations in voltage and current that occur between two identical coated electrodes in electrolyte immersion.[14] The electrode configuration for these experiments is show in Figure 4. This configuration is that used by Eden, et al.,[13] in their studies, and utilizes a reference electrode to monitor the potential fluctuations. The electronics, plus a multiplexer, for running semi-continuous monitoring of up to eight samples in shown in Figure 3. The data from ENM is most commonly analyzed by calculating the noise resistance, R_n,[15,16] from the data. Other parameters can be extracted from ENM measurements by spectral analysis of the time series data.[17,18]

One manner in which the R_n data is presented in shown in Figure 5, where R_n values vs. time of immersion exposure are given for aircraft primers and topcoat systems under consideration in our laboratory. The data shown has been fitted with a straight line in this log (R_n) vs. time plot, indicating an approximately exponential decay of the coating resistance with immersion time. We have been recently analyzing our data by such fitting, and have found that many of our R_n vs. time of exposure results can be fitted in this manner. This would indicate a first order decay of resistance, or barrier properties, with time, and implies that there is a single physical or chemical change causing the fall in coating performance. This type of analysis can be performed also on low frequency |Z| vs. time data, and similar results are seen.[19] We will discuss the use of this type of data analysis for lifetime prediction later in this paper.

3. Hygrothermal Effects

Sometimes these methods coupled to simple panel immersion are insufficient to differentiate between high performance coatings, or the results take excessive amounts of time to achieve proper results. In many cases, the decay of performance of coatings may be accelerated by raising the temperature of immersion in the test electrolyte on a continuous basis or to cycle the temperature from room temperature to a higher temperature and then cool it back to room temperature. The

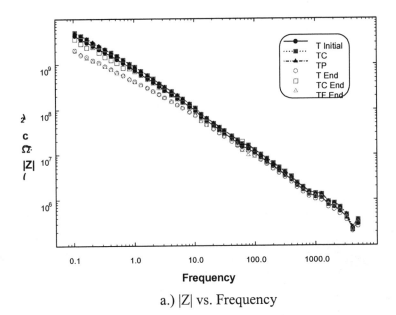

a.) |Z| vs. Frequency

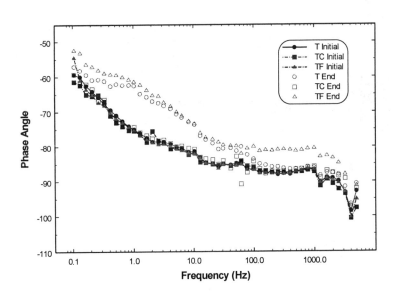

b.)

Figure 3. Bode Plots for Coated Metal Substrates:
a. |Z| vs. Frequency for Various Panels
b.   Phase angle for vs. Frequency for Various Panels

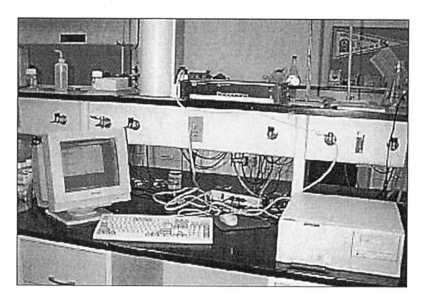

Figure 4. Electronics Configuration for ENM Measurements

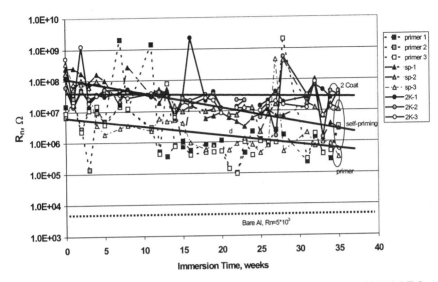

Figure 5. R_n versus Immersion Time for Aircraft Coatings over Al 2024 T-3

cycling may also be performed several times to achieve further differentiation between samples that are quite close in performance, or to analyze the quality of a very good sample. Initial analyses of the effects of heating and cooling of an epoxy powder coating (sometimes known as a fusion bonded epoxy coating, or FBE) on pipeline steel has been presented in prior publications from this laboratory.[20,21] Granata and co-workers have also considered related effects,[22] as have Kuwano, *et al.*[23] One of the major features of thermal cycling in immersion is that it accelerates the water uptake in the coatings, and in the case of many coatings this water plasticizes the film and decreases the barrier properties of the film considerably. In ref. 20, we have shown that we can use ENM and EIS to measure the glass transition temperature, T_g, of the immersed, plasticized, coating, and also showed that the water uptake is irreversible as long as the coating is kept immersed.

More recently, we have considered thermal cycling to analyze plasma polymer pretreatments + a chrome-free cathodic electrodeposition primer over Al 2024 T-3 as a chrome-free replacement for traditional pretreatments and primers for this aerospace alloy. The system under study give exceptional performance in an undamaged mode, and to differentiate between these good samples we have been utilizing thermal cycling. The thermal cycle we have found useful is outlined in Figure 6.

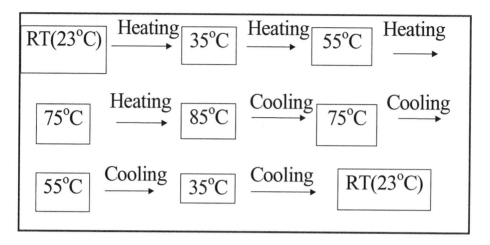

Figure 6. Thermal Test Cycle Used to Differentiate "Good" Samples

III. Consideration of Film Thickness Effects in Corrosion Protection

A. Introduction

In organic coatings used for corrosion protection, it is widely known, in the form of collective experience, that thicker coatings generally provide better protection than thinner coatings, that there may exist a thickness below which the coating has significant performance degradation[24] and that there may be a second thickness above which no further increase in performance occurs. Further, it also seems to be widely known that multiple coating layers provide better corrosion protection than one-coat systems. In discussions with many users and researchers in the field of corrosion protective organic coatings[25], this information is well known, but the specifics of film thickness dependence in electrochemical measurements, and the possible asymptotic nature of the electrochemical response of films is not well documented. Further, we cannot find numerical documentation of single vs. multiple layers examined at equal total film thicknesses. We believe this situation exists for several reasons:

1. The past difficulties with, and the time consuming nature of, multiple electrochemical measurements
2. The difficulties in casting with controllable accuracy a wide range of film thicknesses

3. The presence of intuitive, common knowledge - no perceived need for study.

In the work, described below, we describe our investigation of the film thickness effects on electrochemical properties of several types of coatings as measured in immersion in electrolyte. These studies were performed on cold-rolled steel Q-panel samples using marine two component epoxy-polyamide coatings. The films were cast in "draw-down" form, and multiple samples of each film thickness were examined.

B. Field Knowledge on Film Thickness Effects

Let us call what is known about film thickness effect in the direct use of corrosion protective coatings that exists in non-peer reviewed form "field knowledge." Below we will try to give a very abbreviated form of this field knowledge appropriate to this study. It has been known and observed by many corrosion protective coatings users that "thicker is better" and that multiple coats of paint are better than a single coat. This has been discussed in several books,[24, 26,27] but only qualitative observations are given. There seem to be hints of descriptions the improved performance of multiple coats vs. single coats, but no direct comparisons at constant total thickness. Deliberately thick coatings are known to provide extra corrosion protection on pipelines and the stone impact areas on automobiles and trucks. In these cases, the thick coating provides abrasion and mechanical damage protection resistance supplemental to the corrosion protection of the coating system. This points out that there are often many components to the protection a coating provides that is often summarized only as "corrosion protection." There is documentation that there may exist a minimum film thickness for adequate corrosion protection, but the performance data described is mainly qualitative in nature, and involves no electrochemical measurements.[28] No specifics are provided by observers on this issue, and little has been done to document the minimum film thickness for performance. This may be for several reasons, among them the problem of film application defects which are often judged to be eliminated by a "sufficiently thick coating," a lack of control of applied film thickness, and a concern about insuring more than adequate film thickness by recommending a user film thickness in excess of the performance minimum. Many suppliers of corrosion protective coatings provide qualitative thickness guidelines for use of their materials and may often recommend that the coatings be applied in two-coats, for example, but offer little in the way of numerical performance data to validate their suggested film thicknesses of use. In short, it is difficult and time consuming to accumulate corrosion protection performance data as a function of film thickness, and using the more subjective tests such as continuous salt fog exposure (e.g. ASTM B117), there was not even good numerical data to utilize for differentiation between samples.

C. Literature On Film Thickness Effects

A literature search on this problem revealed very few references that explicitly dealt with film thickness and included numerical, electrochemical characterization of coatings. Di Sarli, et al.[29], examine film thickness effects on coating properties vs. time in immersion in electrolyte for several types of coatings, mainly chlorinated rubbers, phenolics, and mixtures thereof. In a given set of measurements, at most, four film thicknesses values of one coating were studied. No data was analyzed directly vs. film thickness, and a single statement that the relationship between the resistance and capacitance values calculated from impedance measurements and film thickness was non-linear. The film thickness range examined was from 55-270 µm. The authors did observe that thinner films degraded faster than the thick, but used the explanation of Mills and Mayne[30] that a thicker coating meant less chance of weak points in the coating. Amirudin & Thierry[31] examined some film thickness dependence in study of coatings using Electrochemical Impedance Spectroscopy (EIS) methods of coated steel panels in immersion in electrolyte, and used a simple Randles-type equivalent circuit for analysis of the results. They present results as film resistance and film capacitance vs. time at three different thickness each for an alkyd, an epoxy and a bituminous coatings. In all cases, they observe that the resistance of the thicker coatings was greater than thinner coatings, the rate of decay of the coating resistance vs. time was larger with thinner coatings, and the increase of the capacitance with time of the film, a measure of water uptake, decreased with increasing film thickness. They examined only three film thickness for their systems, and no reproducibility data was provided. Scully[32] examined a range of film thicknesses in a study of marine epoxy-polyamide coatings in immersion in sea water by impedance and visual rankings. His results very closely parallel those of ref. 31, and again, no explicit analysis of film thickness is presented. Mertens et al.[33] confirmed that the film thickness of organic coatings has severe influence on the metal substrate. In this case, EIS measurements were completed on alloy pre-coated steel substrates with extreme thin unpigmented coating films (0.5 ~2.0 µm). The electrochemical parameters (film resistance, effective pore area, and active surface area fraction) were plotted against film thickness. The inflection points manifested in all these plots, and the authors related these points to the critical film thicknesses under various immersion conditions considered. Since only 3 film thicknesses for each system were checked in this study, more detail and complete information about the film thickness effect could not be gained. Babic, et al.[34], consider film thickness in EIS studies of "thick" vs. "thin" coatings, and make conclusions similar to the previous references, the resistance of thicker films is higher, and tends to decay more slowly than thinner coatings. Again, no data is plotted vs. film thickness, and the reproducibility vs. thickness is not shown. Murray[35] presents the most complete set of data on the examination of film thickness effects that we have been able to find. The electrochemical methods he used were EIS and dc resistance measurements vs. time on polyurethane films over steel at thickness values ranging from 930-7010 µm. For dc resistance values vs. film thickness, he saw a linear relationship of the form, $R_{dc} = kh$, where h is film thickness. Using the simple relationship to analyze the capacitance, C, data for a dielectric film vs.

thickness, $C = \varepsilon\varepsilon_0/h$, where ε and ε_0 are the dielectric constants of the film and free space, respectively, he showed a linear inverse dependence of the dry coating capacitance with film thickness. He also observed a linear relationship between film thickness and the $\sqrt{t_b}$, where t_b is the time at which the corrosion potential of the metal in immersion begins to change. The latter suggests a diffusion control process. Again, the rate of decay of film resistance values was greatest with thicker films. Haryama, et al [36]., examined the effects of film thickness with respect to film delamination by cathodic charge transfer, and showed that the area of disbonding and charge transfer, independent of film thickness.

D. Basic Theory of Film Thickness Effects

If one measures the resistance, R (area corrected, in units of ohm-cm^2), of organic coating films electrochemically at various film thicknesses, h, then one can determine a film resistivity, ρ, from these data as

(1) $$\rho = \frac{R}{h}$$

This follows directly from the basic examination of the resistivity, where the total resistance is the length times the resistivity.[37] This implies that there should be a continuous increase of film resistance with film build as measured by electrochemical means, and that there should be no upper limit to the film resistance due to a thickness increase for uniform coating films. We have discussed how this has been examined by past workers, and that there appears significant data to show that in thin films and in very thin films there is a divergence from this expected behavior. Analysis of the resistance behavior of organic coating films immersed in electrolyte has shown that films do not very often show simple behavior as a pure resistor (purely resistive behavior in EIS studies, for example)[38]. The study of film thickness behavior on electrochemical measurements of film performance has been very often based on the use of equivalent circuit modeling.[17] The coating resistance can also be characterized by Electrochemical Noise Methods (ENM) [20,21,39]. In this case, two nominally identical coated metal panels are usually used as the work electrodes and joined together during measurement via a zero resistance ammeter, and the spontaneous naturally occurring voltage and current fluctuations are measured without imposing a external signal, i.e., voltage or current. Then, the electrical resistance, now called noise resistance R_n, can be calculated as[40]

(2) $$R_n = \frac{\sigma_v}{\sigma_i}$$

where, the σ_v and σ_i are the standard deviations of the voltage and current noise signals, respectively.

In a simple analysis of a single film, as described immediately above, this is the sum of the film resistivity and the film thickness plus a term dependent on only the metal and its oxide surface. Thus we could write simply for a single coat system of resistivity ρ and thickness h,

(3) $\qquad R = \rho h + R_m^{surface}$

By this simple model, a plot of resistance vs. film thickness should yield a straight line with a slope ρ and intercept $R_m^{surface}$ if we could extrapolate to zero thickness. However, very thin paint films fail and begin to delaminate almost immediately upon immersion in electrolyte, and the limiting low film thickness behavior of thin films as analyzed by this method is not readily measured by EIS or other resistance methods depending upon immersion. Further, as films become very thin, it is very difficult to cast traditional paint films without out defects or non-uniformities, and the measurement of R becomes dominated by film defect effects, and the resistance measured is the resistance of substrate/electrolyte interface. If we have multiple paint coatings, by these arguments we should be able to calculate the resistance of our simple circuit in Fig. A as

(4) $\qquad R = \sum_{i=1}^{n}(\rho_i h_i) + \sum_{j=1}^{n} R_j^{surface} + R_m^{surface}$

where we have n layers of coatings and n surfaces between layers plus the metal surface. When a coating over a metal fails, there is always a R value characterizing the "failed" system, usually somewhat higher that the bare metal because it includes effects of a corrosion grown metal oxide. However, we are attempting here to generate a model for the interpretation of results that will allow us to separately examine the ρ_i terms, and thus terms that are distinctive to the coating material itself. It may also allow us to isolate and analyze the interfacial resistance term for the metal/substrate interface. Figure 7 illustrates schematically the film layers and interfaces.

E. Experimental Studies

In general, the experimental methods follow the methods as described in previous publications from this laboratory.[14,39,41] The substrate used for film thickness studies was cold rolled steel test panel from Parker-Amchem Co., Madison Heights, Michigan. The 76mm×228mm×0.7mm (3"× 9" × 0.026") panel was polished and pretreated with Bonderite 1000 and Parcolene 60, followed by de-ionized water rinse. All the panels were cleaned by acetone prior to use. Red alkyd and green epoxy coatings, supplied by Courtaulds Coatings, Inc., Houston, Texas, were used. They are commonly applied as primers in marine environments for their good corrosion resistance in seawater. The alkyd paint is a one-pack system while the epoxy one is a two-pack system consisting of epoxy and polyamide resins with the ratio of approximately 1 to 1 by volume. As proprietary products, very little technical information could be obtained about these paints.

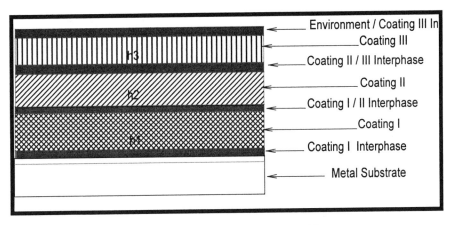

Figure 7. Schematic of Multilayer film

The red alkyd paint was applied as one-coat primer on the steel panels via spray. As far as possible, the spray conditions used were similar to what might be achieved in the actual painting of a ship although on a smaller scale. The wet coating films were allowed to dry for one week under ambient condition, then their dry film thickness was measured. The green epoxy paint was applied as single and double coat films via an 8-path wet film applicator by hand. The wet one-coat films were allowed to air dry under ambient condition for three days before their dry film thickness was measured. Some of the one-coat panels were given a second coat of paint right after the measurement of film thickness. For each two-coat panel, the second coat thickness was controlled as the same as the first one. After one week of air drying, film thickness of the two-coat panels was measured.

F. Film Thickness Measurements

The dry film thickness of coated steel panels was measured by an electromagnetic digital coating thickness gauge, Elcometer 345 FS, made by Elcometer Instruments Ltd., England. This is a ferrous type, used to measure the dry film thickness of nonmagnetic coatings on ferrous substrates, and is based on the principles of electromagnetic induction.[42, 43] The voltage controlled oscillator in the gauge produces a low level alternating voltage to induce a magnetic field by energizing one of the coils in the probe. The magnetic field then interacts with the ferrous substrate, and the strength of substrate influence is sensed by other probe coils in the form of a voltage. The probe is always kept contacting with the coating film surface on the substrate. As the magnitude of the reflective voltage depends on the distance between the probe and the ferromagnetic substrate, the resulting voltage measurement is converted to a thickness reading on the liquid crystal display of the gauge. Electromagnetic gauge may provide accurate readings and the ability to make many measurements in a relatively short time. The accuracy of our used gauge is ± 3% of the reading within the range from 0 to 1250 μm with the resolution of 0.1 μm. This gauge's statistics mode can show the number of readings, mean value, and standard deviation.

In the measurement of the dry film thickness of painted panels, to ensure the accurate readings, gauge was first calibrated by a bare steel panel identical to that being coated as well as a series of plastic shim standards covering the expected range of coating film thickness. For statistical accuracy, six different places in the central region of each panel were measured. The average and standard deviation of the film thickness for each of the tested groups were calculated and listed in the following tables. The coating film thicknesses were highly reproducible, while all of them being within ±10% range of the respective mean value, most of them within ± 5% range.

Table I. Panels with Red Alkyd One-Coat Films

Group	1	2	3	4	5
Pairs Number	3	3	3	3	3
Thickness (μm)	33.8 ± 2.0	65.5 ± 3.4	74.1 ± 2.8	119.2 ± 2.5	145.7 ± 4.3

Table II. Panels with Green Epoxy One-Coat Films

Group	1	2	3	4	5	6	7	8
Pairs Number	6	3	6	3	3	3	3	6
Thickness (μm)	24.2 ± 2.6	53.2 ± 3.6	76.4 ± 2.0	87.5 ± 2.7	104.0 ± 1.7	121.0 ± 3.5	141.2 ± 1.5	155.4 ± 6.5

Table III. Panels with Green Epoxy Two-Coat Films

Group	1	2	3	4
Pairs Number	6	6	6	6
Thickness (μm)	50.7 ± 2.0	89.1 ± 2.9	140.8 ± 3.7	157.5 ± 5.3

G. Experimental Details

In ENM configuration, two nominally identical tested panels were used as working electrodes, and a saturated calomel electrode (SCE) as reference electrode. To each of the tested panels, using a Goop® marine grade glue (Eclectic Products,

Inc., Carson, CA), a cross section of PVC pipe with 2.0 cm inner diameter and 5.2 cm height was attached to act as the electrochemical cell for holding the 3 wt.% NaCl electrolytic medium. The area of painted panel left to expose to the solution was 12.56 cm². Salt bridge was used as a conductive path connecting the electrolytes in the cells of each working electrode pair. The setup for EIS measurements was very similar to that for ENM test, except it consisted of one tested panel as working electrode and an additional platinum net electrode as counter electrode without using salt bridge. Both ENM and EIS measurements were performed through Gamry electrochemical measurement systems (Gamry Instruments, Inc., Willow Grove, PA): CMS 120 system was used for ENM while CMS 300 system for EIS. ENM and EIS experiments were run on a weekly/biweekly basis, and when the panels were measured, ENM measurement was first made, then followed by EIS. In the ENM measurement, readings were taken every 2 seconds over a 512-second interval (block time). During each measurement, 10 intervals of noise data were acquired on each panel. For EIS experiment, the scanning frequency range was 0.02~ 65,500 Hz with 10 mV (rms) applied AC voltage. The two panels of each pair were wired together when the panels were set aside between the measurements to ensure their equal potential before next measurement.

IV. Lifetime Prediction Studies

A. Introduction

In many of our studies of the electrochemical properties of organic coatings over metals during exposure in a Prohesion cyclic test cabinet or immersion in an electrolyte solution, we find that both the noise resistance, R_n, or the low frequency impedance modulus, $|Z|$, show an almost exponential decay a function of time. Further, the "infinite time" limits of these coating resistance values have to be these values for the un-coated, bare metal substrate. We discussed this above, and now consider this in more detail, and pursue the implications of these results for lifetime prediction.

B. Data Analysis

If we fit our resistance data by supposing that both $|Z|$ and R_n are exponential decreasing functions of exposure time with limiting bare metal values of $|Z|_m$ and R_n^m, respectively, we have the following expressions for the low frequency impedance and noise resistance values

$$(5) \quad |Z|(t) = |Z|_m + \left(|Z|_0 - |Z|_m\right)\exp\left(-t/\theta\right),$$

and

$$(6) \quad R_n(t) = R_n^m + \left(R_n^0 - R_n^m\right)\exp\left(-t/\vartheta\right),$$

where $|Z|_0$ and R_n^0 are the film resistance values at $t = 0$. The form in which the data is analyzed is

$$(7) \quad \ln\left(\frac{|Z|(t) - |Z|_m}{|Z|_0 - |Z|_m}\right) = -\frac{t}{\theta},$$

and

$$(8) \quad \ln\left(\frac{R_n(t) - R_n^m}{R_n^0 - R_n^m}\right) = -\frac{t}{\vartheta}.$$

The decay constant θ (the inverse of the slope of the exponential fit line in the graph of $\ln\left(|Z|(t) - |Z|_m\right)$ vs. exposure time) and ϑ (the inverse of the slope of the exponential fit line in the graph of $\ln\left(R_n(t) - R_n^m\right)$ vs. exposure time) can be used to quantify the decay of $|Z|$ and R_n for different samples. These constants, θ and ϑ, have the dimension of time and can be considered characteristic decay times for the coatings. If EIS and impedance and ENM are tracking basically the same changes in the coatings, θ and ϑ should be almost equal. The constants characterizing the initial and final EIS data, $|Z|_0$ and $|Z|_m$, and the initial and final ENM data, R_n^0 and R_n^m, along with the decay constants can provide a quantitative basis for estimating the corrosion rates and performance lifetimes of our samples. For example, a numerical corrosion resistance ranking of samples can be determined from these constants. Such an analysis also shows the agreement in the predictive value this type of ENM and EIS data. By defining failure times as the values for $|Z|(t)$ and $R_n(t)$ to arbitrary failure values, $|Z|_{fail}$ and R_n^{fail}, we can calculate a time to failure as the time at which the resistance has decayed to these values. These time-to failure values are given by

$$(9) \quad t_{fail} = \theta\left[\ln\left(\frac{\left(|Z|_0 - |Z|_m\right)}{|Z|_{fail} - |Z|_m}\right)\right],$$

and

$$(10) \quad t_n^{fail} = \vartheta \left[\ln \left(\frac{\left(R_n^o - R_n^m \right)}{R_n^{fail} - R_n^m} \right) \right].$$

The mathematical description of the decay of the coating resistance written in this way seems to treat many of the features that we see in our almost every sample that we have thus far examined. The initial decay rate is quite large, and then the samples settle into almost a steady state of slow decay. We are not yet giving a theoretical interpretation to the form or values of the decay constants, but we are just using it as an empirical tool in our ranking of the samples. As we go further, these decay rates may prove useful in lifetime estimates. For example, we can rank coatings based on their decay constants or we can rank them on the basis of their time-to-failure estimate, as given above or in another form. In all of the above, we can calculate the resistivity from EIS, ρ, and the resistivity from ENM, ρ_n, according to equation (3) as

$$(11) \quad \rho(t) = \frac{|Z|(t) - |Z|_m}{h},$$

and

$$(12) \quad \rho_n(t) = \frac{R_n(t) - R_m^m}{h}.$$

Data in this form can be also analyzed as discussed above.

C. Results and Discussion

The electrical resistance, low frequency modulus ($|Z|$ at 0.05Hz or 0.08Hz) and noise resistance (R_n), were calculated from the raw data acquired from the EIS and ENM data on the samples vs. exposure time. The resistances and resistivities (ρ) results were graphed as function of testing time when coated panels were constantly immersed in the salt solution during the experimental period of time. In all of our studies of the time dependence of the electrochemical properties $|Z|$ and R_n, we have observed there is a very distinct early time period immediately after test panels are immersed where these properties change very rapidly. The system is changing faster at this point than data acquisition can proceed, and the system is definitely not "stationary" in the manner required by the definitions of $|Z|$ and R_n.[44,45] We often get very high readings for resistances measured during short time periods from initial immersion. After 1-5 hours, the system is now changing slowly enough that the system is locally stationary (the mean of the data does not change with the measurement period) and the measured values of $|Z|$ and R_n are now physically meaningful. This is the point that we choose as time $t=0$ in analyzing our data according the eq. 5-10, defining the $|Z|_0$ and R_n^0 values used in these equations. We often find that values of $|Z|_0$ and R_n^0 measured in this manner are the same as the $|Z|_0$ and R_n^0 values one would get from including all data, even questionable short term values, but extrapolating the linear portion of the plots of $\log|Z|$ and $\log R_n$ vs. time back to $t=0$. In other words, the data fitting we suggest in equations 5-14, is valid only outside the early immersion times. We do not know exactly what is happening during these short times of coated metal immersion, but it appears to be related to early stages of diffusion of the immersion electrolyte into the film. One may imagine that in these short times, non-linear processes related to polymer/water interactions, surface wetting and interaction, local polymer relaxation, etc., are occurring. Measurement techniques that have an appropriately short time scale are needed to detail the events that take place in this early immersion period. We assume in our analysis of long term data that the details of these short term events are not crucial to our data analysis. Our data implies that on the long term that first order decay processes that can be described by a single time constant, θ or ϑ, accurately enough to give reasonably accurate forecasting of coating lifetime.

In our data analysis, equations (3) and (4) were employed to fit our time series data of resistance against immersion time of the panels. Through the least squares fit method, the first order exponential fitting equations of our experimental time series data can be obtained. The corresponding fitting curves show as straight lines in the graphs of logarithmic resistance versus immersion weeks. From the fitting equation, the decay constant as well as decay amplitude values can be derived. Further, given the resistance values of a bare metal panel, coating service times = times-to-failure may simply be estimated by the equation (5) and (6).

1. Alkyd Marine Coating Systems

In our film thickness studies, the alkyd marine coatings with five film thickness groups were first investigated by ENM and EIS electrochemical techniques. The experimental results plotted against immersion time or film thickness were the averaged data from each group consisting of three pair of coated panels.

Figure 8 shows the time dependent low frequency $|Z|$ values at different film thickness. $|Z|$ increases with increasing film thickness, and this trend is more clear after 30 days in immersion. At the end of experiment, the $|Z|$ of the panels with thinnest films dropped by 8 orders of magnitude to 3×10^2 $\Omega.cm^2$, essentially the same value oas the steel substrate, 2×10^2 $\Omega.cm^2$, while the $|Z|(t)$ data for the panels with the thickest films maintained around 5×10^8 $\Omega.cm^2$, just 1 order less than their initial values.

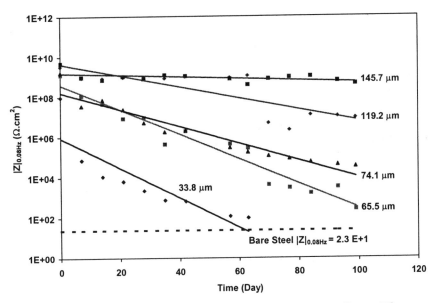

Figure 8. $|Z|_{Low\ Frequency}$ *vs. Time for Alkyd Marine Coatings at Different Film Thickness*

The same data is shown in resistance units, impedance resistance, in Figure 8 for the alkyd marine coatings. The variation in film impedance resistiveity ρ_Z with immersion time at different values of film thickness is illustrated in Figure 9. In figure 10, the noise resistance. R_n, vs. time is shown. Although the trend is not obvious during the early immersion period of time, R_n tends to be smaller with longer exposure time due to the solution penetration into the film. Also, thicker films possess higher R_n values. By the end of the experiment, the coating films in the thinner film thickness groups (33.8μm, 65.5μm,) show total failures with an R_n around 6×10^4 $\Omega.cm^2$, which is equal to that of the steel substrate, 6×10^4 $\Omega.cm^2$. On the other hand, the films from other groups (119.2 μm,145.7 μm) exhibited more stable during the whole testing period of time.

A large number of rusty spots were all observed on the surface of 33.8μm and 65.5μm panels, while little rusty holes were seen on the surface of 74.1 μm panels. On the contrary, for the panels with thickest films, no rust appeared, and the solutions were clear by the end of experimental. Although, no rusty area was found on the panels with 119.2-μm films, test solution in the cell became unclear, resulting from the coating degradation process.

In both measures of electrochemical properties of alkyd marine coatings, $|Z|$ and R_n, we see film thickness dependence of properties, and we observe two different decay rates, a large one for thinner coatings (<100μm), and another smaller rate for

336

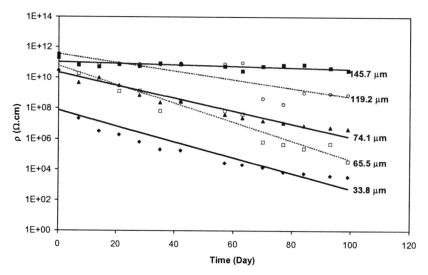

Figure 9. Impedance Resistivitiy vs. Time for Alkyd Marine Coatings

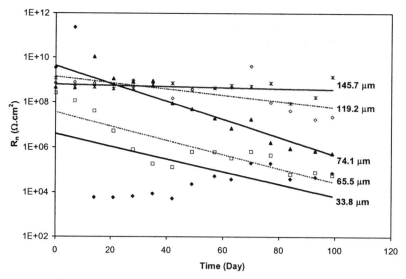

Figure 10. R_n vs. Time for Alkyd Marine Coatings at Different Film Thickness

the thicker coatings. Not only are the thinner coatings lower in protection, but the rate of decay of resistance properties is greater than for thicker coatings of the same paint.

The decay rates calculated from $|Z|$ and R_n are almost equal, once again showing that $|Z|$ and R_n are tracking the same changes in coatings[46], especially in stationary systems and with smoothed data. The same is true for the predicted "lifetime to failure using both sets of data. These data are shown in Figure 11, as well as their sensitivity to the choice of $|Z|_m$ and $R_n{}^m$, the "failure values for the bare metal resistances.

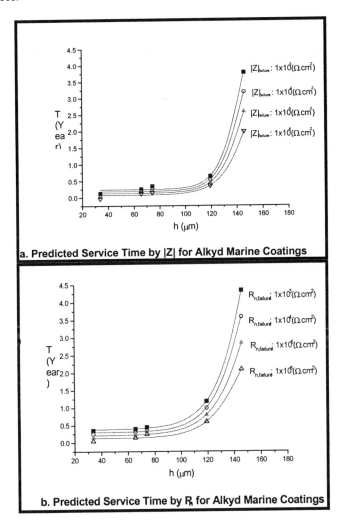

Figure 11. Lifetime to Failure Estimations for Alkyd Marine Coatings

2. Marine Epoxy Coatings

In our studies of film thickness effects in marine epoxy coatings, we included in our experimental studies the consideration of one- and two-coat systems at equivalent film thickness to consider the effects of multi-coat application on coatings properties. The data for impedance resistivity data vs. exposure time is given in Figure 12 for one-coat systems, and in Figure 13 for 2-coat systems. In both figures, we again note that there are two distinct groups of data, one from lower film thicknesses with a high rate of decay vs. exposure time, and a group from higher thicknesses with a much lower decay rate. The data of Figures 12 and 13 presented from the point of view of 2-coat vs. 1-coat systems at equivalent film thicknesses is shown in Figures 14 and 15.

As can be seen, in thin films of about 51 microns in thickness, 2-coat systems have significantly higher impedance resistivity values than 1-coat systems of the same thickness, and these decay at a much lower rate than the one coat systems. The ratio of the 2-coat to 1-coat system values decreases as the films get thicker so that at 141 microns, the 2-coat systems are only slightly higher that the 1-coat systems, and their decay rates are nearly equal.

Figures 16-19 show the same presentation of the data as measured by ENM. In these data also, a clear distinction between thicker and thinner films, especially in the rate of

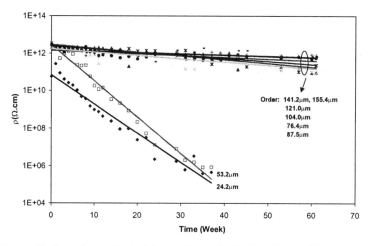

Figure 12. Impedance Resistivity vs. Time for 1-Coat Epoxy Marine Coatings

decay of noise resistivity vs. exposure time, as well as the higher values for 2-coat systems than 1-coat systems at equivalent film thicknesses. The separation between 1-coat and 2-coat behavior also diminishes as total film thickness increases. The most straight forward explanation of this behavior is that multi-coat systems have few

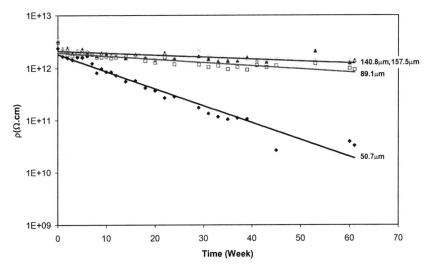

Figure 13. Impedance Resistivity vs. Time for 2-Coat Epoxy Marine Coatings

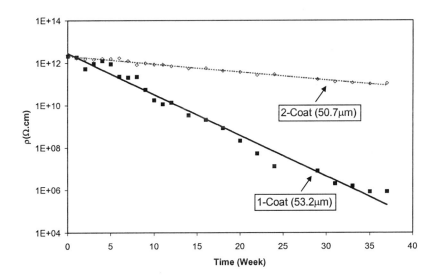

Figure 14. Impedance Resistivity vs. Time for 1&2-Coat Epoxy Marine Coatings

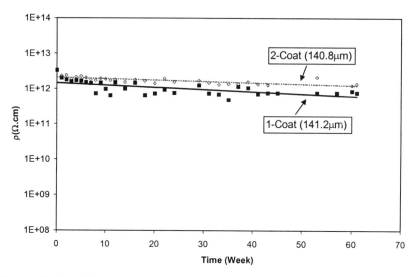

Figure 15. Impedance Resistivity vs. Time for 1&2-Coat Epoxy Marine Coatings

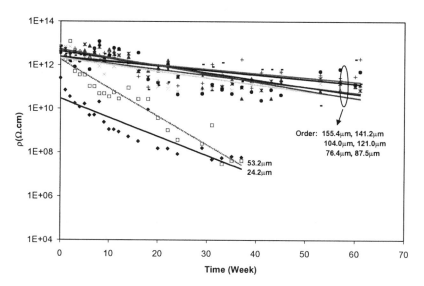

Figure 16. Noise Resistivity vs. Time for 1-Coat Epoxy Marine Coatings

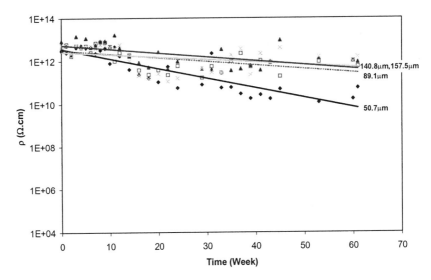

Figure 17. Noise Resistivity vs. Time for 2-Coat Epoxy Marine Coatings

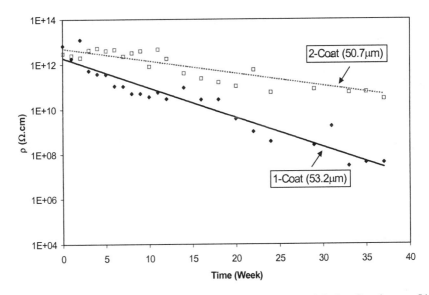

Figure 18. Noise Resistivity vs. Time for 1&2-Coat Epoxy Marine Coatings at 51 microns

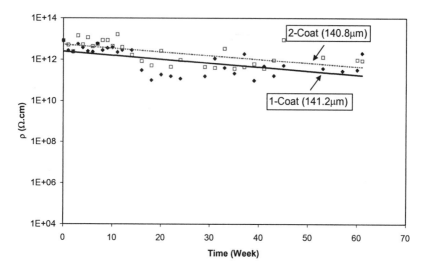

Figure 19. Noise Resistivity vs. Time for 1&2-Coat Epoxy Marine Coatings at 141 microns

through film spanning imperfections than 1-coat systems, and that as paint films become thicker, the probability of film spanning imperfections decreases drastically.

In Figure 20, the decay constants for the $|Z|$ data in terms of both film thickness and 1- vs. 2-coat systems are compared. A distinct break is noted in both curves between thinner and thicker films both for 1- and 1-coat systems. In the epoxy coatings, another effect can be noted not seen in the alkyd films – a constant slope region at higher film thicknesses. There seems to be a critical thickness above which decay is much slower than in thinner film. If this is true of most corrosion resistant coatings, it is an important design parameter of this class of coatings to characterize. Figure 21 shows the lifetime prediction estimates based on impedance data for the 1-coat and 2-coat systems at four different $|Z|_m$ "failure" values. The $|Z|_m$ values we examined range from 10^4 to 10^7 ohms-cm^2, a range which covers almost unnoticeable corrosion to "bare-metal" corrosion failure. The 10^7 ohms-cm^2 value is of the range that might be used by a company that markets corrosion protective coating might wish to use for choosing a warranty time for a coating. It would be a conservative estimate, and one that covers a small amount of imperfections in coating application. The lowest $|Z|_m$ would be the lifetime to complete failure, with poor appearance and considerable substrate attack.

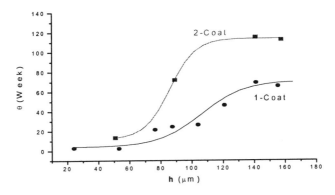

Figure 20. Impedance Decay Constant vs. Film Thickness for 1&2-Coat Epoxy Marine Coatings

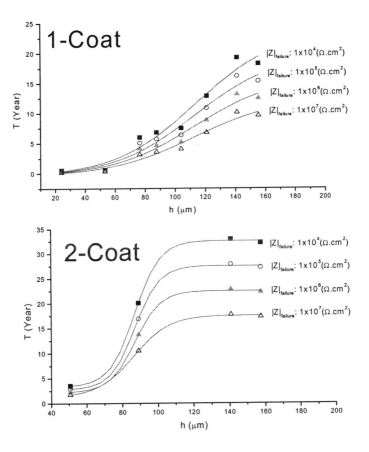

Figure 21. Lifetime –to-Failure Estimations from |Z| Data for Marine Epoxy Coatings

344

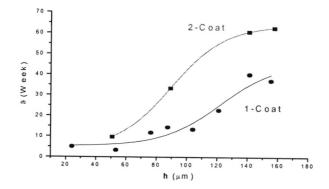

Figure 22. Noise Decay Constant vs. Film Thickness for 1&2-Coat Epoxy MarineCoatings

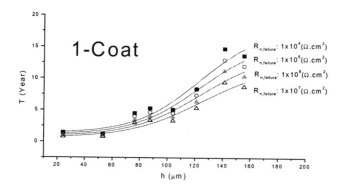

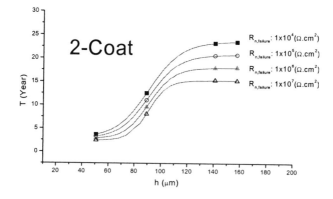

Figure 23. Lifetime –to-Failure Estimations from R_n Data for Marine Epoxy Coatings

Figure 22 portrays the same presentation of results for R_n data., and show the strong similarity in values the derived parameters such as decay constants and lifetime-to-failure values are for EIS and ENM data.

2. Aircraft Coatings

Recently we analyzed some of the data we have been acquiring on aircraft coatings subjected to Prohesion exposure in dilute Harrison's Solution. The time series data is not presented here for the sake of brevity, but suffice it to say, the $|Z|$ and R_n data plotted vs exposure time showed the same exponential decay in time that the other systems that we discussed above. We present these results in Figures 24 and 25.

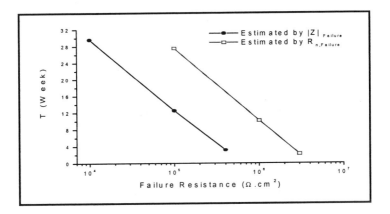

Figure 24. Estimated Time to Failure vs. Resistance at Failure for MIL-P-23377 Aircraft Primer

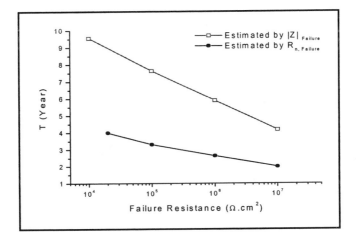

Figure 25. Estimated Time to Failure vs. Failure Resistance for (MIL-P-23377 +Glossy MIL-C-85285) Aircraft Coating Systems

V. Summary and Conclusions from Performance vs. Time and Film Thickness Studies

The following summary and conclusion statements may be made about these results:

1.   The electrochemical methods, ENM and EIS, have all exhibited usefulness by providing numerical, objective data for corrosion resistant rating in our systematic film thickness studies. Although the noise data are more scattered than those from EIS, the results from both EIS and ENM techniques are in very good agreement. The resistance values of coatings over poorly prepared substrates was always lower than the resistance values of the same coatings over well prepared substrates at equal film thickness values. We cannot yet say whether the rate of film resistance decrease is pre-treatment dependent.

2.   Corrosion protection performance of coatings increases with film thickness in a nonlinear way: it changes slowly in the range of thin film thickness, then increases rapidly beyond a certain film thickness, and, after a critical film thickness, again seems to achieve an asymptotic value that is almost constant with increasing film thickness.

3.   The studies indicate that increasing film thickness makes little contribution to corrosion protection below a minimal film thickness and above a maximal limiting film thickness. Those two kinds of critical film thickness are of the coating characteristics: they are low for high quality coatings. Also, in practice, there is a film thickness for "adequate" corrosion protection, which is determined by the coating users. This film thickness is neither necessary beyond the maximal limiting film thickness nor necessary to be the most efficient film thickness.

4.   The methodology for predicting coating lifetimes, the use of exponential decay fit to time series data of resistance from EIS and ENM vs. exposure time works for all coatings systems tested and applied to acceleration of failure by both immersion and Prohesion cyclic testing exposure.

5.   At equal film thickness, films applied in two coats were always higher in resistance values than films applied in a single coat. At the same applied film thickness, 2-coat systems provide better corrosion protection than 1-coat ones. The relative difference between 2-coat and 1-coat films decreases with increasing total film thickness.

6.   Resistivity increases nonlinearly with film thickness. These results imply more efficiency of corrosion protection with thicker films above a critical film thickness than with thinner films

7.   Initial (dry) film resistance or resistivity at very short times is not very good predictor of coating corrosion protective performance.

8. The use of first order exponential decay function requires long term data: (1) The decay constant within a short period of time may quite different from that got from a long observed term; (2) If the time series data changes only very slightly in the time period of your immersion/exposure testing, i.e., the slope of straight line in the graph of logarithmic data versus testing time is close to zero, there may give a large error in estimating the decay constant obtained from the fitting curve. This implies that appropriately accelerated corrosion tests are helpful to accurate prediction of coating service time. These acceleration test protocols give significant changes of coating resistance values in a shorter time period of exposure than possible in direct exposure, and give large enough chances to achieve an accurate measure of the decay constants under the accelerated exposure conditions. One may have trouble doing accurate lifetime predictions u with the accelerated exposure data, but a rapid ranking of systems based on decay constants is possible if the accelerated exposure did not change the true mechanism of failure, but only "accelerated" it.

Acknowledgements

This work was performed with the support of the Office of Naval Research under Grant # N00014-95-10507, Dr. A.J. Sedriks, Program Manager; with the support of the Air Force Office of Scientific Research under Grant F49620-96-1-0284, Maj. H. DeLong and Maj. P. Trulove, Program Managers; with the support from a sub-contract from Boeing, Dr. J. Osborne, PI, the prime contractor for DARPA under contract # F33615-96-C-5078; and with the support from a subcontract with U. Missouri-Columbia, Prof. H Yasuda, PI, the prime contractor for DARPA under Contract # F33615-96-C-5055.

References

[1] Simpson, T.C., "Electrochemical Methods to Monitor Degradation of Organic and Metallic Coatings." **Proc. 12th International Corrosion Congress**, Vol. 1, p.157-170, Houston, TX (Sept 1993)
[2] Appleman, B.R., "Survey of Accelerated Test Methods for Anti-Corrosive Coating Performance", **J. Coatings Tech., 62, 787**, 57-67 (1990).
[3] Appleman, B.R., "Cyclic Accelerated Testing: The Prospects for Improved Coating Performance Evaluation", **J. Protective Coatings & Linings**, 71-79 (Nov. 1989).
[4] Reference needed – use P. Elliott work as example.
[5] Skerry B. S., D. A. Eden, "Electrochemical Testing to Assess Corrosion Protective Coatings", **Progress in Organic Coatings**, 15, 269 (1987).
[6] G.P. Bierwagen, "Reflections on Corrosion Control by Coatings," **Prog. Organic Coatings, 28 (1996)** 43-48
[7] J.R. Macdonald, Ed., *Impedance Spectroscopy*, Wiley, NY (1989)

[8] Scully, J.R., "Electrochemical Impedance of Organic-Coated Steel: Correlation of Impedance Parameters with Long-term Coating Deterioration", **J. Electrochem. Soc., 136(4)**, 979-990 (1989)

[9] Scully J.R.; Silverman D.C.; Kendig M.W., Eds., *Electrochemical Impedance*, Special Tech. Publ. 1188, ASTM, Philadelphia, PA, **1993**

[10] Murray, J.N., "Electrochemical test methods for evaluating organic coatings on metals: an update. Part II: single test parameter measurements," **Progress in Organic Coatings, 31** (1997) 255-264

[11] J..H.W. de Wit, "Inorganic and Organic Coatings," Ch. 16 in P. Marcus & J. Odar, ed. , *Corrosion Mechanisms in Theory and Practice,* Marcel Dekker, NY (1995) 581-627

[12] M. van Westing, G. M. Ferrari, F.M. Geenen, & J.H.W. de Wit, **Prog. Organic Coatings, 23** (1993) 89-103

[13] Eden, D.A, M.Hoffman, and B.S.Skerry, "Application of Electrochemical Noise Measurements to Coated Systems," **ACS Symposium Series, 322,** *Polymeric Materials for Corrosion Control*, R.A.Dicke & F.L.Floyd, eds., American Chemical Society, Washington, D.C (1986)

[14] D. J. Mills, G. P. Bierwagen, D.E. Tallman, & B.S.Skerry, "Charaterization of Corroison under Marine Coatings by Electrochemical Noise Methods," **Proc. 12th International Corrosion Congress**, Vol. 1, p. 182-193, (paper 486), Houston,TX (Sept. 1993)

[15] Bierwagen, G.P. "Calculation of Noise Resistance from Simultaneous Electrochemical Voltage & Current Noise Data," **J. Electrochemical Soc., 141** (1994) L155-L157

[16] J. Chen & W.F.Bogaerts, **Corrosion Sci., 37** (1995) 1839-1842

[17] G. P Bierwagen, V. Balbyshev, D. J. Mills, and D. E. Tallman, "Fundamental Considerations on Electrochemical Noise Methods to Examine Corrosion under Organic Coatings," *Proceedings of the Symposium on Advances in Corrosion Protection by Organic Coatings* II, The Electrochemical Society, Proceedings Volume 95-13, 1995

[18] U. Bertocci, C. Gabrielli, F. Huet, and M. Keddam, "Noise Resistance Applied to Corrosion Measuerements, Part I: Theoretical Analysis" **J. Electrochem. Soc., 144** (1997) 31-37

[19] G.P. Bierwagen, D.E.Tallman, S. Touzain, A. Smith, R. Twite V. Balbyshev & Y. Pae, "Electrochemical Noise Methods Applied to the Study of Organic Coatings and Pretreatment," *Paper 380, Corrosion 98,* 1998 Annual Meeting of the Nat. Assoc. Corrosion Eng. (NACE), NACE Int., Houston TX, March 1998

[20] J.Li, C.S.Jeffcoate, G.P.,Biewagen, D.J.Mills & D.E.Tallman, "Thermal Transition Effects and Electrochemical Properties in Organic Coatings: I. Initial Studies on Corrosion Protective Organic Coatings," **Corrosion, 54** (1998) 763-771

[21] G.P.Bierwagen*, C.Jeffcoate, D.J.Mills, J. Li, S. Balbyshev & D.E.Tallman, "The Use of Electrochemical Noise Methods to Study Thick, High Impedance Coatings," **Prog. Organic Coatings, 29** (1996) 21-30

[22] R.Granata, & K.Kovaleski, "Hydrothermal Properties of Protective Polymer Coatings on Steel," *Proc. 13th Int. Congress on Corrosion*, Houston, TX. Sept 1993, Vol. 1 (Coatings) p. 24; and R.Granata, R.McQueen & K.Kovaleski, "Polymer

Coatings Degradation Mechanisms Related to Hot Production," *Proc. 13th Int. Congress on Corrosion*, Houston, TX. Sept 1993, Vol.5 (Oil & Gas Production) p. 2612

[23] E.Kuwano, T.Fujitani & T.Satoh, "A New Approach to the Anti-Corrosion Function of Coatings," *Extended Abstracts of the Int. Symposium on Advances in Corrosion Protection by Organic Coatings,* Oct.29-31, Noda, Japan, Abstract 111

[24] Z. Wicks, S.P.Pappas, & F.N.Jones, *Organic Coatings*, Vol II. Wiley & Sons, New York(1994) Ch. 27, p. 181

[25] Private communications, D. Scantlebury, UMIST, Manchester, UK April 1996; W. Funke, U. Stuttgart, Stuttgart, Germany, April 1995; D. Mills, U. Nene, Nene, UK, April 1996

[26] R. A. Fraser, "Painting of Railroad Bridges and Structures," ch 10 in J. Keane, ed., *Good Painting Practice, Vol. 1,* 3rd Ed., Steel Structures Painting Council (SSPC), Pittsburgh, PA (1994) pp263-279

[27] C. H. Hare, *Protective Coatings – Fundamentals of Chemistry and Composition*, Technology Publishing Company, Pittsburgh, PA, 1994.

[28] J.D.Keane, W. Wettach, & W. Bousch, "Minimum Paint Film Thickness for Economical Protection of hot-Rolled Steel Against Corrosion," **J. Paint Tech.**, 41(#533)(1969) 372-382

[29] A.R. Di Sarli, E.E. Schwiderke & J.J. Podesta, "Evaluation of Anticorrosive Paint Binders by Means of AC Techniques - Influence of Coat Thickness," *Proc. 10th Int. Cong. Metallic Corr., Madras, India* (1987) Section 6.9, p. 1091-1100 Karaikudi, CERI

[30] D.J.Mills & J.E.O.Mayne, in H. Leidheiser, ed., *Corrosion Control by Organic Coatings*, NACE. Houston, TX (1981) p 12

[31] A.Amirudin & D. Thierry, "Comparative Evaluation of Alkyd, Bituminous, and Epoxy Paints on Steel in Chloride Media by Impedance Spectroscopy," **British. Corr. J.**, 26 (1991) 195-201

[32] J.R.Scully, "Electrochemical Impedance of Organic-Coated Steel: Correlation of Impedance Parameters with Long -Term Coating Deterioration," **J. Electrochem. Soc.**, 136 (1989) 979-990

[33] S. F. Mertens, C. Xhoffer, B. C. De Cooman, & E. Temmerman, "Short-Term Deterioration of Polymer-Coated 55% Al-Zn – Part I: Behavior of Thin Polymer Films," **Corrosion, 53**, 5 (1997).

[34] R. Babic, M. Metikos-Hukovic, & H Radovcic, "The Study of Coal Tar Epoxy Protective Coatings by Impedance Spectroscopy," **Prog. Organic Coatings, 23** (1994) 275-286

[35] J.N.Murray, "Evaluation of Thick Cast Polyurethane coatings on Steel using EIS and DC Electrochemical Techniques," in *Proc. of the Symposium on Advances in Corrosion Protection by Organic Coatings II*, Special Publication of The Electrochemical Society, Proceedings Volume 95-13, D.Scantlebury & M.Kendig, editors, (1995) 69-81) pp.132-150

[36] S. Haryuma, M. Asari, & T. Tsuru, in M. Kendig & H. Leidheiser, Jr., eds., *Corrosion Protection by Organic Coatings,* The Electrochemical Society, Proceedings, (1987) pp. 11-22

[37] D. Halliday & R. Resnick, *Physics,* Wiley, New York (1960) Ch. 31, p.658

[38] see for example J.R. Macdonald, ed., *Impedance Spectroscopy,* Wiley, NY (1989)

[39] G. P. Bierwagen, D. J. Mills, D. E. Tallman, and B. S. Skerry, "Reproducibility Analysis of Electrochemical Noise Data for Coated Metal Systems," *ASTM Special Technical Publication STP 1277, Electrochemical Noise Measurement for Corrosion Applications,* ASTM, 1996.

[40] G. P.Bierwagen, "Calculation of Noise Resistance from Simultaneous Electrochemical Voltage and Current Noise Data," **J. Electrochem. Soc.,** 141 (1994).

[41] a. D.J.. Mills, S. Berg, & G.P. Bierwagen, "Characterization of the Corrosion Control Properties of Organic Electrodeposition Coatings," in *Proc. of the Symposium on Advances in Corrosion Protection by Organic Coatings II,* Special Publication of The Electrochemical Society, Proceedings Volume 95-13, D.Scantlebury & M.Kendig, editors, (1995) 69-81

b. D. J. Mills, G. P. Bierwagen, D.E. Tallman, & B.S.Skerry, "Investigation of Anticorrosive Coatings by the Electrochemical Noise Method," **Material Perf., 34,** (1995) 33

[42] Elcometer, Inc., *Electromagnetic Induction Principle (Ferrous Gauges),* Rochester Hills, Michigan, 1997.

[43] T. Cunningham, "Measuring Dry Film Thickness Using Electromagnetic and Eddy Current Gauges," *Material Performance,* 5, 39 (1995).

[44] G. P. Bierwagen, "Calculation of Noise Resistance from Simultaneous onElectrochemical Voltage & Current Noise Data," **J. Electrochem. Soc.,**141(1994) L155-L1957

[45] B. H. Dougherty & S. I. Smedley, "Validation of Experimental Data from High Impedance Systems Using The Kramers-Kronig Transforms," in *Electrochemical Impedance – Analysis and Insterpretation,* J.C. Scully, D.C Silverman & M. W.Kendig, eds., STP 1188, ASTM, Philadelphia (1993) p. 154

[46] G.P. Bierwagen, D.E.Tallman, S. Touzain, A. Smith, R. Twite V. Balbyshev & Y. Pae, "Electrochemical Noise Methods Applied to the Study of Organic Coatings and Pretreatment," *Paper 380, Corrosion 98,* Reviewed paper for the 1998 Annual Meeting of the Nat. Assoc. Corrosion Eng. (NACE), NACE Int., Houston TX, March 1998

Chapter 17

Modeling and Laboratory Simulation of Grain Cracking of Latex Paints Applied to Exterior Wood Substrates

F. Louis Floyd

PRA Laboratories, Inc., 430 West Forest Avenue, Ypsilanti, MI 48197

Introduction and Background

All architectural coatings on exterior wood substrates will eventually crack. If they have deficient adhesion, they will subsequently exhibit peeling and flaking. If adhesion is adequate, no further failures will be noticed, and cracking may only be noticed under close inspection.

There are several kinds of cracking exhibited by architectural coatings applied to exterior wood substrates:

Initial (seen within a day of application):

- If paints are formulated at too high solids, they can exhibit mud-cracking, which results from premature surface drying and splitting of the surface "skin." Such cracks do not usually extend through the thickness of the paint.
- If a latex is too high in Tg, has a deficient coalescent level, or uses an inappropriate coalescent type, cracking due to deficient film-formation can occur. These cracks do extend through the thickness of the film

Long term (seen months to years after application):

- Shrinkage cracking (also called checking, aligatoring), caused by residual unreacted crosslinking capability that continues to react, albeit slowly; characterized by cellular pattern in the cracking;
- Grain cracking, which mimics the grain pattern of the wood substrate.

Only grain cracking has an external cause. All other forms of cracking have internal (to the paint) causes. This paper focuses on grain cracking.

Why does grain cracking occur?

Wood is anisotropic. It expands and contracts as the environment goes through wet/dry, freeze/thaw, and hot/cold cycles. But unlike metals, or other homogeneous materials, wood does not swell and contract uniformly in all three dimensions. That which we call "grain" is actually the pairing of summer and winter growth rings in the wood. The summer grain is typically wider, lighter in color, lower in density, and more porous than the winter grain. The summer grain also swells and contracts more than the winter grain. This sets up a differential dimensional change in 3 dimensions, that create cyclic stresses in paint films applied to the wood substrate. Grain cracking is primarily a physical (mechanical) property of paint films, rather than a chemical or photo-chemical property, and is generally believed to be related to the initial mechanical integrity of a paint film. [1]

Early workers believed that tensile properties controlled cracking. Jaffe & Fickenscher[2] showed data suggesting that elongation-at-break was the main property controlling cracking. Schurr, Hay, and van Loo[3] on the other hand, suggested that changes in tensile strength during aging were influencing the process. Work-at-break (area under the stress-stain curve) was implied to be significant, but not explicitly claimed by either group.

Yamasaki[4] thought that permeability was an important factor, at least for alkyds, and conducted wet/dry cycling tests that produced results that seemed visually similar to cracking failures observed in the field. However, no mathematical correlations were attempted; no attempt was made to relate cracking to material properties; and the acceleration factor he obtained relative to a actual exterior exposure was small.

It is commonly believed that adhesion promoters enhance the crack resistance of latex systems, although the author is not aware of any studies that have been published in support of this belief. It is possible that any perceived improvement may relate more to the visual impact of peeling, rather than enhanced cracking resistance per se. It is also possible that adhesion promoters promote transfer-to-polymer reactions during polymerization, which in turn might account for the perception that adhesion promoters enhance crack resistance. (see following)

It is well-documented by latex suppliers that latex systems have substantially better crack resistance than alkyds, with pure acrylics being better than styrene acrylics and vinyl acrylics. This follows a general molecular weight order of acrylics >> styrene acrylics ~ vinyl acrylics >> alkyds. Moon and Baker[5] (among others) have shown a strong molecular weight effect on the fracture resistance of polymers, even well above the critical chain entanglement molecular weight.

However, the reader should exercise caution with such broad conclusions, since individual examples of the various classes of latex can perform well outside the range normally assumed for its class, and can also have quite different molecular weight distributions than assumed for its class. For example, the present work does not show the presumed clear-cut advantage of pure acrylics over styrene/acrylic and vinyl/acrylic latexes.

Reid[6] applied fracture mechanics (see next section) to brittle furniture lacquers, while seeking an alternative to the Weibull distribution to describe observed cracking behavior. He related cold checking (hot/cold cycling) to fracture energy by visual inspection of data plots. He stopped short of claiming a cause-effect relationship. Instead, he focused on the shortcomings of various laboratory temperature cycling tests.

Fracture mechanics

Griffith[7] was the first to view cracking as a material process, although his work was not picked up in other technological areas for decades. His interest was in the several orders of magnitude discrepancy between theoretical strength of materials based on chemical bond energies and their measured strengths. Prior to Griffith, the prevailing assumption was that all bonds support an equal load, and that all bonds rupture simultaneously, thus requiring that the crack velocity be infinite. In fact, Griffith pointed out that crack velocity is finite and fracture is sequential through a material.

His explanation was that an applied stress will concentrate at the random flaw sites that exist in all materials, where it can easily exceed bond energies, resulting in crack propagation. The wide data scatter normally seen in stress-strain testing is due to the random distribution of initial flaw sizes. This random distribution of flaws also accounts for the inverse dependence of strength on sample size; i.e. the larger the sample cross-section, the more likely that it will contain a critical (large) flaw.

One implication of this flaw-induced stress concentration model is that the largest random flaw dominates the process, by becoming the critical flaw that ultimately leads to fracture. Griffith showed that by intentionally introducing flaws into materials, he was able to control and study this process as a function of initial flaw size.

Griffith determined that the force required to cause crack propagation is proportional to the size of the initial flaw and governed by the following equation:

$$\sigma_B = [\, 2\, E\, \gamma\, /\, \pi\, A\,]^{\frac{1}{2}} \qquad (1)$$

where σ_B = tensile strength
$\quad$ E $\;$ = Young's modulus
$\quad$ γ $\;$ = fracture energy (FE)
$\quad$ A $\;$ = ½ initial flaw size (induced manually).

To correct for edge effects of narrow samples used in tensile testing, Brown & Strawley[8] introduced the following correction factor:

$$Y = 1.99 - .41\,(A/W) + 18.7\,(A/W)^2 - 38.48\,(A/W)^3 + 53.85\,(A/W)^4 \quad (2)$$

where $\;$ Y = compensation factor for edge effects
$\quad$ W = specimen width.

This correction factor is valid over the range of $0 < (A/W) < 0.6$ or for initial flaws that range up to 60% of the sample width.

This resulted in the following final form of the Griffith equation which is used experimentally:

$$\sigma_B = [\, 2\, E\, \gamma \, / \, A\, Y^2 \,]^{1/2} \qquad (3)$$

For elastic materials (with negligible viscous component),

$$W_B = \tfrac{1}{2}\, \sigma_B\, \lambda_B \qquad (4)$$

Where W_B = work-at-break
     σ_B = stress-at-break (force)
     λ_B = strain-at-break (elongation)

Rearranging and substituting into the Griffith equation (3), results in the following expression:

$$W_B = \gamma \, / \, A\, Y^2 \qquad (5)$$

Thus, Griffith showed that a plot of work-at-break vs. reciprocal initial flaw size resulted in a straight line, the slope of which was the fracture energy. Fracture energy is a property of the material itself, independent of geometry, and has the units of energy required to generate a unit of new surface area.

Focus of Present Work

The goals of the present work were: (1) Develop a laboratory test that simulates exterior cracking with sufficient accuracy to permit its use in lieu of exterior exposure testing. (2) Determine whether fracture mechanics can be used to predict the grain-cracking behavior of latex paints, based on tensile testing as described by Griffith.

Experimental

Test panel preparation:
- 8 in. x 18 in. Douglas fir plywood substrate (for lab testing)
    - 2 paints per panel: 1 experimental plus control
- 6 in. x 36 in. Douglas fir plywood (for exterior testing)
    - 5 paints per panel: 4 experimental plus one control
- control: 1st line exterior latex flat from major manufacturer

- locations on panel are randomized
- applied by brush, 2 coats (i.e. self-primed)
- 24 hr dry between coats
- backs and edges sealed with an alkyd barrier primer
- 1 week final dry before commencing testing
- replication error is derived from control paint present on each panel

Laboratory simulation test:
- Soak painted panels face down in water for 1 hour
- Remove panels from water and place in panel rack
- Freeze ($<0°$ F) wet panels overnight (16 hours)
- Thaw and dry panels at ambient temperature (7 hours)
- One cycle takes 24 hours; panels are left at ambient temperature over weekends
- Run test for 30 cycles
- Evaluate for cracking (ASTM D-661-93)
- Normalize for panel control (see following discussion)
- Report mean and standard deviation of control paint as error of test.

Exterior exposure
- South $45°$, northern Ohio location
- 6 months' duration, initiated in September
- Evaluate for cracking (ASTM D-661-93)
- Normalize for panel control (see discussion for explanation)
- Report mean and standard deviation of control paint as error of test

Tensile testing
- Cast 10 mil dry paint films on polypropylene sheet
- Detach and cut into strips 0.75 in. wide by 2 in. long
- Cut initial flaw into one side of specimen with scalpel, and measure actual flaw length under low power magnification (~20X). Actual initial flaws cut ranged from 0.02" to 0.30".
- Pull to failure in Instron tensile tester (5 in./min; 1 in. initial gauge length)
- Record stress and strain at break, and calculate work (area under stress/strain curve – a planimeter was used in this work).
- Note that replicates *per se* cannot be run, since it is impossible to perfectly reproduce the initial flaw cut length. Instead, 6 different initial flaw lengths were run to produce the desired plot from which fracture energy is calculated (or read).

Permeability

- Soak filter paper in a 3% aqueous solution of $CoCl_2$. Dry filter paper in 105°C oven, till it changes from pink (wet) to blue (dry).
- Place blue indicator paper on plate glass. Add 2 in. x 2 in. section of paint film cast as for tensile testing. Add 6 in. x 6 in. piece of electroplating tape into which has been cut a ½ in. x ½ in. hole. Press tape firmly to rest of assembly to assure complete bonding.
- Place assembly into tray containing ambient tap water with glass side up. Suspend glass plate on thin shims (e.g. bottle caps) so sample assembly is not touching bottom of tray.
- Measure the time (stopwatch) it takes from first immersion till the indicator paper changes from blue to pink.
- Run 6 replicates, and report mean value. Note that the time-to-change (permeability time) is inverse to most measures of permeability: larger numbers represent lower permeability.

Paint samples employed. All paints were prepared as PVC (pigment volume concentration) ladders, with a constant 20% PVC in TiO_2, constant 36% volume solids, and varying amounts of extender to achieve the indicated PVC's for each group. Coalescent level (Texanol ®) was adjusted to achieve a knife point minimum film formation temperature (MFT) of 10°C for all cases.

- Binder variations (24, 32, 40, 48, 56 PVC)
  - o BA/MMA, single mode, HEC stabilized
  - o BA/MMA, bimodal, high solids, surfactant stabilized
  - o VA/BA, broad distribution, colloid stabilized
  - o BA/S, single mode, surfactant stabilized
  - o Extender: calcined clay, silica
- Pigment variations (40, 48, 56, 64 PVC; 2 variations: 30% and 36% volume solids)
  - o Talc, 1 – 2 microns
  - o Talc, 10 – 30 microns
  - o Clay, 0.5 microns
  - o Mica, 40 microns
  - o Binder latex: BA/MMA, single mode, HEC stabilized

Weather-Ometer

- 3 in. x 6 in. polypropylene (pp) panels
- carbon arc with Corex D filters
- 1000 hours, standard cycle
- films cut to size, detached from pp substrate, and tested for fracture energy

Results and Discussion

Lab Simulation of Exterior Cracking

Normalization. Grain-cracking of paints applied to wood substrates is an inherently variable process, due to the variability in both the wood substrate and the painting process. As the USDA Forest Products Laboratory reminds us, wood itself is a highly variable material, even within the same lot of boards. [9] This variability is compounded by application techniques such as brushing, rolling, and spraying that generate less than uniform and perfect films. Not only can average film thickness vary over a panel, but there can also be variations on a small scale sufficient to influence cracking behavior (thinner sections crack more readily than thick sections).

One solution to this variability problem is to employ the technique of normalization of test results, based on the presence of a control paint on each board being rated. Normalization is a relatively simple process, and can be expressed mathematically as

| individual test paint rating | − | control paint rating on same panel | + | mean of all control paint ratings | = | normalized test rating | (6) |

By subtracting the control paint rating from the test paint section, one obtains a relative value for cracking on that particular piece of wood. By adding back the mean of the control paint ratings over all panels in the study, one obtains a non-relative value once again, which can be used to compare to other test paints in the study, including those on different pieces of wood. In short, this filters out the effect of the variability of the wood test substrate from the cracking results.

The reader should realize that implicit in this process is the assumption that the paint application technique is the same for each test section and for each control section. Given that such studies are typically conducted by skilled professionals, that is probably a reasonable assumption, provided that the area painted is large enough to be representative, as was the case in the present study.

Validity of laboratory simulation of cracking. A comparison of the normalized cracking behavior obtained by the accelerated laboratory test (described above) and that obtained by exterior testing shows excellent correlation ($r = .85$ on average), regardless of data sets employed. [see Tables I and II, and figures 1 and 2] Visual inspection of the panels showed that the type, magnitude, and relative degree of cracking were strikingly similar for the two test conditions. It was therefore concluded that it would be adequate in the future to perform the laboratory test as an accurate prediction of actual exterior performance. The time required for lab testing is approximately 1/5 the time required for a South 45° exterior exposure test.

Table I. Binder Variations

Binder latex	PVC	Fracture Energy	Lab test Cracking	Exterior Cracking
acrylic	24	8.59	9	9
	32	5.07	9	9
	40	2.55	7	7
	48	1.3	4	4
	56	0.97	4	4
high solids acrylic	24	16.06	7	8
	32	5.56	7	7
	40	3.16	3	4
	48	1.61	2	3
	56	0.95	2	2
vinyl acrylic	24	16.38	9	9
	32	11.04	9	7
	40	6.41	5	7
	48	1.78	3	3
	56	0.51	2	2
styrene acrylic	24	15.21	10	7
	32	5.53	9	6
	40	2.69	4	6
	48	1.28	3	4
	56	0.79	2	5

correlations:	r
PVC vs. fracture energy	(0.87)
PVC vs. lab cracking	(0.89)
PVC vs. exterior cracking	(0.89)
Fracture energy vs. lab cracking	0.77
Fracture energy vs. exterior cracking	0.73
Lab cracking vs. exterior cracking	**0.86**

Table II. Pigment Variations

Pigment	% Volume Solids	PVC	Fracture Energy	Lab Test Cracking	Exterior Cracking	Permeability Time (minutes)
talc, 1-2 micron	30	40	31.58	7	6	14
		48	13.2	5	5	17
		56	6.3	3	4	14
		64	2.3	3	3	9
	36	40	9.96	8	9	7
		48	4.08	4	7	7
		56	3.54	2	4	4
		64	1.05	2	3	nd
talc, 10-30 microns	30	40	14.7	5	7	12.5
		48	8.8	5	6	7.1
		56	3.6	3	2	4
		64	2.2	3	2	2
	36	40	13.7	8	9	9
		48	5.4	6	8	7.5
		56	2.2	5	4	5
		64	1.1	4	3	nd
clay 0.5 microns	30	40	20.51	2	4	11
		48	10.03	2	3	7
		56	3	2	3	2
		64	1.25	2	2	1.3
	36	40	5.74	6	2	6
		48	3.38	2	2	9
		56	1.89	2	2	1.2
		64	1.35	2	2	nd
mica, 40 micons	30	40	6.82	8	9	23
		48	4.59	7	9	23
		56	2.2	7	7	15
		64	1.62	4	6	4
	36	40	6.01	7	10	21
		48	2.46	7	9	22
		56	1.77	6	7	7
		64	1.02	3	5	2

correlations:
PVC vs. fracture energy (0.68)
PVC vs. lab cracking (0.61)
PVC vs. permeability (0.54)
fracture energy vs. lab cracking 0.30
fracture energy vs. permeability 0.29
lab cracking vs. permeability 0.62
lab cracking vs. exterior cracking **0.84**

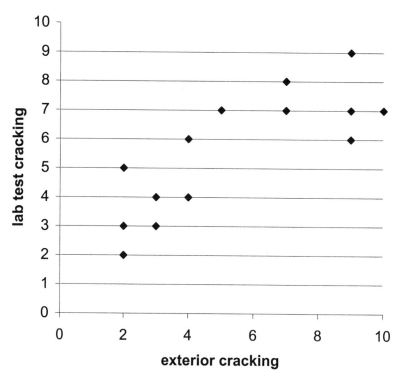

Figure 1
Lab vs. Exterior Cracking
Binder Variations

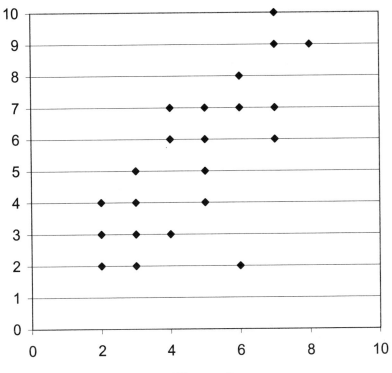

Figure 2
Lab vs. Exterior Cracking
Pigment Variations

The error of *normalized* cracking results is on the order of +/- 1 rating unit, which means that differences in test systems less than this amount are not statistically significant. Without normalization, the error is more on the order of +/- 3 units, which would nullify any attempt to relate causes and effects.

Predicting Grain Cracking from Material Properties

Binder variations. The first series of paints examined contained variations in the binder latex and pigment loading (PVC), the assumption being that fracture energy would be most influenced by these two variables. As one can see from Table I, as PVC increases, fracture energy decreases and cracking increases. The calculated correlations between PVC and cracking and fracture energy are highly negatively correlated (r = -0.87 to -0.89).[10] This provides a useful reality check for this experiment.

Also strongly significant is the correlation between fracture energy and cracking: r = 0.77 for lab cracking and 0.73 for exterior cracking [figures 3 and 5}. Meanwhile, the correlation between lab cracking and tensile strength was only 0.16, and between lab cracking and elongation-at-break was only 0.12, neither of which is statistically significant. The author therefore concludes that contrary to the findings of earlier workers, tensile properties are not adequate predictors of cracking tendencies of latex paints, while fracture energy is (at least for this data set).

Pigment variations. Table II contains the data for a study that kept binder constant and varied pigmentation, in an attempt to discover the effect of different sized and shaped extender pigments on cracking behavior. As the reader can see, different extenders do indeed produce quite different results, whereas the binder variations discussed above show a relatively smaller effect. Again, the PVC effect was quite noticeable, although the correlations were a bit lower.

Contrary to the finding of the first series, the crack ratings were *not* predicted by fracture energy [figures 4 and 6]. The regression analysis provided very low correlations of 0.3 and 0.24, with non-significant T values on 32 data points.

However, it was observed that exterior crack ratings were correlated to permeability as demonstrated in Table III and figure 7 with a correlation coefficient of 0.62, with a T value of 4.12 on 29 data points (confidence level: 99 %). This suggested that a multiple-parameter model may well be more appropriate than the single-parameter model.

Additional calculations bore out this speculation. In Table 3, it can be seen that the best correlation to cracking was the combination of permeability and log [fracture energy]. This yielded a correlation of 0.73, with a confidence level in excess of 99%, based on 29 data points.

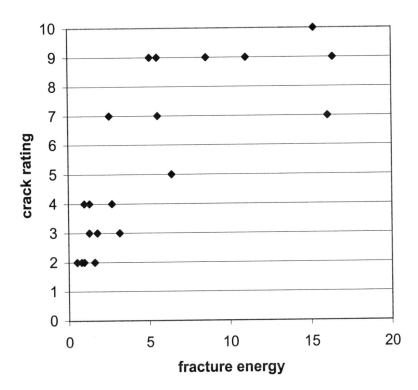

Figure 3
FE vs. Lab Cracking
(binder variations)

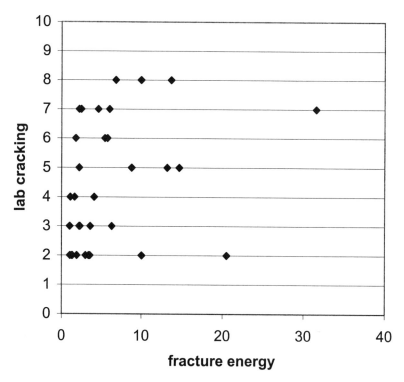

Figure 4
FE vs. Lab Cracking
(pigment variations)

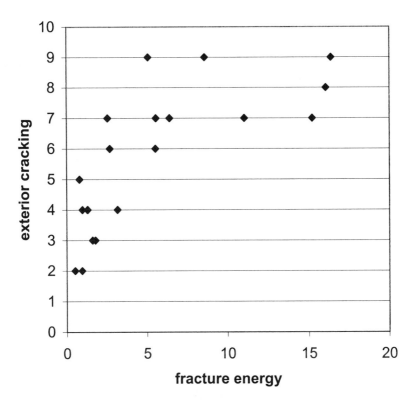

Figure 5
FE vs. Exterior Cracking
(binder variations)

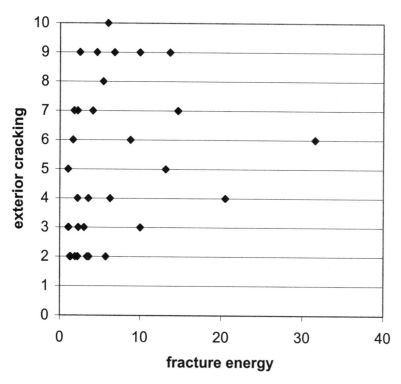

Figure 6
FE vs. Exterior Cracking
(pigment variations)

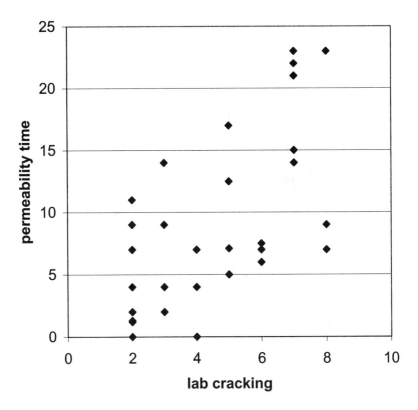

Figure 7
Cracking vs. Permeability

Table III. Summary of Correlations

Data Set	Sample Size	Independent Variable	Dependent Variable(s)	Correlation Coefficient	Significant at 99% Conf ?
Binder Variations	17	lab cracking	exterior cracking	0.86	yes
			fracture energy	0.77	yes
			tensile strength	0.16	no
			elongation-at-break	0.12	no
Pigment Variations	32	lab cracking	exterior cracking	0.84	yes
			fracture energy	0.30	no
			permeability	0.62	no
			log [fracture energy], permeability	0.73	yes

Note: confidence = 1 - significance

Permeability measurements were subsequently made for the first study, but this parameter was not significant. It therefore appears that permeability is largely controlled by pigment type, while fracture energy is jointly determined by binder type and pigment level. It is such behavior which leads to multiple parameter models. It also suggests why earlier workers drew different conclusions: their range of variables was much smaller.

A reasonable conclusion at this point is that while the binder choice strongly influences the fracture energy of a paint, the choice of pigmentation type strongly influences the permeability of a paint. Thus, a 2-factor model adequately covers both issues, without needing to resort to any information about the actual composition of the paint.

Effect of Artificial Weathering on Fracture Energy

In most of this work, the free film samples evaluated for fracture energy (FE) were aged six days at ambient conditions and one day at 70°F and 50% RH prior to testing. In order to evaluate the effect of weathering on the relative values of fracture energy, a series of paints were tested for FE after 500 and 1000 hours of exposure in a carbon arc Weather-Ometer with Corex D filters.

The paints employed were the all acrylic paints from the binder variation study. Figure 8 illustrates the surprisingly minimal effect that 1000 hours of exposure had on the intrinsic fracture energy of the samples. While there does appear to be some fall-off in FE after 1000 hours, the rank ordering of the samples was preserved. This lack of variation may well be due to the choice of the relatively durable acrylic binder, and may not hold true for less durable systems, or for systems having residual unreacted functionality, such as alkyds.

This result may help to explain why fracture energies measured on unexposed samples can correlate with the cracking of samples exposed either artificially or naturally. It is certainly tempting to reassert once again that cracking is a function of the mechanical integrity of a paint film, rather than its resistance to UV degradation. However, it must be acknowledged that a similar experiment performed with less durable alkyd paints may have produced different results.

Suggestions for future work

- For anyone following on from this work, it is suggested that logical extensions would be to check for validity for (1) alkyd paints, (2) unsealed wood substrates (no edge or back sealer), and (3) primer/topcoat pairs including alkyd/latex.

- Although elastic behavior was assumed in the mathematical derivations, the samples evaluated in this work were viscoelastic in behavior. One

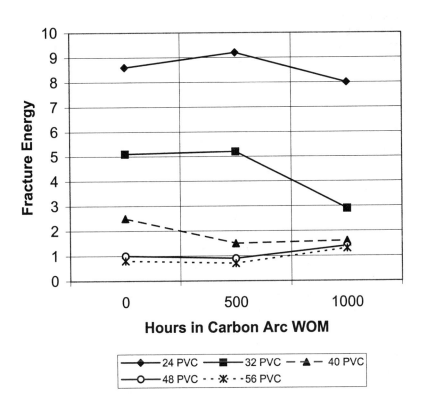

Figure 8
Effect of UV Exposure on FE

alternative approach would be to *not* apply the elastic behavior assumptions of equations 1 to 5, but instead to obtain the fracture energy directly from the slope of a graph of "σ^2" vs. "$2E/AY^2$" as flaw size is varied [see eqn. 1]. With today's equipment and computers, this is a simple matter. At the time this work was originally done, it was not.

- It would be interesting to learn if cracking behavior is influenced by residual internal residual stress in paint films. The work of Perera[11] of the Belgium Coatings Research Institute and Croll[12] of the National Research Council of Canada (now at North Dakota State University) is worth reviewing by anyone desiring to follow this line of reasoning.

Proposed Mechanism for Grain Cracking

As a result of the accumulated data and observations, the following view of the grain cracking process is offered:
- Differences in coefficients of expansion along grains of the wood substrate generate cracks in the wood itself, during exposure to weather cycles.
- As cracks in the wood propagate or widen, paint films bridging the cracks become highly stressed.
- Inclusions in the paint film generate different stress distributions in different parts of the film. These inclusions may consist of voids, pigments, microcracks, etc. They become stress-concentrators when the film is stretched across the wood crack. Cracks are initiated at these sites of stress-concentration.
- As further stress is applied to the film, these cracks will propagate until the strain energy is consumed in the generation of new surface area.
- The rate of *initiation* of cracks is related to the permeability of the paint film to water. A film with low permeability to water will protect the wood substrate from swelling, hence minimizing the initiation step in cracking.
- The rate of crack *propagation* (generation of new surfaces) is related to the surface energy of the composite; i.e. the fracture energy. A composite with low fracture energy propagates cracks more easily.
- Overall cracking behavior of a given paint film will depend on its intrinsic permeability and fracture energy.

Summary

- An accelerated laboratory test has been designed that reasonably predicts the cracking behavior of latex paints applied to exterior wood surfaces.

- Statistical analysis is required of such data since experimental error tends to be large, and correlations of multiple parameter models are not obvious by inspection. Normalization procedures were introduced to minimize the impact of substrate variation in cracking tests.
- Single parameter models are inadequate for describing cracking behavior. In the present case a 2-parameter model involving fracture energy and permeability was required to account for cracking behavior.

Acknowledgments

This work is adapted from an earlier publication by the same author in the *Journal of Coatings Technology*.[13] The author thanks the Federation of Societies for Coatings Technology, publishers of the *JCT*, for their kind permission to publish this adaptation of that earlier work.

The author gratefully acknowledges the contributions of Dr. C. L. Leung and Messrs. J. J. Beauregard, F. A. Wickert, S. S. Shepler, and R. L. Groseclose for their experimental contributions to this work.

The author also thanks the former Glidden Division of SCM Corp. (now known as ICI Paints North America) for its support of this work and permission to publish these results.

Abbreviations Used

Aka	also known as
BA	butyl acrylate
MMA	methyl methacrylate
S	styrene
VA	vinyl acetate
MFT	minimum film formation test
FE	fracture energy
PVC	pigment volume concentration
VS	percent solids by volume
pp	polypropylene

References

[1]Williams, Knaebe, and Feist, **Finishes for Exterior Wood,** Forest Products Society, 1996. ISBN 0-935018-83-2

372

[2] Jaffe & Fickenscher, "Stress Strain Measurements of PVA Films as Correlated with Natural Exterior Exposure," *Official Digest*, **33**(434), 1961, p. 331.

[3] Schurr, Hay, and van Loo, "Possibility of Predicting Exterior Durability by Stress/Strain Measurements," *J. Paint Technology*, **38**(501), 1966, p. 591.

[4] Yamasaki, "Cracking and Flaking of Exterior Oil House Paint on Wood: Laboratory Simulation to Support a Mechanism," *J. Paint Technology*, **44**(568), May 1972, p. 68.

[5] Moon and Baker, *J. Polymer Science: Polymer Physics*, **11**, 1973, p. 909.

[6] Reid, "A Fracture Mechanics Approach to Lacquer Cracking," *J. Oil & Colour Chemists Association*, **59**, 1976, p. 278.

[7] Griffith, *Philosophical Transactions*, A **221**, 1920, p. 163.

[8] Brown and Strawley, *ASTM STP 410*, 1966.

[9] **Wood Handbook: Wood as an Engineering Material**, a compendium of papers generated by the Forest Products Laboratory of the USDA Forest Service; printed by the Forest Products Society in 1999. ISBN 1-892-529-02-5.

[10] Terminology note: confidence = 1 − significance. Statisticians usually speak in terms of significance, while experimentalists tend to speak in terms of confidence. Thus, a .01 (or 1%) significance level is the same as a .99 (or 99%) confidence level.

[11] Perera and Eynde, *J Coatings Technology* :
  (a) **53** (677), June 1981, p.39;
  (b) **53** (678), July, 1981, p. 40;
  (c) **55** (699), April 1983, p. 37;
  (d) **56** (716), September 1984, p. 111;
  (e) **56** (717), October 1984, p. 47;
  (f) **56** (718), November 1984, p. 69;
  (g) **59** (748), May 1987, p. 55.

[12] Croll, *J. Coatings Technology* :
  (a) **53** (672), January 1981, P. 85;
  (b) **52** (665), June 1980, p. 35;
  (c) **51** (659), December 1979, p. 49;
  (d) **51** (648), January 1979 p. 64;
  (e) **50**, (638), March 1978, p. 33.

[13] Floyd, "Predictive Model for Cracking of Latex Paints Applied to Exterior Wood Surfaces," *J. Coatings Technology*, **55**(696), January, 1983, p. 73.

Chapter 18

The Effect of Weathering on the Fracture Resistance of Clearcoats

M. E. Nichols and J. L. Tardiff

Materials Science Department, Ford Research Laboratory, 2101 Village Road, Dearborn, MI 48121

The effect of weathering on the fracture energy of eight automotive clearcoats, both isolated clearcoat and as clearcoats in complete paint systems, was studied. All the clearcoats underwent photooxidation, as measured by infrared spectroscopy, as weathering progressed. In addition, all of the clearcoats embrittled, as measured by changes in their fracture energy. For many of the coatings, the changes in the fracture energy mirrored the chemical composition changes brought on by weathering. However, some clearcoats embrittled at rates that were independent of the rate of photooxidation. The presence of stabilizers, ultraviolet light absorbers and hindered amine light stabilzers, also effected the fracture energy of the clearcoats, both initially and as weathering progressed. Weathering-induced changes in the fracture energy of the clearcoats were related to weathering-induced changes in the chemical composition.

The long-term weathering behavior of automotive topcoats is of prime concern to both automotive manufacturers and consumers. Consumers appreciate coatings with better long-term appearance retention, while automotive manufacturers demand equal or superior weathering performance in addition to better processability, initial appearance, and economic value. These often conflicting requirements must be balanced against the paramount concern of increasingly stringent environmental emissions regulations, which may force dramatic changes in future topcoat systems technology. Reliable, rapid weatherability testing will be crucial to the timely adoption of new technologies, as long-term weathering testing is by definition a time consuming process. Traditional weatherability testing protocols are well suited to monitor the gradual changes in appearance that were the hallmark of monocoat paint systems. Monocoats tended to lose gloss and distinctness of image (DOI) gradually. Superior coatings were those that lost gloss and DOI most slowly. Thus, monitoring gloss and DOI were simple and useful means for assessing the weatherability of monocoat paint systems. The transition to basecoat/clearcoat systems fundamentally changed the manner in which weathering was manifested in automotive coatings and subsequently the tests needed to differentiate superior from inferior coating systems. In basecoat/clearcoat systems, gloss and DOI can remain high for long periods of time, after which failure can occur quickly and with little warning by either cracking, delamination, or a combination of the two. Thus, using gloss and DOI measurements alone to assess the long-term weathering behavior of basecoat/clearcoat systems is insufficient.

Much of the recent innovation in paint weathering research has focused on detecting small changes in the chemical composition of clearcoats in the early stages of weathering, either natural or accelerated exposure. These chemical changes are mainly due to photooxidation and to a lesser extent hydrolysis. The ability to measure these small changes allows a degree of anticipation: those clearcoats undergoing a higher rate of chemical composition change early in the weathering process are likely to have the shortest useful life. The methods by which these small chemical changes have been measured are quite varied, and have recently been reviewed by Gerlock et. al.[1] For example, infrared spectroscopy can be used to monitor both the accumulation of photooxidation products and the rate of film erosion.[2,3] In addition, microscopic infrared spectroscopy can be used to measure the spatial distribution of degradation through the thickness of the paint system.[4] Time-of-flight secondary ion mass spectrometry (TOF-SIMS) can also be used to map the locus of photooxidation in automotive paint systems. The TOF-SIMS technique relies on weathering the paint system in the presence of $^{18}O_2$ to label the photooxidation products and make them detectable in the TOF-SIMS.[5-7]

Because the effects of additives, ultraviolet light absorbers (UVAs) and hindered amine light stabilizers (HALS), can be important to the long-term weathering performance of paint systems, much work has been directed at measuring their longevity and effectiveness. UV spectroscopy has proved useful in measuring UVA loss rates.[8] Recently, the distribution of UVAs in weathered clearcoats has been measured by UV micro-spectroscopy, thus showing that UVAs are not depleted uniformly in clearcoats and both posing and answering new questions about additive migration in paint systems.[9] HALS longevity has also been recently assessed by peracid oxidation using electron spin resonance (ESR) to quantify nitroxyl radical concentration.[10-11] While powerful, the underlying assumption that is made when using any of these techniques to anticipate the weathering performance of coatings is that the amount of chemical change that occurs during weathering is directly proportional to the changes in the governing mechanical properties of the clearcoat.

While the chemical composition changes that occur during weathering are the most fundamental manifestation of weathering, most of the potential catastrophic failure mechanisms are mechanical in nature: cracking, peeling, and microchecking. These failures are due to the changing mechanical behavior of the coating system and the stresses to which it is exposed. The size and spatial distribution of these stresses have recently been investigated and shown to be on the order of 15-25 MPa for a weathered clearcoat.[12] The stresses arise from a number of sources including: thermal expansion coefficient mismatch, humidity expansion coefficient mismatch, densification, and solvent loss. Perera and coworkers have studied these stresses extensively.[13-15] The stresses due to environmental changes are, of course, cyclic in nature. Some finite mean stress is maintained upon which a cyclic stress is superimposed. Thus, the paint system undergoes fatigue loading.

In addition to the changing stress state, the fundamental mechanical properties of coatings can change as weathering progresses. These changing mechanical properties are largely a result of the chemical composition changes - no chemical composition changes translate into no (or very little) mechanical property changes. There has been little previous research on the weathering-induced changes in the mechanical properties of high solids coatings. Hill and coworkers have reported on the changing dynamic mechanical behavior of coatings as the coatings weathered, with their attention focused on the changes in the plateau region of the modulus curve.[16,17] Depending on the details of the degradation, the rubbery modulus, and therefore the crosslink density, can either increase or decrease on exposure. Detailed studied of other weathering-induced changes in the mechanical properties of automotive clearcoats are non-existent.

For any mechanical failure, a priori identification of the failure modes and critical mechanical properties is crucial in designing a robust testing program. We have proposed that for one of the most likely possible catastrophic basecoat/clearcoat failure modes, clearcoat cracking, the fracture energy (G_c) is the most relevant mechanical property.[18,19] The fracture energy is the amount of mechanical energy required to unstably propagate an already existing crack in a material or at an interface. Thus, for clearcoat cracking the germane fracture energy would be that of the clearcoat, while for delamination of the clearcoat from the basecoat the fracture energy of that interface would be most important. Meth and coworkers have reported on a method for measuring the fracture energy of the basecoat/clearcoat interface using Laser Induced Decohesion Spectroscopy (LIDS). This method appears to be the most quantitative for measuring the fracture energy of the basecoat/clearcoat interface.[20]

We have previously measured the fracture energy of a number of isolated automotive clearcoats using techniques previously developed for measuring the fracture energy of thin films of interest in microelectronic devices.[18,19] These measurements should have a direct bearing on the ability of these clearcoats to resist cracking. Our previous measurements were made on isolated clearcoats, not clearcoat in a complete paint system. However, this was the first step in quantifying the brittleness of these materials. In this report we have continued our work and report on the fracture energy of both isolated clearcoats and clearcoats in complete paint systems. In addition, the fracture energy of these clearcoats has been measured as a function of weathering time, both accelerated and outdoor Florida exposure. Finally, these results are compared to chemical composition change measurements in the same clearcoats to clarify the mechanical tolerance of the clearcoats to chemical composition changes.

Experimental

Materials

Both isolated clearcoats (clearcoats adhering directly to metal substrates) and clearcoats in complete paint systems were studied. The commercial clearcoats represented a wide variety of chemistries including: acrylic/melamine, epoxy-acid, polyurethane-polyol/melamine, acrylic-silane/melamine, and 2K polyurethane. All of the isolated clearcoats were tested in two forms: the commercial formulation containing all the stabilizers (both UVA and HALS) and an additive-free formulation containing neither UVA or HALS. Clearcoats on a variety of metallic substrates were prepared using Bird applicators and were cured for the recommended times at recommended temperatures. The thickness of all the cured films was measured with a micrometer. Typical films were 50 μm thick. Details of these sample preparation techniques have been described elsewhere.[18]

Full paint systems were obtained from paint suppliers in the form of standard 4" x 12" paint panels. All panels for accelerated weathering testing were dark metallic red basecoat/clearcoat systems. They were obtained as either the commercial paint system containing all the stabilizers or as a system with no UVA in any of the layers. Additional panels of the same formulations that had already been exposed in Florida were also obtained, but were typically of different basecoat colors.

Exposure

Isolated clearcoats were exposed to accelerated "light only" weathering in a xenon arc weatherometer.[21] The weatherometer provided borosilicate S/borosilicate S filtered xenon arc light at 0.45 W/m^2 irradiance with no water spray or dark cycle included. Black panel temperature was 60°C and the dew point was 25°C. Samples were exposed for up to 4000 hours. Full paint systems were exposed to the standard SAE J1960 June '89 weathering protocol (70°C black panel during irradiance) modified with borosilicate S inner and outer filters. Samples were exposed for up to 3000 hours. Additionally, some panels (see above) were exposed in south Florida for outdoor natural weathering at 5° south. These panels were exposed for up to 5 years.

Infrared Spectroscopy

Fourier transform infrared (FTIR) spectra were obtained using a Mattson FTIR 5000 at 4 cm^{-1} resolution. Isolated clearcoats ~8 μm thick were coated and cured on 1cm silicon discs and placed in the same weatherometer as the clearcoats on metal substrates. The discs were periodically removed and the spectra of the clearcoat recorded. The amount of photooxidation was generically quantified using the Δ(-OH,-

NH)/-CH] method, which ratios the absorbance in the 3800-2000 cm^{-1} region to that in the 3100-2800 cm^{-1} region.[2,3] For the complete paint systems photoacoustic (PAS) infrared spectroscopy was used to quantify the chemical changes taking place during weathering. All PAS data was obtained on a Mattson Cygnus 100 rapid scan FTIR system equipped with a water cooled source, a variable source aperture at 50% and a MTEC 2000 cell. All spectra were transformed using 4000 data points and one order of zero filling to give spectral resolution of 8 cm^{-1} and a digital resolution of 4 cm^{-1}. A scan velocity of 3.6 kHz and an assumed coating thermal diffusivity of 1x10-3 cm^2/sec yielded a thermal diffusion length at 4000 cm^{-1} of 6 μm and at 2000 cm^{-1} of 8 μm. This procedure led to a sampling depth of between 12 and 16 μm into the coating.[22] Samples for PAS experiments were taken from paint panels that had been exposed to the same accelerated weathering conditions as those for fracture energy measurements. The amount of photooxidation was again quantified using the Δ[(-OH,-NH)/-CH] method.

Fracture Energy

Details of the fracture energy measurements have been reported previously along with a complete summary of the theoretical background.[18,19] In brief, the fracture energy (G$_c$) is the amount of mechanical energy necessary to propagate a crack in a material. For coatings on substrates this quantity can be measured in a straightforward manner. Thin strips of the coating/metal composite were cut using a hand shear. The cutting process introduced a multitude of sharp edge cracks in the specimens, which is a necessary condition for the fracture energy testing. Typical strips are 10 cm x 1 cm. The strips were then pulled in tension in a mechanical testing machine at 20 mm/min. The strain at which cracks propagated across the clearcoat was recorded. Typically 8-10 samples were tested for each material at a given amount of weathering time. This reduced the uncertainty on any fracture energy value to ± 25%. The fracture energy was then calculated as

$$G_c = \frac{h\sigma^2 \pi g(\alpha,\beta)}{2\bar{E}} \qquad (1)$$

where h was the film thickness, σ was the stress, Ebar=E/(1-v^2) (E is the modulus of the clearcoat, v is Poisson's ratio taken to be 0.35), and g(α,β) was a constant relating to the modulus mismatch between the coating and the underlying layer. For a clearcoat on steel or aluminum g(α,β) = 0.78 while for a clearcoat on top of a basecoat in a full paint system g(α,β) = 1.26. The stress, σ, is calculated assuming linear elasticity, that is σ = Eε, where ε is the strain at cracking. To be rigorously correct the stress would be determined from the actual stress strain curve of the material, and we have previously termed this the true or actual fracture energy, while that calculated assuming linear elasticity we have termed the apparent fracture energy. We have choosen to concentrate on the apparent fracture energy for two reasons. First, it is simpler to calculate requiring only a knowledge of the modulus of the coating, which can be measured using dynamic mechanical testing or can be estimated. Second, the true fracture energy discriminates against coatings with low yield stresses as the stress during cracking is quite low, while the strain can be quite high. These low yield stress coatings perform well in the field, with regards to cracking, but would posses low true fracture energies. Additionally, as weathering progresses the amount of anelastic deformation a coating undergoes before fracture decreases. Thus, the difference between the true and apparent fracture energy becomes negligible at long weathering times. For the remainder of this paper the fracture energy will refer to the apparent fracture energy.

Results

The fracture energy of eight unweathered isolated commercial clearcoats is shown in Figure 1. The fracture energies ranged from 230 J/m^2 to 14 J/m^2, similar to what has previously been reported.[19] For reference, the fracture energy of window glass is approximately 6 J/m^2, while polymers can range from a less than 100 J/m^2 to well over 1000 J/m^2. On weathering, all of the coatings underwent photooxidation. The extent of photooxidation was measured from the FTIR spectra of the clearcoats that had been deposited on silicon discs. The Δ(-OH,-NH)/-CH] ratio at 4000 hours of accelerated weathering exposure is shown in Figure 2 for the fully formulated and additive-free formulation of each clearcoat. The larger the Δ(-OH,-NH)/-CH] ratio the greater the extent of photooxidation in that clearcoat. For most coatings the additive-free formulation photooxidized more quickly, as expected. Clearcoat F was a notable exception as its rate of chemical composition change was low and appeared to be unrelated to the presence of additives.

For most of the clearcoats, the change in the fracture energy mirrored the chemical composition change. Clearcoat B in Figure 3 exemplifies this behavior, where the fracture energy is shown as a function of weathering. Clearcoat B photooxidized quickly when formulated without stabilizers, but can be stabilized nicely by the addition of HALS and UVA (Figure 2). It's fracture energy decreased slowly when fully stabilized, leveling off at approximately 70 J/m^2, but the fracture energy of the additive-free formulation decreased rapidly with weathering and was near zero after 3000 hours of exposure. Also of note is the difference in the initial fracture energy of the two formulation of clearcoat B. The presence of the additives alone increased the fracture energy by a factor of two. Clearcoat F's fracture energy behavior also mirrored it's photooxidation behavior (Figure 4). Clearcoat F is the one clearcoat studied whose photooxidation rate was nearly uneffected by the presence of HALS and

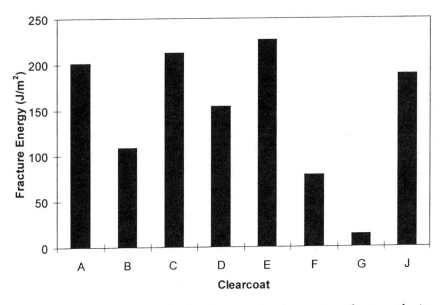

Figure 1. Fracture energy of isolated automotive clearcoats without weathering. Smaller values of fracture energy indicate more brittle coatings.

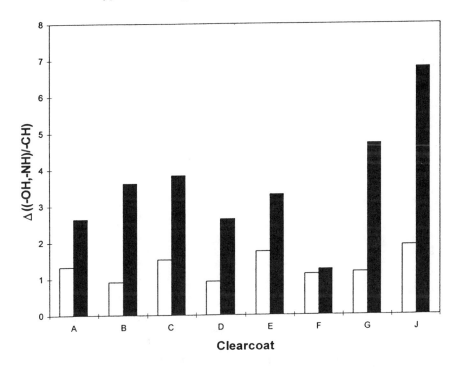

Figure 2. Change in the [(-OH,-NH)/-CH] ratio of clearcoats after 4000 hours of accelerated weathering. Filled bars are additive-free formulations. Unfilled bars are the clearcoats containing HALS and UVA. FTIR measurements made in transmissions on coated Si discs.

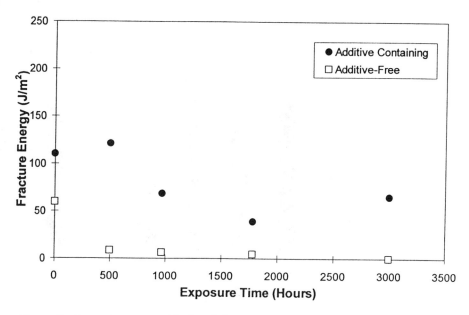

Figure 3. Fracture energy of isolated clearcoat B as a function of accelerated weathering exposure. Filled circles are the formulation containing additives, open squares are the additive-free formulation.

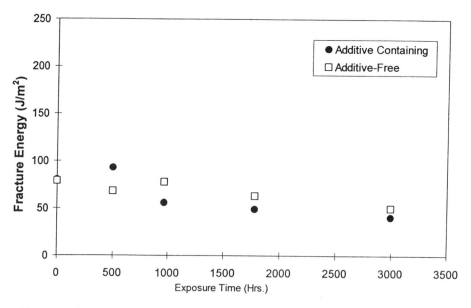

Figure 4. Fracture energy of isolated clearcoat F as a function of accelerated weathering exposure. Filled circles are the formulation containing additives, open squares are the additive-free formulation.

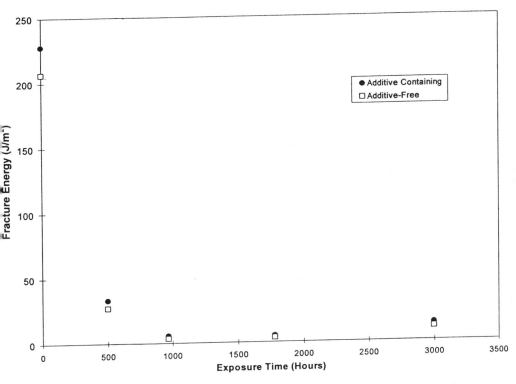

Figure 5. Fracture energy of isolated clearcoat E as a function of accelerated weathering exposure. Filled circles are the formulation containing additives, open squares are the additive-free formulation.

UVA. While the coating started out with a relatively low fracture energy, it decreased only slowly with weathering. Both the fully formulated and additive-free formulations changed at the same rate. Clearcoats E behaves fundamentally differently (Figure 5). Its fracture energy decreased rapidly with weathering in either the fully stabilized or additive-free formulations. However, the additive-free formulation photooxidized much more rapidly than the formulation containing the additives.(Figure 2).

When testing the fracture energy of the clearcoat in the full paint system, one significant complication arises. While most coating systems fail during testing by brittle crack growth in the clearcoat, some coating systems do not. Two other failure modes were observed: slow tearing of the paint system from the edges, and brittle crack propagation through the complete paint system followed by some delamination at the e-coat/phosphate interface. While values of the fracture energy for the clearcoats in these systems were calculated they are not rigorously correct. Typically, those systems that start out with one of these failure modes will progress to the more brittle failure as weathering progresses. Those coating systems that fail during fracture testing in a manner other than clearcoat cracking are noted in the data. For these clearcoats the fracture energy values are only qualitatively correct and serve only to gauge trends in the early portion of exposures.

Regardless of the type of fracture that occurs during testing, the values of the fracture energy of the clearcoat in the full paint system were generally quite different than that of the isolated clearcoat. In all the cases the fracture energy is higher in the full system. Figure 6 shows the fracture energy of the nonweathered fully formulated clearcoats in a variety of full paint systems. Figure 7 shows the fracture energy of three fully formulated clearcoats in full paint systems as a function of accelerated weathering time. All of the clearcoats become more brittle as weathering progresses, with some clearcoats embrittling quickly and others changing only slowly.

Clearcoats also embrittled due to outdoor exposure in Florida. Figure 8 shows the fracture energy of three clearcoats in full paint systems exposed in Florida for one, two, and three years. Again natural weathering decreased the fracture energy of the clearcoat. For most of the clearcoats in Fig. 6 the fracture energy of the Florida weathered panels similarly decreased to the same level as the 3000 hour accelerated weathered panels after approximately 2-3 years.

The sensitivity of the clearcoat's fracture energy to the extension rate was also examined. Figure 9 shows the fracture energy of the clearcoats B, E, and G in complete paint systems as a function of testing rate, at both 2 mm/min and 20 mm/min. All of the panels had been exposed for 1000 of accelerated weathering before testing. Clearcoat E's fracture energy was quite dependent on rate, clearcoat B's was somewhat dependent, and clearcoat G's was independent of rate over the range tested.

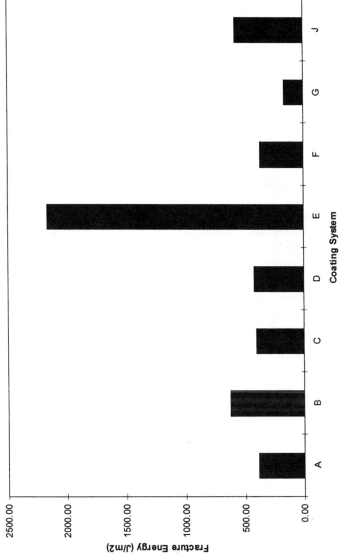

Figure 6. Fracture energy of non-weathered clearcoats in complete paint systems. Clearcoats A and C failed by brittle cracking to the substrate. Clearcoats D and E failed by slow crack tearing. Remaining clearcoats cracked by brittle channeling in the clearcoat only.

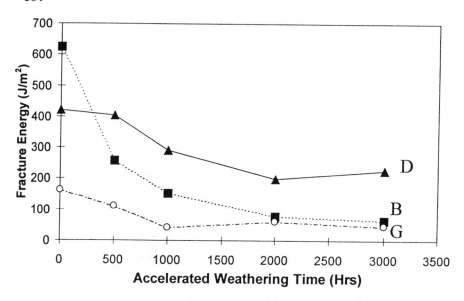

Figure 7. Fracture energy as a function of accelerated weathering exposure for three clearcoats in complete paint systems. Clearcoats B and G failed by brittle clearcoat cracking, clearcoat D failed by slow crack tearing initially, but after 500 hours of weathering failed by brittle clearcoat cracking.

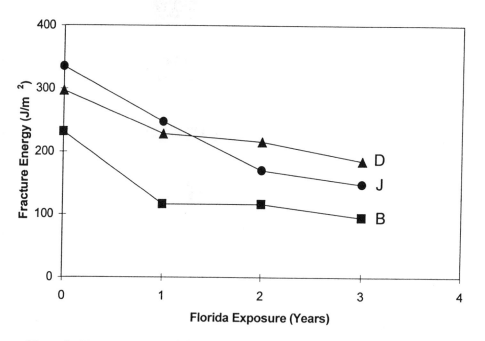

Figure 8. Fracture energy of clearcoats in complete paint systems as a function of Florida exposure time

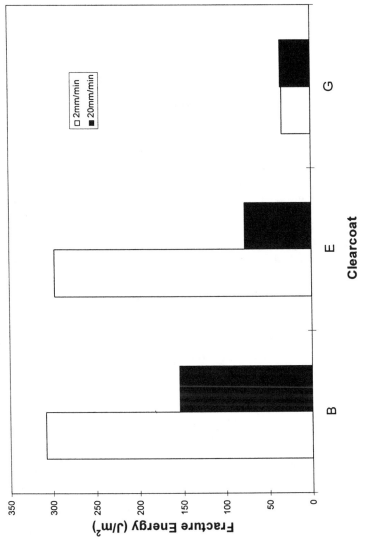

Figure 9. Effect of extension rate on the fracture energy of clearcoats in complete paint systems. All clearcoats had 1000 hours of accelerated weathering before testing. Open bars are 2mm/min extension rate, closed bars are 20mm/min extension rate.

Discussion

Isolated Clearcoats vs. Clearcoats in Full Paint Systems

As weathering progresses, all of the clearcoats, whether isolated or as part of a full paint system, became more brittle, meaning their fracture energy decreased. However, not all of the clearcoats embrittled at equal rates. The embrittlement was, of course, intimately tied to the chemical composition changes taking place in the clearcoat. Embrittlement can result from at least two different mechanisms: chain/crosslink scission and additional crosslinking. New crosslinks can form due to additional thermal cure taking place and also due to the free radical driven photooxidation chemistry that occurs during exposure.[23] Often the new crosslinks will be more rigid than the original crosslinks, such as the melamine-melamine crosslinks that can form during the weathering of acrylic/melamine clearcoats. Decreasing the crosslink density through chain or crosslink scission can also lead to embrittlement. These scission events introduce defects into the polymer network, which can ease crack propagation. Carried too a much greater extent, however, a loss of crosslink density will sufficiently lower the T_g of the coating as to make it unusably soft and even tacky. The exact details of the chemical changes that occurred during photooxidation are unknown for each of the clearcoats studied. Most of the clearcoats likely underwent both crosslink/chain scission and additional crosslink formation, with the ratio between the two being different for each coating and coating formulation. We have broadly measured their rates of photooxidation using the $\Delta(-OH, -NH)/-CH]$ technique, either by transmission FTIR for isolated clearcoats, or by PAS-IR for the clearcoat in full paint systems. However, like the fracture energy, regardless of whether the clearcoat was an isolated film or part of a complete paint system, the $\Delta(-OH, -NH)/-CH]$ ratio grew as weathering progressed.

There were, however, important differences between the behavior of the isolated clearcoats and the clearcoats in complete paint systems. For example, the fracture energy of the clearcoats in complete paint systems was significantly higher than the fracture energy of the same clearcoat as an isolated film. This can be explained by two factors. First, all of these clearcoats have been formulated to perform as part of a full paint system. Thus, their chemical composition and mechanical performance have been selected to perform best when they are cured on top of basecoat, not bare metal. Because all the basecoat/clearcoat system are processed wet-on-wet, that is the basecoat and clearcoat are cured together, some degree of mixing is expected between the basecoat and clearcoat. Hakke has shown that UVAs can migrate between the basecoat and clearcoat during curing, and it is likely that other species of similar molecular weights, such as crosslinkers and HALS, can also migrate.[24] Thus, it would be unexpected if the initial fracture energy of each clearcoat was the same when measured as an isolated clearcoat or as a clearcoat in a full paint system. The difference in composition between the isolated clearcoats and the clearcoats in complete paint systems has the potential to effect the fracture energy in another manner. To calculate the fracture energy, the modulus of the clearcoat must be known. This is typically measured using DMA on isolated clearcoat films. However, due to mixing during cure, the clearcoat in complete paint systems will not have the same composition as the isolated clearcoat. This is unlikely to have introduced large errors, though, as the modulus of most crosslinked glassy polymers does not vary by more than 10-20%. These results, showing a difference in the behavior between isolated clearcoats and clearcoats in complete paint systems, are in contrast to previous results that showed little difference.[19] This is likely due to the highly weathered and very brittle nature of the clearcoats previously studied.

The second reason that the fracture energy of the clearcoat in the complete paint system should not match that of the isolated clearcoat is the that the mechanics of the fracture process are somewhat different when the clearcoat is on top of metal or on top of a ductile substrate like a basecoat. While differences in the modulus between basecoat and metal were accounted for by adjusting the value of $g(\alpha,\beta)$, this does not account for plasticity of the underlying layer.[25] When a clearcoat is on top of a basecoat, some of the strain energy in the clearcoat can be relieved by the low yield stress basecoat underneath. This may in turn induce some plasticity in the clearcoat and allow the clearcoats to tolerate higher strains before cracking.

The change in the fracture energy of a clearcoat as it weathers appears to be a sensitive measure of the clearcoat's long-term weathering resistance that provides additional information not provided by chemical composition change or appearance measurements. For the isolated clearcoats in Figures 3 and 4, the change in the fracture energy closely mirrored the rate and amount of chemical composition change. Clearcoat B in Figure 3 photooxidized quickly in the absence of UVA and HALS, but when fortified with these stabilizers its photooxidation rate was much slower. Its fracture energy behavior was similar. Clearcoat B embrittled quickly in the unstabilized form, but retained a higher fracture energy for a longer time when fully stabilized. Clearcoat F's fracture energy behavior was also similar to its chemical composition change behavior. Whether fortified or not, clearcoat F photooxidized at a very slow rate and its fracture energy also decreased at a very slow rate. The clearcoat is inherently photooxidation stable.

Clearcoat E, however, shows a marked difference between the photooxidation behavior and the fracture energy behavior and points to the importance of measuring both chemical and physical property changes. When unfortified, Clearcoat E photooxidized quickly, but when fully stabilized it photooxidized slowly. The fracture energy, however, decreased at the same rapid rate regardless of whether the clearcoat was stabilized or not. The fracture energy of clearcoat E in the full paint system also changed rapidly even though PAS-IR results show that its Δ(-OH,-NH)/-CH] change was low when fully stabilized. Thus, for this clearcoat there does not appear to be a direct link between the weathering-induced chemical composition changes and the changes in the mechanical behavior. However, more likely is that the chemical composition changes that drive the changing mechanical behavior of clearcoat E were not captured by the Δ(-OH,-NH)/-CH] method and indeed may not be directly related to photooxidation but rather hydrolysis or secondary curing reactions. Further experiments were not attempted to clarify this result. The fact remains, however, that photooxidation rates need not be directly proportional to mechanical property change rates.

Like the isolated clearcoats, the clearcoats in full paint systems can be either quite tough or quite brittle. Clearcoat E stood out as being extraordinarily tough (much tougher than as an isolated clearcoat), but quickly embrittled on weathering. The rates of embrittlement were different for each clearcoat. As shown in Figure 7, Clearcoat B started out the toughest but embrittled quickly. Clearcoat G started out brittle and became more brittle, while clearcoat D started out intermediate and lost fracture energy gradually. Clearcoat D, therefore, has the best chance of resisting cracking during long-term exposure.

The accelerated weathering results were confirmed by fracture energy testing of panels returned from Florida exposure. Three systems, which are quite similar to systems B, D, and J, are shown in Figure 9. These paint systems were nominally the same as in Figure 6, but the basecoats were different colors and the panels were produced and cured at a different time than those panels used to produce the data in Figure 6. Again all embrittled as weathering progressed, but system D again maintained the highest level of fracture energy after two years of Florida exposure. Based on the rate of change of the fracture energy, one year of Florida exposure induced the same change in the fracture energy as between 1000 and 1500 hours of accelerated weathering. While the rate of embrittlement between Florida and accelerated exposure may be similar based on the energy dose, the actual times to failure could be quite different due to the different stresses a coating system experiences outdoors versus in the weatherometer.

Non-ideal cracking

The cracks that occurred during fracture testing of the clearcoat in the complete paint systems can be classified into three distinct types. The first type was by far the most common type and is classified as "ideal", that is, during testing the cracks propagate unstably across the clearcoat and down to the basecoat. The cracks typically arrest at the basecoat/clearcoat interface due to the higher toughness of the basecoat and then spread some of the energy out over that interface. Determining the fracture energy of these coatings is straightforward and analogous to the calculation for the isolated clearcoats, requiring only a change in the value of $g(\alpha,\beta)$ in eq. 1 from 0.78 to 1.26 to account for the modulus of the basecoat under the clearcoat. Eq. 1 is only strictly applicable to this mode of cracking, as it has been specifically derived for the case of channeling of cracks through the full thickness of the coating, in this case the clearcoat over the basecoat.

The second and third types of cracking occured only at high strains and allow for only a qualitative determination of the fracture energy. The second type of cracking was unstable (high speed) crack propagation across the specimen and down to the e-coat/phosphate interface, and the third type was slow, stable crack growth in the clearcoat alone. These two modes of cracking are observed in only in a few nonweathered systems. Weathering typically embrittled these systems such that after limited exposure they displayed brittle crack growth in the clearcoat alone. For both the second and third types of cracking behavior, eq. 1 cannot be rigorously applied. Because the cracks propagate all the way to the e-coat/phosphate interface in the second type of cracking, they move through a non-uniform medium (clearcoat, basecoat, primer, and e-coat), the constitutive properties of which are also non-uniform. The fracture mechanics from which eq. 1 is derived are strictly assume homogeneous isotropic materials.[26] One could assume some average constitutive behavior for the entire paint system. However, this does not describe the brittleness of the clearcoat alone, which is the goal of this work. An alternative method of looking at these fractures is to proceed with the calculation on the basis of the clearcoat thickness and constitutive properties alone, implicitly assuming that the cracks begin in the clearcoat and then propagate down through the other layers. This is the likely scenario, as we have typically observed the cracks to start at the edge and surface of the paint system and then propagate across and then down through the system at high rates. Because these types of failures were typically seen only at high strains it was assumed that the high driving force for cracking that was required to propagate cracks, also kept the cracks from arresting at the basecoat. In a practical sense, the fracture energy of clearcoats that failed during testing in this manner are inherently tough enough to belay any worries about mechanical weakness. When the materials embrittle enough due to weathering, they typically fail by clearcoat cracking and their fracture energy numbers can be calculated with confidence.

When the clearcoat undergoes stable crack propagation (cracking type number 3), that is the slow growth of cracks in the clearcoat which grow in length as the strain increases, the calculation of the fracture energy is also not truly appropriate. Eq.1 has been developed for the unstable cracking of brittle films, where plasticity is a minor component. Clearcoats that exhibit slow crack growth do not meet this criterion and therefore are not amenable to this type of analysis and would be better analyzed using J-integral techniques or the method of essential work. Methods for calculating J-integral toughness and the essential work of fracture for thin films on substrates have not yet been developed. However, again, these tearing-type of cracks were only seen at high strains and would not present any potential cracking problems until the clearcoats have weathered and embrittled. For consistency the few materials that failed during testing in this manner are noted and their fracture energy values were calculated based on the strain at which cracks appeared regardless of the type of crack and using the thickness of the clearcoat alone. While these numbers are quantitatively incorrect, they give a feeling for the high toughness of these clearcoats, and provide an initial point at which to evaluate their weathering performance.

Tolerance of Clearcoats to Chemical Change

The central question that can be answered by combining fracture energy measurements with chemical composition change measurements is: how much weathering-induced chemical composition change can a clearcoat tolerate without cracking? The simpler question is, of course, how much weathering can a clearcoat tolerate without cracking? This second question assumes that the chemical change data (photooxidation rate, UVA effectiveness and loss rate, HALS longevity and effectiveness) is not necessary for anticipating the long-term cracking behavior of clearcoats. While this may be true, we believe that chemical composition change data is at least as valuable as the fracture energy results for at least three reasons. First, chemical composition change data can be obtained quickly from clearcoats applied on silicon discs or using PAS-IR on clearcoats in full paint systems. Changes can be observed in as little as 250 hours of accelerated weathering with small samples. 1000 hours of accelerated weathering can provide solid direction as to the rate of photooxidation. These experimental conditions enable these techniques to be used for both forumlation screening and comparatively fast trouble shooting. Second, not all potential clearcoat failures are due to cracking. For every mechanical or non-mechanical failure a new phenomenological test would have to be developed. Chemical change measurements are generic. In the absence of a mechanical test that is predictive for every failure, chemical change data can be the first indication that weathering performance is suspect. Third, from the coatings producer vantage, if the exact chemical composition of the coating is known, the details of the chemical changes that occur during weathering can be monitored and deciphered, not just the generic amount of photooxidation. Thus, while fracture energy measurements directly probe one of the most relevant potential failure mechanisms, chemical composition change measurements provide additional information on other potential failures in a rapid and inexpensive manner. Thus, the two types of measurements are complimentary and synergistic in nature.

One way of determining the tolerance of coatings to chemical change is shown in Figure 10 where the change in the fracture energy is plotted against the change in the $\Delta[(-OH,-NH)/-CH]$ value for paint systems B and D. Clearcoat D underwent roughly 50% more photooxidation than clearcoat B, but its fracture energy change was only 1/3 that of clearcoat B. Thus, clearcoat D is very tolerant to chemical change and clearcoat B is intolerant. Chemical composition changes alone would lead to the selection of clearcoat B over clearcoat D as a superior clearcoat for weathering resistance, an erroneous conclusion with regards to cracking resistance. Measuring fracture energy along, however, could lead to the selection of a coating that is unsatisfactory in other aspects, such as gloss loss or long-term adhesion.

Application to other Failure Modes

The highly crosslinked nature of the clearcoats typically minimizes the viscoelastic effects during crack propagation unless the polymer is near T_g or contains secondary phases. However, some clearcoats do show substantial rate effects. These rate effects can have a direct impact on field performance as crack growth during exposure is likely to be slow and driven by fatigue loading. The effects of varying the rate of extension during fracture testing are shown in Figure 9. Clearcoat E, which was very tough initially and then embrittled rapidly on weathering, was almost three times as tough when tested at a slower rate. Clearcoat G, which was inherently brittle showed

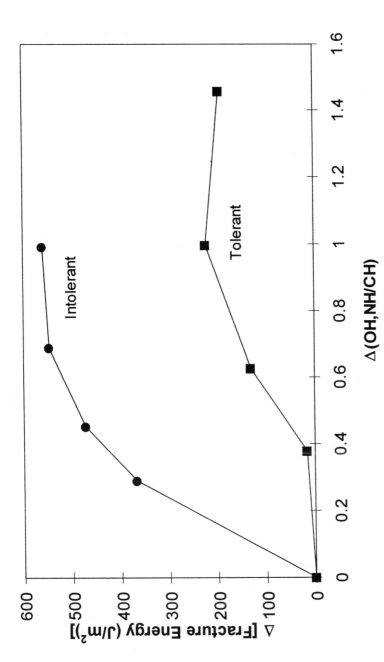

Figure 10. Chemical composition changes and fracture energy changes for two clearcoats in complete paint systems. Both clearcoats started at approximately the same initial fracture energy values.

no rate effects, while clearcoat B was intermediate. Thus, building in mechanisms for viscoelastic processes to absorb energy at lower strain rates, can significantly improve the low strain rate toughness of materials, and consequently improve their resistance to long-term cracking. Conversely, those systems that are indeed quite brittle at high strain rates would likely be more susceptible to cracking during high strain rate events such as stone impact.[27]

As a final note, fracture energy measurements may also provide insight into subtle formulation changes. For example, clearcoat B showed a marked difference in the initial fracture energy when fully formulated as opposed to when it was formulated without HALS or UVA. The fully formulated version has a fracture energy of 120 J/m^2 versus 60 J/m^2 for the additive-free version. This difference is greater than the typical uncertainty on the measurements. The higher fracture energy of the fully formulated clearcoat could have been due to a different state of cure. Most of the clearcoats' curing chemistry is acid catalyzed. While the HALS the coating contains is supposed to be non-basic, there could have been some residual impurities that interfere with the curing reaction, leading to different network structures in the two formulations. Furthermore, the addition of a few percent of soluble small molecules may have plasticized the network increasing its fracture energy. Thus, small formulation changes such as changing the amount or type of HALS or UVA can have significant effects on both initial and long-term performance.

Conclusions

The catastrophic coating failures that can potentially occur in basecoat/clearcoat paint systems are the result of the changing mechanical properties of the coatings and of the stresses to which they are subjected. The changes in the mechanical properties are clearly driven by the weathering-induced chemical composition changes in the paint system. While the potential failures (cracking and peeling) can be dramatic and sudden, the mechanical properties of the coatings and interfaces do not change discontinuously, but rather in a gradual fashion. When the stress to which the coating is subjected provides enough driving force to overcome the fracture energy of the coating, cracking will occur. By measuring the fracture energy of automotive clearcoats as a function of weathering time, particularly in complete paint systems, these cracking events can be anticipated. Those systems whose fracture energy values are highest after exposure stand the greatest chance of not cracking during long-term in service exposure. By linking the fracture energy changes with a knowledge of the stress state in coatings it should be possible to make quantitative predictions on the lifetime of coating systems. Coupling fracture energy measurements with the indispensable chemical composition change measurements, will certainly reduce the risk of choosing coating systems with inferior long-term weathering performance.

Acknowledgements

The authors would like to thank PPG, BASF, and DuPont for their generous donation of materials used in this research. The authors would also like to thank Dr. John Gerlock for stimulating discussions of the results of this research.

Literature Cited

1. Gerlock, J. L., Smith, C. A., Nichols, M. E., Tardiff, J.L., Kaberline, S.L., Prater, T. J., Carter III, R.O., Dusbiber, T.G., Cooper, V. A., and Misovski, T., to be published in proceeding of 2nd Conference on Service Life Prediction of Coatings.
2. Gerlock, J. L, Smith, C. A., Nunez, E. M., Cooper, V. A., Liscombe, P., Cummings, D. R., and Dusibiber, T. G., in Polymer Durability, eds. R. L. Clough, N. C. Billingham, and K. T. Gillen, ACS Advances in Chemistry Series 249, Wash. D. C., 1996, p 335.
3. Gerlock, J. L., Smith, C. A., Cooper, V. A., Dusbiber, T. G., and Webber, W.H., *Polym. Deg. and Stab.,* accepted for publication.
4. Lemaire, J. and Siampiringue, Proc. 1st Conf. on Service Life Prediction of Coatings, Breckenridge CO., ACS, Sept. 1997.
5. Gerlock, J. L., Prater, T. J., Kaberline, S. L., and deVries, J. E., *Polym. Deg. and Stab.,* **47**, (1995), 405.
6. Kaberline, S. L. and Gerlock, J. L, Jones, J., and Prater, T. J., to be published in proceeding of 2nd Conference on Service Life Prediction of Coatings
7. Gerlock, J. L., Prater, T. J., Kaberline, S. L., Dupuie, J. L., Blais, E. J.; Rardon, D. E., **65**, *Polym. Degrad. Stab.,* (1999), 37.
8. Gerlock, J. L., Smith, C. A., Cooper, Dusbiber, T. G., Prater, T. J., Kaberline, S. L., deVries, J. E., and Carter, R. O., Nichols, M. E., Dearth, M. A., and Dupuie, J. L., *Surf. Coat. Austral.,* **34**, No. 7, 14 (1997).
9. Smith, C. A. and Gerlock, J. L., *Poly. Deg. and Stab.,* in press.
10. Kucherov, A. V., Gerlock, J. L., and Matheson Jr., R. R., *Polym. Deg. and Stab.,* **69**, (2000), 1.
11. Kucherov, A. V., Gerlock, J. L., and Matheson Jr., R. R., *Polym. Deg. and Stab.,* in press.
12. Nichols, M. E. and Darr, C. A., *J. Coat. Tech.,* **70**, (1998), 141.
13. Perera, D. Y. and Schutyser, P., *Prog. Org. Coat.,* **24**, (1994), 299.
14. Perera, D. Y. and Schutyser, P., *Proc. of 22nd FATIPEC Congress,* Budapest, (1994).
15. Perera, D. Y. and Vanden Eynde, D., *J. Coat. Tech.,* **59**, (1987), 55.
16. Hill, L. W., Korzeniowski, H. M., Ojunga-Andrew, M., and Wilson, R. C., *Prog. Org.Coat.,* **24**, (1994), 147.
17. Hill, L. W., *J. Coat. Tech.,* **64**, (1992), 29.
18. Nichols, M. E., Darr, C. A., Smith, C. A., Thouless, M. D., and Fischer, E. R., *Polym. Deg. and Stab.,* **60**, (1998), 291.
19. Nichols, M. E., Gerlock, J. L., Smith, C. A., and Darr, C. A., *Prog. Org. Coat.,* **35**, (1999), 153.

20. Meth, J. S.; Sanderson, D.; Mutchler, C.; Bennison, S. J., *J. Adhes.*, **68,** (1998), 117

21. Bauer, D. R., Mielewski, D. F., and Gerlock, J. L., *Polym. Deg. and Stab.*, **38,** (1992), 57

22. Carter III, R. O., Gerlock, G. L., and Smith, C. A., *Proceedings of the SPE ANTEC*, New York, (1999).

23. Nichols, M. E., Gerlock, J. L., and Smith, C. A., *Polym. Deg. Stab.*, **56,** 81, (1997).

24. Haacke, G., Andrawes, F. F., and Campbell, B. H., *J. Coat. Tech.*, **68,** (1996), 57.

25. Beuth, J. L., *Int. J. Solids Struct.*, **29,** (1992), 1657

26. Hutchinson, J. W. and Suo, Z., *in Advan. in Appl. Mech.*, **29,** Academic Press, (1992), 63.

27. Ramamurthy, A. C., Ahmed, T., Favro, L. D., Thomas, R. L., Hohnke, D. K., and Cooper, R. L., *SAE Preprint*, 930051, (1995).

Mathematical Model

Chapter 19

Using Accelerated Tests to Predict Service Life in Highly Variable Environments

William Q. Meeker[1], Luis A. Escobar[2], and Victor Chan[1]

[1]Department of Statistics, Iowa State University, Ames, IA 50011-1210
[2]Department of Experimental Statistics, Louisiana State University, Baton Rouge, LA 70803

Today's manufacturers need to develop newer, higher technology products in record time while improving productivity, reliability, and quality. This requires improved accelerated test (AT) methods that can usefully predict service life. For example, automobile manufacturers would like to develop a 3-month test to predict 5 or 10-year field reliability of a coating system. Such estimation/prediction from ATs involves *extrapolation*. Seriously inadequate predictions will result unless adequate models and methods are used. This paper describes a general framework within which one can use laboratory test results to predict product field service life performance of certain products in a highly-variable environment.

Difficulty establishing correlation between laboratory tests and outdoor weathering tests for paints and coatings

Manufacturers of paints and coatings, for example, have had difficulty in establishing adequate correlation between their laboratory tests and field experience. Most of the laboratory tests attempt to accelerate time by "speeding up the clock." This is done by increasing average level of experimental factors like UV radiation, temperature, and humidity and cycling these experimental factors more rapidly than what is seen in actual use, in an attempt to simulate and accelerate outdoor aging. Such experiments violate some of the rules of good experimental design (e.g., by varying important factors together in a manner that confounds the effects of the factors) and the high levels of the accelerating variables can induce new

396

failure modes. Thus such accelerated tests provide little fundamental understanding of the underlying degradation mechanisms. Because experience has shown that the results of these tests are unreliable, standard product evaluation for paints and coatings still requires outdoor testing in places like Florida and Arizona (hot humid and hot dry environments, respectively). Such testing, however, is costly and takes too much time.

Other possible reasons for differences between laboratory tests and outdoor weathering tests include

- Inadequate control/monitoring of laboratory accelerated test conditions [e.g.,
  $\xi = (UV, temperature, humidity)$].

- Inadequate control/monitoring of field testing environmental conditions at outdoor exposure sites.

- Physical/chemical models that do not provide an adequate description of the relationship between degradation rates and experimental/environmental variables.

- Prediction models and methods that do not properly account for temporal environmental variability.

See [4] and [3] for a detailed description of issues relating to prediction of service life for paints and coatings.

Traditional Accelerated Tests

Reference [6] describes traditional accelerated life tests that are often used to provide timely information about life characteristics of components and materials. Other useful references for accelerated testing include [8] and Chapters 18 and 19 of [5]. This section briefly reviews these methods and provides an introduction to methods for using more powerful accelerated degradation tests.

An accelerated life test

Figure 1 shows the data from an accelerated life test on an electronic device (which we call Device A). These data were originally analyzed in [1]. The response was failure time for those devices that failed and the amount of running time for the others. The purpose of the experiment was to predict the early part of the failure-time distribution of the devices at an ambient use temperature of 10°C, presumably for a system to be installed under-sea. Superimposed on Figure 1 is a lognormal-Arrhenius model that

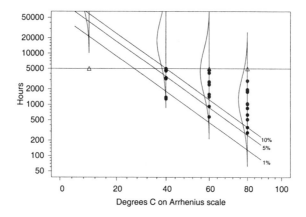

Figure 1: The Arrhenius-lognormal log-linear regression model fit to the Device-A ALT data.

describes the failure time distribution for the devices as a function of ambient temperature.

The Arrhenius-lognormal regression model is

$$\Pr[T(\text{temp}) \leq t] = \Phi_{\text{nor}} \left[\frac{\log(t) - \mu}{\sigma} \right]$$

where Φ_{nor} is the standard normal cumulative distribution function, $\mu = \beta_0 + \beta_1 x$,

$$x = \frac{11605}{\text{temp K}} = \frac{11605}{\text{temp} \,^\circ\text{C} + 273.15},$$

temp K is temperature Kelvin, and $\beta_1 = E_a$ is the effective activation energy.

For this application, the use-environment conditions were stable and well characterized. See [1] and Chapter 19 of [5] for additional details on the model and the analyses of these data.

An accelerated degradation test

Degradation data, when they are available, provide more information than the more common failure-time data. Degradation data also offer advantages for developing mechanistic models that are important for accelerated testing and other applications requiring extrapolation outside the

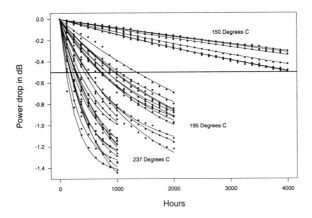

Figure 2: Device-B power drop accelerated degradation test data and fitted curves. Reproduced from Reference 7. Copyright 1998 American Statistical Association.

range of one's data. Chapters 13 and 21 of [5] describe methods for analyzing such data. We outline some of the basic concepts here.

Figure 2 shows the data from an accelerated degradation test on an RF power amplifier (which we call Device B). These data were originally analyzed in [7]. The Device B test is called a degradation test because the response was degradation in performance, in this case power drop, as a function of operating time and temperature. The temperatures shown are the junction temperatures corresponding to the environment in which the devices were operating. A device was defined to have failed after its power had dropped by 0.5 decibels (dB). The purpose of the degradation test was to estimate the fraction of units that would fail after 10 years at an operating junction temperature of 80°C.

The degradation in power was believed to have been caused by a mechanism that could be adequately described by single-step chemical reaction with rate constant $\mathcal{R}$. Diagrammatically,

$$A_1 \xrightarrow{\mathcal{R}} A_2$$

The rate equations for this reaction are

$$\frac{dA_1}{dt} = -\mathcal{R}A_1 \quad \text{and} \quad \frac{dA_2}{dt} = \mathcal{R}A_1, \quad \mathcal{R} > 0 \tag{1}$$

and the power drop was believed to be proportional to the level of A_2. Because power drop is observable, we use A_2 to denote this quantity. The solution of the system of differential equations in (1) gives

$$
\begin{aligned}
A_1(t) &= A_1(0)\exp(-\mathcal{R}t) \\
A_2(t) &= A_2(0) + A_1(0)[1 - \exp(-\mathcal{R}t)]
\end{aligned}
\tag{2}
$$

where $A_1(0)$ and $A_2(0)$ are initial conditions. If $A_2(0) = 0$, letting $A_2(\infty) = \lim_{t\to\infty} A_2(t) = A_1(0)$, the solution for $A_2(t)$ is

$$A_2(t) = A_2(\infty)[1 - \exp(-\mathcal{R}t)]. \tag{3}$$

The asymptote $A_2(\infty)$ reflects the limited amount of a material that was available for reaction to the harmful compounds.

Reference [7] describes Device B degradation with the model in (3), assuming that random distributions on the asymptote $A_2(\infty)$ and the rate constant $\mathcal{R}$ reflect unit-to-unit variability. The lines superimposed on Figure 2 are fitted regression curves, for each device, using the model in (3).

Simple temperature acceleration

The Arrhenius model describing the effect that temperature has on the rate of a simple first-order chemical reaction is

$$
\begin{aligned}
\mathcal{R}(\texttt{temp}) &= \gamma_0 \exp\left[\frac{-E_a}{k_{\mathrm{B}} \times (\texttt{temp} + 273.15)}\right] \\
&= \gamma_0 \exp\left(\frac{-E_a \times 11605}{\texttt{temp} + 273.15}\right)
\end{aligned}
\tag{4}
$$

where $\texttt{temp}$ is temperature in $°C$ and $k_{\mathrm{B}} = 1/11605$ is Boltzmann's constant in units of electron volts per $°C$. The pre-exponential factor γ_0 and the reaction activation energy E_a in units of electron volts are characteristics of the particular chemical reaction. Taking the ratio of the reaction rates at temperatures $\texttt{temp}$ and $\texttt{temp}_U$ cancels γ_0 giving an Acceleration Factor

$$
\begin{aligned}
\mathcal{AF}(\texttt{temp}, \texttt{temp}_U, E_a) &= \frac{\mathcal{R}(\texttt{temp})}{\mathcal{R}(\texttt{temp}_U)} \\
&= \exp\left[E_a\left(\frac{11605}{\texttt{temp}_U + 273.15} - \frac{11605}{\texttt{temp} + 273.15}\right)\right]
\end{aligned}
\tag{5}
$$

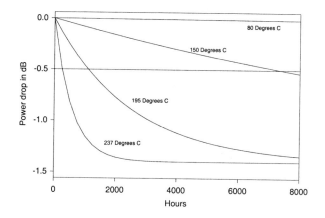

Figure 3: Device-B power drop predictions as a function of temperature. Reproduced from Reference 7. Copyright 1998 American Statistical Association.

that depends only on the two temperature levels and the activation energy. If $\text{temp} > \text{temp}_U$, then $\mathcal{AF}(\text{temp}, \text{temp}_U, E_a) > 1$. For simplicity, we use the notation $\mathcal{AF}(\text{temp}) = \mathcal{AF}(\text{temp}, \text{temp}_U, E_a)$ when temp_U and E_a are understood to be, respectively, product use (or other specified base-line) temperature and a reaction-specific activation energy.

The lines in Figure 3 were obtained by evaluating (3) at the estimates of the means of the distributions of the random asymptote and rate constant, as a function of temperature, using the Arrhenius model in (4).

General time transformation model

The Arrhenius model and other simple acceleration models result in what Meeker and Escobar [5] (Chapter 18) call a scale accelerated failure time (SAFT) model, in which the time to failure $T(\boldsymbol{\xi})$ at environment conditions $\boldsymbol{\xi}$ is related to the time to failure time $T(\boldsymbol{\xi}_0)$ at environment conditions $\boldsymbol{\xi}_0$ through the relationship

$$T(\boldsymbol{\xi}) = T(\boldsymbol{\xi}_0)/\mathcal{AF}(\boldsymbol{\xi}).$$

Although this model is adequate for some simple situations, more complicated acceleration models are often needed (e.g., when a failure mechanism involves two important rate constants with different activation energies).

A more general time-transformation function can be expressed as

$$T(\xi) = \Upsilon\left[T(\xi_0), \xi\right].$$

When the function $\Upsilon(t; \xi)$ is monotone increasing in t, the quantiles of the life distribution at ξ and ξ_0 are related by

$$t_p(\xi) = \Upsilon\left[t_p(\xi_0); \xi\right], \quad 0 < p < 1.$$

The cdfs at ξ and ξ_0 are related by

$$\begin{aligned} \Pr\left[T \le t; \xi\right] &= \Pr\left[\Upsilon(T; \xi) \le t; \xi_0\right] \\ &= \Pr\left[T \le \Upsilon^{-1}(t; \xi); \xi_0\right]. \end{aligned}$$

A General Approach to Service Life Prediction in Complicated Environments

As described above, tests that simply "speed up the clock" have not provided adequate predictions of field performance of organic paints and coatings. In general, the mechanistic modeling of failure and their dependency on accelerating variables is important for the successful application of accelerated testing. In order to do realistic service life prediction for complicated environments based on accelerated laboratory tests, we propose the following general approach, generalizing the traditional methods described in the previous section.

1. Use understanding of the physical/chemical mechanisms underlying product degradation and failure along with the experimental results to develop a deterministic product degradation model. To be workable, it will be necessary to develop a relatively simple model that, for example, identifies and focuses on the rate-limiting steps in the overall failure model.

2. Conduct laboratory experiments, using standard principles of experimental design, to gain fundamental understanding of the mechanisms leading to failure. Factors studied in the experiment should correspond to the environmental variables that affect service life.

3. Iterate between these first two steps in order to find and refine a model that will give the rate of degradation as a function of environmental variables and other important factors.

4. Use manufacturing process or experimental data to model and quantify product variability (e.g., unit-to-unit variability due to differences in raw materials and processing).

5. Use environmental time series data on the important factors that affect degradation (e.g., UV radiation, temperature, humidity) to characterize the environment. This could be done, for example, by identifying multivariate stochastic process models.

6. Use the variabilities in steps 4. and 5. along with the physical/chemical mechanism model identified in step 1. to define a stochastic process model for product degradation.

7. Use available data to estimate the unknown model parameters.

8. Use the product degradation model to generate a product service life prediction model.

9. Use statistical inference methods to quantify uncertainty in the service life distribution predicted from available data.

The remainder of this paper will develop and illustrate the basic modeling and prediction methods.

Degradation Model

Degradation, $\mathcal{D}(t)$, usually depends on environmental variables like UV, temp, and RH, that vary *over time*, say according to a multivariable profile $\xi(t) = [\text{UV}, \text{temp}, \text{RH}, \dots]$. Laboratory tests are conducted in well-controlled environments (usually holding variables like UV, temperature, and humidity constant). Interest often centers, however, on life in a variable environment.

Modeling begins by developing a deterministic physical/chemical models for the failure mechanism. Then random and stochastic process distributions can be added, as needed, to account for important process variabilities (unit-to-unit, stochastic over time, or both). There are three situations to consider:

- The environmental conditions described by the vector ξ are constant over time.

- The environmental profiles $\xi = \xi(t)$ have a variable but *deterministic path* in time (i.e., an experimental step-stress vs. time profile).

- The environmental profiles $\xi = \xi(t)$ are random in time (e.g., outdoor/real-world weather conditions) and the distribution of environmental sample paths can be described by a (multivariate) *stochastic process* model with parameters θ_ξ.

For example, Jorgensen et al. [2] used laboratory tests to identify a model similar to

$$\frac{d\mathcal{D}(t; L_{UV-B}, \texttt{temp}, \texttt{RH})}{dt} = A \times L_{UV-B} \times \exp\left(-\frac{E_a}{k_B \times \texttt{temp K}}\right) \times \exp(C \times \texttt{RH})$$

where L_{UV-B} is the instantaneous UV irradiance in the UV-B band (290-320 nm), $\texttt{temp K}$ is temperature Kelvin, RH is relative humidity (all of the environmental variables are potentially functions of time), k_B is Boltzmann's constant, E_a is an activation energy, and A and C are other parameters characteristic of the material and the degradation process.

Deterministic degradation model

For a given environmental profile $\boldsymbol{\xi}(t)$, the cumulative degradation at time t for a particular unit (specified by a unit parameter vector $\boldsymbol{\beta}$) can be expressed deterministically as

$$\mathcal{D}(t) = \int_0^t d\mathcal{D}[\tau, \boldsymbol{\xi}(\tau)] \tag{6}$$

$$= \int_0^t \frac{d\mathcal{D}[\tau; \texttt{UV}(\tau), \texttt{temp}(\tau), \texttt{RH}(\tau), \dots]}{d\tau} \, d\tau$$

where $d\mathcal{D}[\tau; \boldsymbol{\xi}(\tau)]/d\tau$ is the degradation rate and $\boldsymbol{\xi}(\tau)$ is the vector of environmental conditions at time τ (to simplify notation we suppress, for the moment, the dependence of $\mathcal{D}(t)$ on unit parameters $\boldsymbol{\beta}$). In general, the cumulative degradation paths $\mathcal{D}(t)$ differ from unit to unit due to:

- Intrinsic unit-to-unit differences (raw materials, processing differences).

- Extrinsic differences (e.g., in environmental profiles denoted by $\boldsymbol{\xi}(\tau)$).

Stochastic degradation model

The environmental variables in the profile $\boldsymbol{\xi}(t)$ are controlled in laboratory tests, but will be stochastic over time in actual product use. In order to evaluate the distribution of degradation paths for a stochastic environment, one can still use (6), but the integral becomes a stochastic integral.

A Simple Example

This section presents a simple example to show how to predict the effect of environmental variability on degradation. The example is based on the Device B power drop degradation model in (3). To keep the example simple, we will assume that there is no unit-to-unit variability (i.e., $\beta = (\gamma_0 = 1.59 \times 10^{-13}, A_2(\infty) = -1.42)$ and $E_a = 0.72$ are held constant).

Figure 3 shows the predicted degradation paths for Device B power drop, according to the deterministic degradation model with Arrhenius temperature dependence, but with no unit-to-unit variability.

In order to predict power drop for a unit in which temperature (and thus degradation rate) changes over time, one can use the following generalization of the model in (2):

$$A_2(t) = A_2(0) + A_1(0) \left[1 - \exp\left(-\int_0^t \mathcal{R}[\texttt{temp}(\tau)]\, d\tau \right) \right]. \tag{7}$$

where $\mathcal{R}[\texttt{temp}(\tau)]$ is the degradation rate constant, as a function of temperature, and temperature is allowed to vary with time τ. For example, if the rate is $\mathcal{R}_1$ from 0 to t_1 and $\mathcal{R}_2$ thereafter, then the power drop profile would be

$$
\begin{aligned}
A_2(t) &= A_2(0) + A_1(0)\left[1 - \exp\left(-\mathcal{R}_1 t\right)\right], \ 0 < t \le t_1 \\
A_2(t) &= A_2(t_1) + A_1(t_1)\left[1 - \exp\left(-\mathcal{R}_2(t - t_1)\right)\right], \ t > t_1
\end{aligned}
$$

where

$$A_1(t_1) = A_2(0) + A_1(0) - A_2(t_1).$$

Figure 4 shows the degradation path for a unit run at 150°C for 4000 hours and 237°C thereafter with $A_2(0) = 0$. For a general piece-wise constant temperature profile:

$$A_2(t) = A_2(t_{i-1}) + A_1(t_{i-1})\left[1 - \exp\left(-\mathcal{R}_i(t - t_{i-1})\right)\right], \ t_{i-1} < t \le t_i, \tag{8}$$

where $i = 1, 2, \ldots,$ $\mathcal{R}_i = \mathcal{R}(\texttt{temp}_i)$, $\texttt{temp}_i$ is the temperature between t_{i-1} and t_i, and $t_0 = 0$. Figure 5 shows the power drop path for a unit run at 150°C with a brief excursion to 237°C.

Figure 6 shows a simulated random temperature profile from Gaussian-noise discrete-time (one-hour time increments) first-order autoregressive [AR(1)] stochastic process model with a mean of 150°C, a standard deviation of 40°C, and autocorrelation $\rho_1 = 0.7$. Equation (8) provides a numerical tool for computing power drop for computing power drop for any arbitrary discrete-time temperature profile. The power drop profile corresponding to the temperature profile in Figure 6, computed from (8), is given in Figure 7.

Figure 8 shows simulated sample paths corresponding to five different simulated temperature profiles like that in Figure 6. The stochastic nature of simulated sample paths allows one to visualize the corresponding failure-time distribution. In Figure 8, for example, failure could be defined as

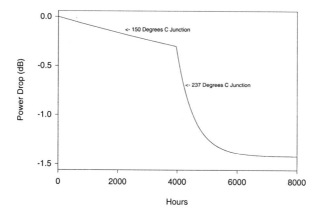

Figure 4: Power drop as a function of time with a change in temperature from 150°C to 237°C after 4000 hours.

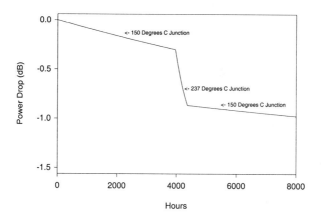

Figure 5: Power drop as a function of time with a brief excursion from 150°C to 237°C.

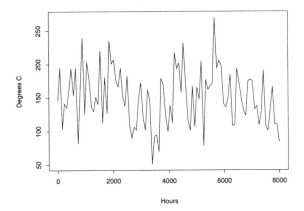

Figure 6: Simulated random temperature profile (AR(1) with $\rho_1 = 0.7$, $\mu = 150°C$ and $\sigma = 40°C$).

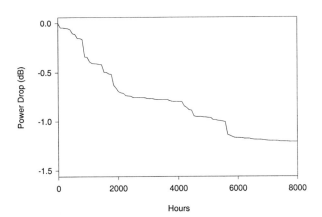

Figure 7: Power drop as a function of time corresponding to the simulated random temperature profile.

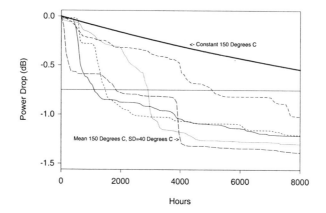

Figure 8: $\mathcal{D}(t|\boldsymbol{\xi})$ Power-drop cumulative degradation paths.

the first time at which the power drop reaches -0.75 dB. Simulating a large number of such curves would provide an evaluation of the failure-time probability distribution.

The smooth curve in Figure 8 was obtained by substituting the average temperature $(150°\text{C})$ into the constant temperature power drop model (3). It is interesting to note that, in this example, the stochastic sample paths tend to be lower than what would be predicted by substituting the average temperature into the constant temperature model. This shows the potential danger of substituting average values of random inputs into a deterministic model when trying to predict a response. The bias in the prediction is due to a combination of nonlinearity in (3) and in the Arrhenius relationship (4).

Service Life Distributions

In this section we return the more general notation used at the beginning of this paper. As described in Chapter 13 of [5], a closed form equation for the failure-time distribution for a degradation model can be obtained only for some very special, simple degradation models, even when environmental conditions are constant over time. Numerical or simulation-based methods are generally used and several general approaches are described there.

This section extends the methods described in [5], providing expressions that apply with both unit-to-unit variability and random environmental profiles. We start by reviewing methods for deterministic environmental profiles.

Deterministic environmental profile

In this section we condition on a fixed environmental profile $\boldsymbol{\xi}$ and use $\mathcal{D}(t|\boldsymbol{\xi},\boldsymbol{\beta}) = \mathcal{D}(t|\boldsymbol{\xi},\beta_1,\ldots,\beta_k)$ to denote the degradation for a unit as a function of $\boldsymbol{\beta}$ and time t. For a specific unit, $\boldsymbol{\beta}$ contains parameters describing unit-to-unit variability. We suppose that in a population of units, $\boldsymbol{\beta}$ has the density $f(\boldsymbol{\beta})$. Note that for a given $\boldsymbol{\xi}$ and $\boldsymbol{\beta}$, the path $\mathcal{D}(t|\boldsymbol{\xi},\boldsymbol{\beta})$ is deterministic. Conditional on the $\boldsymbol{\xi}$, the environmental profile,

$$
\begin{aligned}
\Pr(T \leq t|\boldsymbol{\xi};\boldsymbol{\theta}_\beta) &= F(t|\boldsymbol{\xi};\boldsymbol{\theta}_\beta) = \Pr\left[\mathcal{D}(t|\boldsymbol{\xi};\boldsymbol{\beta}) \leq \mathcal{D}_\mathrm{f};\boldsymbol{\theta}_\beta\right] \qquad (9) \\
&= \int \delta\left[\mathcal{D}(t|\boldsymbol{\xi};\boldsymbol{\beta}) \leq \mathcal{D}_\mathrm{f}\right] f(\boldsymbol{\beta};\boldsymbol{\theta}_\beta)d\boldsymbol{\beta}
\end{aligned}
$$

where the indicator function $\delta[\;]$ is 1 if the argument is true and 0 otherwise. For a fixed environment profile $\boldsymbol{\xi} = \boldsymbol{\xi}_0$, the cdf $F(t|\boldsymbol{\xi}_0;\boldsymbol{\theta}_\beta)$ gives the fraction failing as a function of time, reflecting unit-to-unit variability.

As described in Section 13.5.2 of [5], when $\mathcal{D}(t;\boldsymbol{\beta})$ is a decreasing function of t, depending on just two parameters $\boldsymbol{\beta} = (\beta_1,\beta_2)$ that are bivariate normal distributed, and if $\mathcal{D}(t;\boldsymbol{\beta})$ is decreasing in β_2, then (notationally suppressing the dependency on the fixed $\boldsymbol{\xi}$)

$$
F(t;\boldsymbol{\theta}_\beta) = P(T \leq t;\boldsymbol{\theta}_\beta) = \int_{-\infty}^{\infty} \Phi\left[-\frac{g(\mathcal{D}_\mathrm{f},t,\beta_1) - \mu_{\beta_2|\beta_1}}{\sigma_{\beta_2|\beta_1}}\right] \frac{1}{\sigma_{\beta_1}}\phi\left(\frac{\beta_1 - \mu_{\beta_1}}{\sigma_{\beta_1}}\right) d\beta_1
$$

where $g(\mathcal{D}_\mathrm{f},t,\beta_1)$ is the value of β_2 that gives $\mathcal{D}(t;\boldsymbol{\beta}) = \mathcal{D}_\mathrm{f}$ for specified β_1 and t and where

$$
\begin{aligned}
\mu_{\beta_2|\beta_1} &= \mu_{\beta_2} + \rho\sigma_{\beta_2}\left(\frac{\beta_1 - \mu_{\beta_1}}{\sigma_{\beta_1}}\right) \\
\sigma^2_{\beta_2|\beta_1} &= \sigma^2_{\beta_2}(1 - \rho^2).
\end{aligned}
$$

This approach is easy to adapt to functions $\mathcal{D}(t;\boldsymbol{\beta})$ that increase in β_2 or t. Also, this approach can, in principle, be extended in a straightforward manner when there are more than two continuous random variables. The amount of computational time needed to evaluate the multidimensional integral will, however, increase exponentially with the dimension of the integral.

Stochastic environmental profile

In this section suppose that the variability in the environmental profile $\boldsymbol{\xi}(t)$ can be described by a stochastic process model (say with controlling parameters θ_ξ). Then, because the degradation rate depends on the environment, $\mathcal{D}(t;\beta)$ is also stochastic, as illustrated in Figure 8 with the power drop example. Considering the variability in the environmental profile ξ, $F(t|\xi;\theta_\beta)$ is a random function. For fixed time t_0, $W = F(t_0|\xi;\theta_\beta)$ is a random variable reflecting the variability in the random environmental profiles.

Under the assumption that the random environment in ξ is independent of the unit-to-unit variability in β, the probability distribution

$$F_W(w;\theta_\beta,\theta_\xi) = \Pr(W \le w) = \Pr[F(t_0|\xi;\theta_\beta) \le w], \quad 0 < w < 1$$

allows assessment of the distribution of fraction failing by t_0, relative to an uncertain future environment. With respect to variability in the environment mental profile ξ, the expectation of W is

$$\mathrm{E}(W) = \mathrm{E}[F(t_0|\xi;\theta_\beta)] = F(t_0;\theta_\beta,\theta_\xi)$$

giving the conditional probability of failure before time t_0 for a single unit. More generally,

$$\begin{aligned} F(t;\theta_\beta,\theta_\xi) &= \int \Pr[\mathcal{D}(t) \le \mathcal{D}_f|\xi;\theta_\beta]\,f(\xi;\theta_\xi)d\xi \\ &= \int F(t|\xi;\theta_\beta)f(\xi;\theta_\xi)d\xi \end{aligned}$$

which is, in effect, averaging over all possible environmental profiles in the environment described by the environmental model parameters θ_ξ.

An alternative representation for the unconditional probability of failure

Conditional on a fixed β, the failure-time distribution is

$$F(t|\beta;\theta_\xi) = \Pr(T \le t|\beta;\theta_\xi) = \Pr[\mathcal{D}(t) \le \mathcal{D}_f] = \Pr\left[\int_0^t \frac{d\mathcal{D}[\tau;\xi(\tau)]}{d\tau}\,d\tau \le \mathcal{D}_f\right].$$
(10)

As in (6), computation of $F(t|\beta;\theta_\xi)$ requires the solution of the stochastic integral for the specific environment characterized by θ_ξ. Accounting for unit-to-unit variability in the β parameters gives the following alternative representation for the unconditional probability of failure for a single unit.

$$F(t;\theta_\beta,\theta_\xi) = \int F(t|\beta;\theta_\xi)f(\beta;\theta_\beta)d\beta.$$
(11)

Time transformation function with stochastic environmental profiles

As shown in the previous section, for a model with stochastic process variability, the failure-time distribution can be determined by solving stochastic differential equations or averaging of the possible realizations of a random environment. In principle, it is also possible to determine a time transformation function Υ for any specified failure model, characterized by a degradation model and a failure criterion, $\mathcal{D}_f$ and such a transformation function can be generalized to allow for stochastic environments. Such a time transformation function can be used to relate the failure-time distributions at two different locations with different environments characterized by different set of stochastic process model parameters, say θ_ξ and θ_{ξ_0}. That is,

$$\Pr\left[T \leq t; \theta_\xi, \theta_\beta\right] = \Pr\left[\Upsilon\left(T; \theta_\xi\right) \leq t; \theta_{\xi_0}, \theta_\beta\right]$$
$$= \Pr\left[T \leq \Upsilon^{-1}\left(t; \theta_\xi\right); \theta_{\xi_0}, \theta_\beta\right].$$

The function $\Upsilon\left(t; \theta_\xi\right)$ can be determined by finding the mapping between the failure time quantiles for different $0 < p < 1$ at conditions characterized by a stochastic process with parameters θ_ξ versus those with parameters θ_{ξ_0}.

Concluding Remarks and Areas for Future Research

This paper has outlined a general framework for using accelerated degradation tests to predict the performance in a highly-variable environment. The methods depend importantly on the ability to describe adequately the failure process by a relatively simple model (so that its parameters can be estimated reliably from experimental data) that gives degradation rate as a function of environmental conditions. Such models will have to be provided by scientists working with particular materials and products. Even with such a model in hand, there remain a number of technical challenges that are the subject of current research. These include

- Modeling of environmental variables (e.g., UV and other weather-related variables) with a stochastic process model.

- Methods for quantifying uncertainty in forecasts due to:

  ▶ Uncertain future weather.

▶ Uncertainty in model parameters (for both the random-effect unit-to-unit variability and the weather models) due to limited data.

Prediction intervals for quantities of interest like cumulative degradation or fraction failing would be the natural means of describing the effects of this uncertainty.

• Numerical techniques for implementing the methods.

• Approximations that will allow rapid analyses, at least for some special-case situations.

Acknowledgments

We have benefited from encouragement and helpful discussions with Gary Jorgensen and Jonathan Martin. Computing for the research reported in this paper was done using equipment purchased with funds provided by an NSF SCREMS grant, award DMS 9707740, to Iowa State University.

References

1. Hooper, J. H., and Amster, S. J. (1990), "Analysis and presentation of reliability data," in *Handbook of Statistical Methods for Engineers and Scientists*, Harrison M. Wadsworth, Editor. New York: McGraw Hill.

2. Jorgensen, G. J., Kim, H. M., and Wendelin, T. J. (1996), "Durability studies of solar reflector materials exposed to environmental stresses," Durability Testing of Non-Metallic Materials, Herling, R. J., Ed., ASTM STP 1294, American Society for Testing and Materials, Philadelphia, PA, 121–135.

3. Martin, J. W. (1999), "A Systems Approach to the Service Life Prediction Problem for Coating Systems," in *A Systems Approach to Service Life Prediction of Organic Coatings* Washington: American Chemical Society, D. R. Bauer and J. W. Martin, Editors.

4. Martin, J. W., Saunders, S. C., Floyd, F. L., and Wineburg, J. P. (1996), "Methodologies for predicting the service lives of coating systems," Blue Bell, PA: Federation of Societies for Coatings Technology.

5. Meeker, W. Q., and Escobar, L. A. (1998), *Statistical Methods for Reliability Data*. John Wiley & Sons: New York.

6. Meeker, W.Q. and Escobar, L.A. (1999), "Accelerated Life Tests: Concepts and Data Analysis," in *A Systems Approach to Service Life Prediction of Organic Coatings* Washington: American Chemical Society, D. R. Bauer and J. W. Martin, Editors.

7. Meeker, W.Q., Escobar, L.A., and Lu, C.J. (1998), "Accelerated Degradation Tests: Modeling and Analysis." *Technometrics* 40, 89-99.

8. Nelson, W. (1990), *Accelerated Testing: Statistical Models, Test Plans, and Data Analyses*, John Wiley & Sons: New York.

Chapter 20

Computer Programs for Nearly Real-Time Estimation of Photolytic and Hydrolytic Degradation in Thin Transparent Films

Brian Dickens

c/o National Institute of Science and Technology, 100 Bureau Drive, Building 226, B350, Stop 8621, Gaithersburg, MD 20899

Computer programs are described that allow the entry of data from the Building and Fire Research Laboratory photoreactor at the National Institute of Standards and Technology into specially designed relational database tables, facilitate the checking of the data, and calculate relative quantum yields of photodegradative processes in the presence of hydrolysis by providing ways of subtracting controls. Many processing options are included. The interface is predominantly driven by mouse clicks and is highly automated. The computer programs run under 32-bit Windows. They are partly a production tool and partly a research tool.

A simplistic view divides estimation of product service life into 1) manufacturing, 2) testing, 3) specifying expected service conditions, and 4) extrapolating the test results to the expected service conditions.

Manufacturing is included because, if nominally identical articles have a wide range of service lives under nominally identical service conditions, then the articles are not only not identical but also they are not even very similar. This may be because of inadequate control of conditions or materials during the manufacturing process. It complicates the analysis and significantly increases uncertainty in the results.

Testing is a matter of exposing the articles to the dominant "stress factors", typically considered for coatings to be UV and visible radiation, moisture, and heat because these factors are present at most locations. If other factors are active locally, such as salt near the ocean, they must be included. The testing must quantify the influence of the most important stress factors on the articles and must include measurements of the interactions between factors and of the effects of cycling the values of the factors. The goal is to predict the response of a typical article to specified combinations of factors. The intention is to use these predictions to estimate what will happen to the article under some chosen sequence of stress factor levels representing the expected service environment.

The expected service environment is a sequence of stress factor levels thought likely to be close to or representative of those that will occur in practice. Based on quantitative information from the testing, the response of the article to the expected service life is tallied until changes in the article reach levels designated as constituting failure. The response level calculated for the article will accumulate all the errors in measuring the levels of the component parameters and their effects. It will be difficult to make the calculated response level precise. Attention must be paid to detail at all times. There is no substitute for careful and thoughtful work, well planned and correctly performed.

Experimental Considerations

A photoreactor was built at NIST (1) to expose coating specimens to a wide range of constant levels of the three main stress factors, UV-visible radiation, humidity, and temperature. (External weathering does not provide the separation of factors, the constant levels of factors, and the wide variation in factor levels that are appropriate to characterize the effects of the factors on the specimens.) The design of the photoreactor proved to be well thought out and the few problems encountered during the first exposures were successfully overcome. The result of exposing a group of specimens was an enormous collection of spectra documenting changes in lamp irradiance, filter transmittance, specimen absorbance and specimen damage. With the computer programs described here, one can systematize and correlate these spectra, together with parameters which document the apparatus and the specimens, in relational database tables and within minutes provide a plot of degree of damage versus incident wavelength for various levels of humidity and temperature. Emphasis and effort can now be concentrated more on assessing the meaning of the results than on processing the data. This report will describe some of the capability of the programs which take the user to that stage.

Photolytic damage is caused by radiation absorbed by the specimen from sunlight or similar radiation incident on the specimen. Photolytic damage is expected to be wavelength specific, partly because different wavelengths contain different amounts of energy per photon and partly because which chemical group acts as the chromophore depends on the wavelength of the incident radiation. Photolytic damage is instantaneous following the absorption of radiation.

Hydrolysis is attainment of a temperature-dependent equilibrium and is driven by the concentrations of hydrolysable groups and water in the material (water being typically expressed in moles/liter). Because hydrolysis is a relatively slow process, it can take hours, days or weeks for equilibrium to be reached, depending on temperature and catalysts.

In the experiments carried out to date, damage (the cumulative effects of photolytic degradation and hydrolysis) has been monitored quantitatively by changes in IR spectra. This has the advantage of providing information on the chemical processes involved. Since the procedure currently involves estimation of apparent relative quantum yields, the emphasis in this system of programs is on photodegradation. Different systems of exposure can be envisioned and other quantitative evidence of damage will be used in future investigations as the need arises.

Rationale

The dosage is estimated from UV-visible spectra of the lamps and the interference filters used to isolate a particular wavelength range, and from the UV-visible absorption of the specimens themselves. Some of these changed with time. For example, the lamps aged and the specimens degraded. The experimental procedure is therefore to monitor the process over short ranges of time over which any change is presumed to be linear. Once the spectral output of the lamp and the transmittance of each filter are known, the amount of radiation incident on a specimen can be estimated. From the specimen absorbance, the transmittance of the specimen is known. Radiation incident on the specimen but which does not emerge at the other side of the specimen is presumed to have been absorbed by the specimen and is taken as being the apparent dosage. Whether to consider that all the absorbed radiation leads to degradation or whether some radiation is considered to be "harmlessly" absorbed by degradation products, acting as light stabilizers, rather than by the original coating material is a choice offered by the programs. The above procedure was developed for transparent specimens. For opaque specimens, absorptance will be estimated from a combination of reflectance and transmittance measurements

The programs allow the rapid estimation of dosage and damage from legions of data. In fact, one could say that the programs make possible the processing of so much data. This in turn makes feasible a systematic study of the effects of the "stress" factors over wide ranges of stress, a desirable and probably necessary condition where correlated processes are to be resolved. The programs are also research tools that organize the data so that aspects of the experiments can easily be recalled and examined. Archiving and copying the data merely require copying a few database tables.

The programs have been designed to be complete yet relatively easy to use. They run under 32-bit Windows, i.e., Windows 95, Windows 98, Windows NT and Windows 2000. The file structure of the database tables is MS FoxPro. This structure was chosen because it is widespread and has each table in a physically different file It is quite flexible and allows the easy manipulation of so-called memos or "Blobs" - collections of binary data which can be stored compactly on the computer disk yet can be read into computer memory extremely quickly. The structure of the data tables is described in the program manual (2).

Calculation of Dosage

Radiation from the lamp is directed to the cell containing the specimen, where it passes through a filter and is then incident on the specimen. The specimen absorption spectrum gives the fraction of the incident radiation that emerges from the other side of the specimen.

Components such as the lamps and specimens change with time. These changes are followed by taking UV-visible spectra and, in the case of the specimens, IR spectra, at reasonable intervals.

The procedure for estimating the dosage is:

Interpolate between lamp spectra adjacent in time to get the output, I_0 , of the lamp in watts at the time of interest.

Interpolate between filter spectra adjacent in time to get % transmittance = f. The transmitted radiation, I_f, is given by $I_f = f I_0/100$

Interpolate between specimen absorbance spectra adjacent in time to get A at each wavelength for which the spectra were measured. The transmitted radiation $I_t = I_f 10^{-A}$ where A = absorbance.

The radiation in watts absorbed by the specimen then given by $I_f - I_t = I_f(1-10^{-A}) = f I_0/100 (1-10^{-A})$. The flux I_0 is in watts (Joules/s) and is wavelength-dependent. Multiplying the flux by the elapsed time in seconds gives the dosage, i.e., the energy the specimen has absorbed, in Joules.

Propagation of error Propagation of error

A model must successfully account for effects seen in the experiments before it can be used to generate prophesies. But before a model can be proposed, the effects must be visible, i.e., above the noise. The noise levels in the measurements are very important. They affect whether or not the effects of the experiments are discernible and usable. They also determine the applicable range (in time or radiation level or hydrolysis level) of any prophesy.

The computational procedures involved in estimating damage and dosage include multiplying, dividing and subtracting combinations of spectral absorbances. It is necessary to be aware of the effect of noise and to minimize the noise produced in the measuring processes.

Given that $z = a + bx + cy$, then the propagation of error formula for subtraction and addition of x and y to yield z is

$\sigma^2(z) = b^2\sigma^2(x) + c^2\sigma^2(y)$, where σ is the standard error in the quantity.

For multiplication/division, $z = x^a y^b$, the formula is

$\sigma^2(z)/z^2 = a^2\sigma^2(x)/x^2 + b^2\sigma^2(y)/y^2$

Clearly, noise is never lost. Subtraction and division are especially troublesome in that the new derived quantity is usually smaller than the original quantities but the noise of the derived quantity is always larger than the noise in either of the original quantities.

Damage

In the photolytic part of the damage suffered by the specimen, the damage at a particular wavelength of incident radiation is given by

damage = flux * time * quantum yield (at each wavelength in the spectrum)

and the dosage, D, is given by D = flux * time.

Flux and time are quantities from the exposure in the photoreactor. The flux is presumed to be constant over each time interval.

If the damage is represented in terms of dosage, D, by a polynomial, then
damage $= a_0 + a_1 {}^*D + a_2 {}^*D^2 + a_3 {}^*D^3 + a_4 {}^*D^4$ (1),
where a_0, a_1, a_2, a_3 and a_4 are coefficients of the polynomial.

A polynomial is convenient for interpolation because it allows for any reasonable shape of damage/dosage curve, including those which are non-linear with increasing dosage due to shielding of the matrix by degradation products, using up weak links, formation of skins, etc. A polynomial should not be used for extrapolation.

Quantum yield

Quantum yield (q) is the damage per unit dose and is wavelength-specific.

The quantum yield for a particular wavelength in the incident radiation is given by differentiating equation (1) to give

$$d(damage)/dD = a_1 + 2*a_2*D + 3*a_3*D^2 + 4*a_4*D^3.$$

The coefficients a_1 etc., propagate the *random* errors arising from
1) the experimental data and
2) the assumption that the polynomial describes the damage/dosage relationship.

Systematic errors are not discerned in this treatment.

From the propagation of errors for multiplication,

$$\sigma^2(damage)/\ damage^2 = \sigma^2(flux_2)/\ flux_2^2 + \sigma^2(time_2)/\ time_2^2 + \sigma^2(q)/q^2$$

where $flux_2$ and $time_2$ are quantities characterizing the expected exposure. The uncertainties in the conditions which characterize the expected service environment play a part in the estimation of the expected damage uncertainty of the result, as do the uncertainties in the quantum yields.

Quantum yield is really the fraction of absorbed photons which lead to the event under consideration. Here, we have used the term quantum yield to represent the yield of damage, as shown by the IR spectrum of the specimen, per joule of absorbed radiation. If the IR absorbance decreases with dosage (because the chemical group causing the IR absorbance is being consumed), our quantum yield will be negative. We have kept the sign as additional information which need not be discarded.

Database tables

The programs allow the user to process data for a particular specimen by clicking on a button. To provide that capability, the programs must have ready access to the required data. The data are stored in various database tables which are inter-related so that any particular item of information is only stored in one table. These tables are in MS FoxPro format and can easily be read by any Windows-based database program. However, to do more than look at the contents of the tables requires some specific computer coding.

Two programs are provided to put data into the tables. One program automates entering spectra and the other makes it easy to enter textual data. They are described in the program manual.

At least two programs are provided to allow the user to browse the data tables. One browses through tables of spectra and the other through textual tables. The spectra-

420

browsing program allows a certain amount of spectral manipulation so that trends and effects can be monitored. Some small amount of editing of the data can be carried out in both programs. This is necessary to correct the mistakes that inevitably creep into large sets of data.

Calculation of dosage, damage and quantum yield

Several steps are required to calculate dosage, damage and quantum yield. The database tables to use must be made known to the program, as in the screen below. Tables can be attached to the program by reading in a previously saved initialization file, by clicking on each table button and selecting the table file name in a standard open file dialog, or by dragging and dropping the file name from Windows Explorer onto the table button (Figure 1).

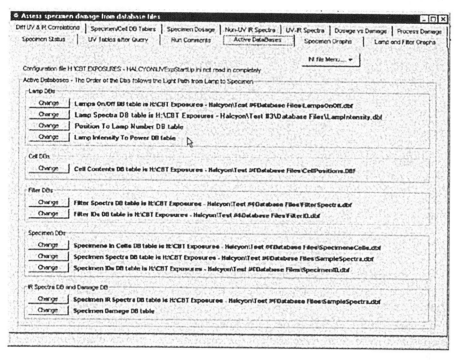

Figure 1. Screen where database tables are assigned.

Specimens can be processed singly, both manually and automatically, or automatically in batches. A specimen is chosen by clicking on the navigator bar at the top of the next screen. With each click, the information for the current specimen is displayed in the lower half of the screen. When first starting, one would select a specimen and click on manual processing (Figure 2).

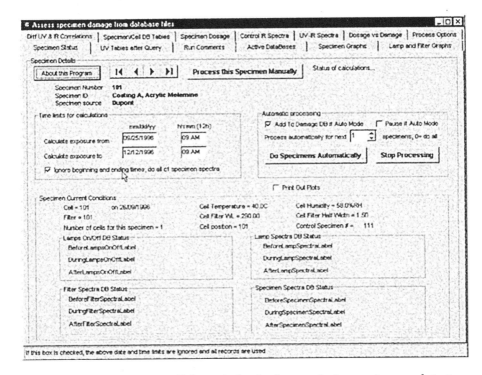

Figure 2. Screen to scroll through the database, select a specimen and start processing .

Control Parameters

The program is very flexible, but this means that choices have to be made. Most of the choices are on the "Process Options", the last page of the tabbed page control (Figure 3).

Zeroing the UV Spectra

Before any curve can be scaled or used in other arithmetic procedures, the data must be corrected for offset from the true zero position. In the case of the spectral curves in this program, the x values are typically wavelengths and are well known, but the measurement of the lamp intensity, for example, includes a dark current which will be non-zero even when the lamp is off. The lamp spectra will be moved along the vertical axis until they are zero at the wavelength specified in the "Zero lamp spectra at" box (Figure 4). The specimen UV-visible spectra will be similarly be moved along the vertical axis until they are zero at the wavelength specified in the "Zero specimen spectra at" box. If the value in a box does NOT fall in the wavelength range of the spectrum, no zeroing will be done, which provides a convenient method of turning the zeroing off.

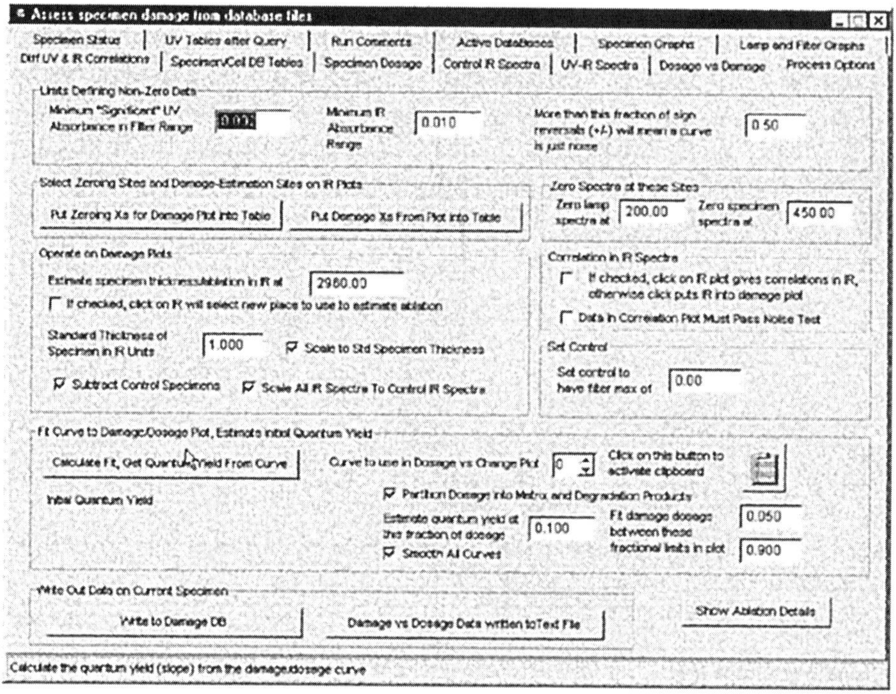

422

Figure 3. The process options screen.

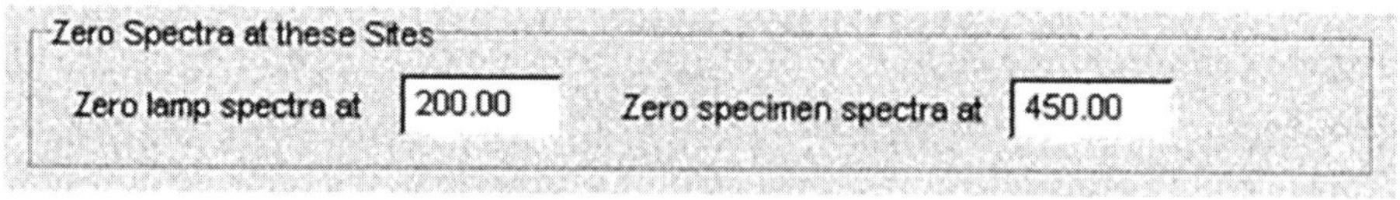

Figure 4. Selecting the zeroing sites for the lamp and specimen UV-visible spectra.

The filter spectra are not zeroed. The lamp and specimen UV-visible spectra are corrected by simple vertical shifts of the data. This removes the dark current equally from all places in the lamp spectra. No corrections for sloping backgrounds are applied to the specimen UV-visible spectra.

Zeroing the Damage (IR) Spectra

The damage spectra (currently IR spectra) can also be zeroed. There are several places in a typical IR spectrum where the absorbance can be expected to be zero. These places are selected by clicking on a plot of the damage spectra. The intersection of the wavelength where the plot was clicked and the absorbance on the curve defines a point which will be made zero. All points on a straight line between any adjacent clicked points will also be made zero. If there are no adjacent clicked points, as for before the first point or after the last point clicked in the plot, all points from the clicked point to the beginning or end of the plot will have the y coordinate of the clicked point subtracted from their y coordinates. This means that a series of straight lines can be defined by clicking on the IR plot and the straight lines between these points will define the new zero baseline.

To specify where to zero the spectra, click on the "Put Zeroing Xs for Damage Plot into Table" button (Figure 5) on the "Process Options" page.

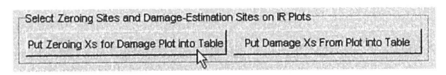

Figure 5. Entry buttons to the select zeroing sites procedure.

The table of sites at which to zero the damage plot is then shown and the damage plot is displayed (Figure 6). Sites can be typed into the table or put into the table by right-clicking on the damage plot. The site produced by clicking is entered into the table of zeroing sites where the cursor is so, if the cursor is at the top of the table, successive clicks will move down the table, replacing whatever is already in the table. To avoid wiping out what is already in the table, one should click on an empty line in the table below any filled lines. When the process is ended, the table will be tidied up by the program.

Zeroing takes place on the data only after they have been read into the program. The original data in the database table are left untouched.

424

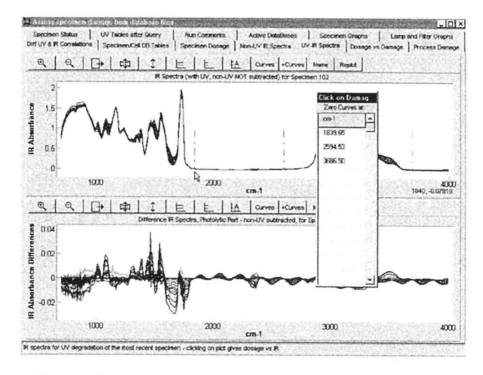

Figure 6. Choosing the zeroing sites in the damage spectra by clicking on the spectra.

The process of adding zeroing sites to the table must be ended so that mouse clicks can revert to their normal usage. The "Put Zeroing Xs for Damage Plot into Table" button on the "Process Damage" page is down during the zero site adding process. Re-clicking on this button ends the process (Figure 7). The program will now use the list of sites to zero all damage spectra before they are processed further.

Selecting the Damage Sites in the Damage Spectra

The spectral sites at which damage is to be assessed are selected in a procedure similar to that used to selecting the zeroing sites. To specify where to monitor the damage spectra, click on the "Put Damage Xs from Plot into Table" button (Figure 8) on the "Process Damage" page.
This brings up the Damage Site table (Figure 9). Right mouse clicks on the damage plot (or, better, on the difference damage plot) will put the abscissa (wave number in the case of IR spectra) values for these sites into the table. The entries begin where

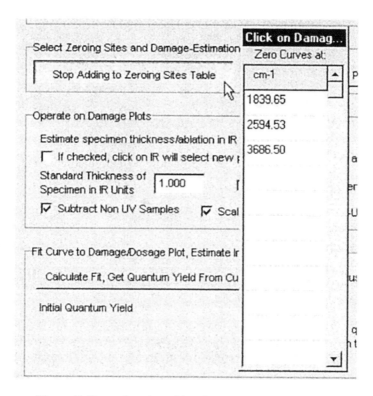

Figure 7. Removing the table of zeroing sites from view.

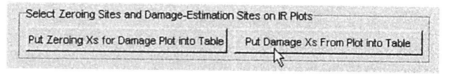

Figure 8. Starting the selection of sites in the damage spectra to use to estimate damage.

426

the cursor is in the table, so before clicking on the damage plots, put the cursor in an empty row of the table below any filled lines.

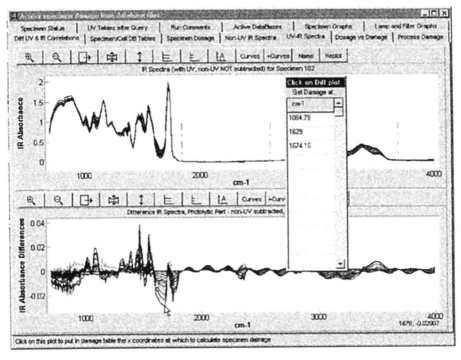

Figure 9. Adding damage sites to the damage site table.

The process is ended by re-clicking on the "Put Damage Xs from Plot into Table" button on the "Process Damage" page. This button is kept down during the process to show that the process is underway. When the table is hidden by the program, the entries in the table are tidied up so that there are no empty rows. When a specimen is processed automatically, damage will be estimated at the sites in this table.

Ablation

Some specimens suffer loss of material during the exposures. For convenience, this type of loss will be referred to here as ablation. The results of ablation must be corrected if effects in specimens of different thicknesses are to be compared or subtracted. This situation occurs when the spectra of controls must be subtracted from those of severely degraded exposed specimens. The program provides a mechanism for correcting for ablation by scaling the IR (damage) spectra so that the absorbance at a user-specified wavelength has a pre-specified value (Figure 10). Alternatively, the thickness of the specimen can be estimated by comparing its absorbance at some wavelength with the absorbance at the same wavelength of a specimen of known thickness. The wavelength in the damage spectrum at which to estimate specimen thickness is specified in the box next to the "Estimate specimen thickness/ablation at" label and should be a place in the spectra not be affected by photolytic or hydrolytic degradation. This box can conveniently be filled in by first checking the check box labeled "If checked, click on IR will select new place to use to estimate ablation", then clicking on the IR plot. Uncheck the check box before proceeding.

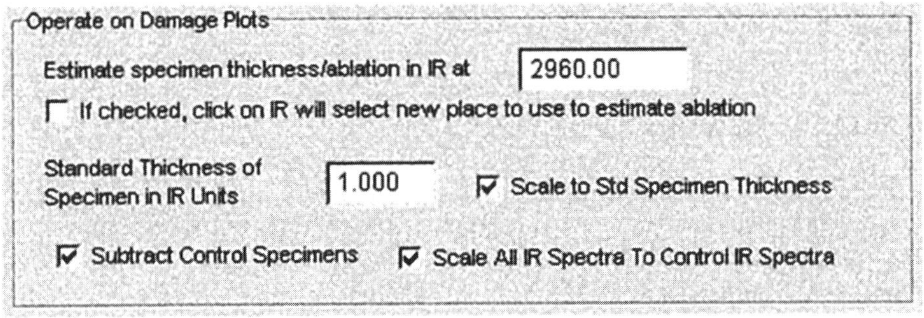

Operate on Damage Plots

Estimate specimen thickness/ablation in IR at   2960.00

☐ If checked, click on IR will select new place to use to estimate ablation

Standard Thickness of
Specimen in IR Units   1.000   ☑ Scale to Std Specimen Thickness

☑ Subtract Control Specimens   ☑ Scale All IR Spectra To Control IR Spectra

Figure 10. The ablation site and spectral scaling options.

Specimen Thickness

To estimate the damage a specimen has sustained, IR spectra from the specimen and a control must be adjusted and subtracted. Typically, the control specimen has been kept in the dark and the "damaged" specimen has been exposed to lamp radiation. These specimens were probably not of exactly the same thickness at the beginning of the exposures and the thickness of the exposed specimens may have changed as a result of ablation during the exposure. Thus, the spectra will probably not have equal intensity where they should have equal intensity. If the "Scale to Std Specimen Thickness" check box (see above figure) is checked (by clicking on it), the spectra are scaled (on the vertical axis) to a standard value, specified in the "Standard Thickness of Specimen in IR units" box, at the wavelength specified in the "Estimate specimen thickness/ablation in IR at" box (see the figure above). Subtractions of the spectra from one another will then be more realistic. If on the other hand the ablation is of major interest, one would not want to correct for ablation in this way because the effect of ablation would then be removed from the data. The check box allows the program to handle both cases.

Subtracting the Control Specimens

The effects in the control specimens are usually subtracted from the exposed specimens - that is why there are controls. None the less, the option is provided not to subtract the controls by unchecking the "Subtract Control Specimens" check box (see the figure above).

Scale all IR Spectra to Control Spectra

If the "Scale to Std Specimen Thickness" check box is not checked, no scaling to the ablation site will be carried out and ablation will be included in the results. Including ablation may override any chemical changes; not including ablation will make the chemical changes the only changes.

Smoothing the spectral curves

If the "Smooth All Curves" check box is checked, the raw data spectra will be smoothed immediately after they are added to the plots (the data on the disk will remain untouched). Subsequent arithmetic operations using these spectra will then use the smoothed values.

Ablation, Dosage and Damage

The dosage of a specimen is calculated from the lamp and filter spectra and from the UV spectra of the specimen. Because the beam traversing the specimen is attenuated by the specimen, degradation is probably not uniform across the specimen. Scaling the dosage as a function of specimen thickness is not a simple matter. Therefore, if ablation has taken place and but corrections for ablation are not to be made, the damage spectra are first scaled so that they can be compared with and corrected by reference to the control damage spectra, then, when making the damage/dosage plot, are rescaled back to the values they should have for a specimen of the thickness that produced the damage spectrum.

Signal versus Noise

The Building and Fire Research Laboratory photoreactor and this system of programs are used to assess the weathering resistance of specimens that are presumably very resistant to the effects of weather. Therefore it may be that some aspects of the test lead to a low signal (result of measurement). Consequently, noise is a problem which must be watched carefully, the experimental conditions must be well-controlled, and the exposures must be long enough for the results to be significantly above the noise.

The **first** test of whether the signal is above the noise is the minimum amount of UV absorbance the specimen is required to have in the region of the filter transmission. This is the quantity specified in the box labeled "Minimum Significant UV Absorbance in Filter Range" in Figure 11. The specimen UV absorbance must be well-determined for the calculations to be meaningful. The specimen UV absorbance enters the dosage calculations as the exponent of 10, i.e., as A in 10^{-A}.

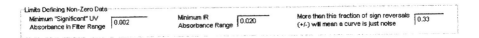

Figure 11. Setting the noise levels and noise test.

The **second** test is that the IR absorbance where the damage is to be monitored must change during the experiments by at least the number of absorbance units specified in the box labeled "Minimum IR Absorbance Range" (Figure 11).

The **third** test is on the fraction of times a damage or dosage curve changes direction when plotted against time of exposure. If the two lines joining three adjacent points in the curve have slopes of different sign, this is counted as a sign of noise. The test then examines all sets of three adjacent points. The fraction of the cases where noise is signaled is compared with the number given in the box labeled "More than this fraction of reversals will mean a curve is just noise" (Figure 11). A curve which first grew and then declined would always have one such point, where it changed from growing to declining, but many more than one such points would signify that the curve changed direction many times and presumably was very noisy. If a curve is found to be noisy and automatic processing is underway, the data in that curve will not be processed further. If manual operation is underway, warning messages will be shown.

Correlations in the Damage Spectra

The damage spectra show the changes which occur as the specimens are exposed to a laboratory environment. It may be that it is not immediately obvious what those changes mean. Therefore an option was provided to assess which in which regions in the damage spectra changes are related to each other.

Chemical groups which appear in the specimen during degradation will have increased IR absorbance. Chemical groups which disappear during the degradation will have decreased IR absorbance. It would simplify the interpretation of the IR spectra if changes arising from a given process (i.e., destruction of a chromophore and creation of degradation products) could be identified and treated together. Secondly, it is futile to examine two regions in the damage spectra which are well correlated because one is then using the same information twice. Thirdly, changes in spectra are best measured at absorbances of 0.4 to 0.7 or perhaps as high as 1.0, so the intra-spectra correlations can be used to select a region with the most-nearly satisfactory absorbance.

A system was therefore devised and implemented to assess correlations among absorbances in the damage spectra. There are typically 10 to 20 spectra in a series of damage spectra. For every wavelength, there are therefore 10 to 20 absorbance values which, if plotted with spectrum number (1,2,3,4...) on the x axis and absorbance on the y axis, constitute a curve. These curves are used to estimate intra-spectral correlations. If the user selects a location in the damage spectra (by clicking on the plot of the spectra) the correlations of that site with all other sites (at all other wavelengths) in the IR spectra is calculated and plotted. This shows which regions of the spectra are correlated, simplifies the interpretation of the spectra, and allows the user to follow damage using the best site, as discussed above.

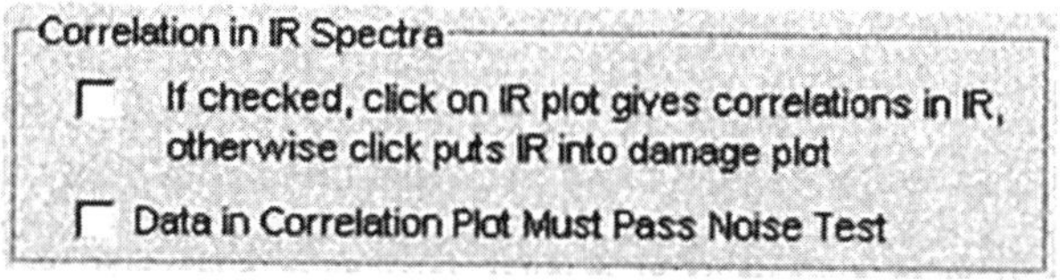

> ┌─ Correlation in IR Spectra ──────────────
> │   ▢   If checked, click on IR plot gives correlations in IR,
> │   otherwise click puts IR into damage plot
> │   ▢ Data in Correlation Plot Must Pass Noise Test

Figure 12. Setting the options to calculate correlations in the damage spectra.

The first check box in Figure 12 sets up the damage plot so that right clicks on it will be interpreted as places for which to calculate the correlation to all other parts of the spectrum. The various correlation curves are plotted in a correlation plot. Uncheck the check box after the tests so that mouse clicks can revert to their normal functions.

If the "Data in Correlation Plot Must Pass Noise Test" check box is checked, the program first checks that the trend with spectrum number (i.e., time) is not noisy according to the fraction of sign reversals permitted for the slope of lines joining adjacent points. If the trend fails the test, no correlation plot will be made. The severity of the test for noise can be reduced by increasing the fraction of allowed reversals (i.e., making the number in the box shown in Figure 13 closer to 1.0).

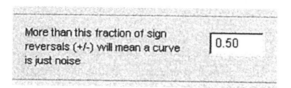

Figure 13. The fraction of sign reversals permitted for non-noisy curves.

When the correlation plot is being made, trends in the spectra which do not pass this same noise test will have their correlation with the trend at the point clicked set to zero.

As an example of a correlation plot, a right click at 1674 cm^{-1} in the spectrum in Figure 14, i.e., on the right-hand peak of the double-peaked envelope near the middle of the plot,

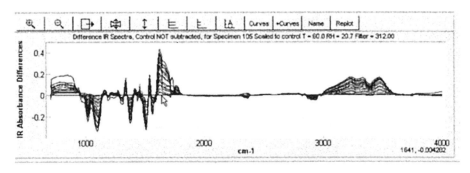

Figure 14. Sample damage plot providing the correlation plot in Figure 15.

gives the correlation plot in Figure 15.

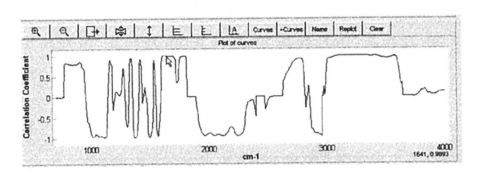

Figure 15. The correlation plot from Figure 14.

Places in the correlation plot where the vertical coordinate is at or near +1 contain the same trend with time (i.e., with damage spectrum number) as the absorbance at 1674 cm^{-1}, although the actual values of the IR absorbances are almost certainly different. Places where the vertical coordinate is at or near -1 have the opposite trend with time as the series of IR absorbances at 1674 cm^{-1} and may be from a group which is consumed when the group absorbing at 1674 cm^{-1} is produced.

Saving the processing parameters

When the control parameters have been set, they can be saved with the names of the database tables in an initialization file which can be used to start the program with the same conditions in the future.

Processing a batch of specimens

For automatic processing, the user clicks on the button on the first page (Figure 16).

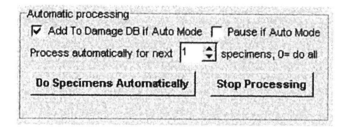

Figure 16. The automatic processing options.

selecting the number of specimens given in the spin box, or, if all specimens from the current one to the end of the data set are to be processed, selecting 0 as the number of specimens to do. The results are shown visually in plots and may be written to a damage database table that can be examined using two other programs. Examples are given in the program manual. Data from most of the plots can be written in textual form to the Windows clipboard for pasting into other packages.

Comparing results from different specimens

One "comparison of results" program provides interpolations of damage/time curves to 0%, 20%, 40%, 60%, 80% and 100% levels of RH and allows other operations on the damage/time curves. The second program compares the quantum yields themselves and allows the user to assess the evidence the quantum yield program amassed for each estimate of quantum yield. Examples are given below.

The user can select to plot quantum yields as a function of the various exposure parameters. Figure 17 is a plot of quantum yield versus filter wavelength for damage at 1630 cm^{-1} in an acrylic melamine exposed to temperatures of 50 and 60C. Data

from all humidities at these temperatures are included. The amount of photolytic
degradation was generally small, especially at the lower wavelengths.

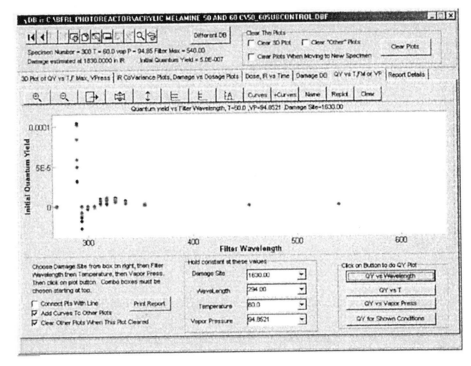

Figure 17. Quantum yield versus filter wavelength .

The data seem to be very consistent with wavelength among all these cases. This
provides confidence in the data collection and in the calculation procedure. The fall-
off in quantum yield below 320 nm is due to little or no perceptible degradation
because at those wavelengths 1) the lamp output is low, 2) the filter range is narrow
(2 nm) and 3) the filter transmission is low (20% for the lowest).

The conditions imposed on the data in the program that wrote the damage database
table are shown in Figure 18, a screen from the dosage-calculating program.

The damage-dosage plot is the lower of the two plots in Figure 19 – there are some
cases which have very low dosage, as shown by the short lines.

In Figure 20, the plot was zoomed (using the first plot button) and the short lines
identified using the plot "Names" button. As is shown, one of the short lines is for
the wavelength of 290 nm.

434

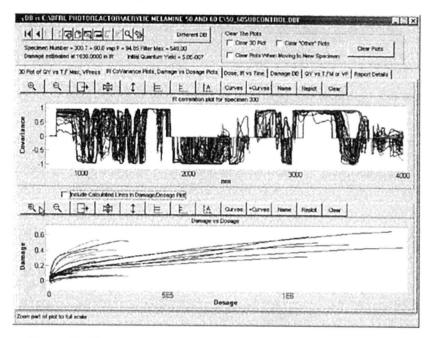

Figure 18. *The conditions used in the data-processing program to give the results shown in Figure 17.*

Figure 19. *Top = damage spectra intra-spectral correlations, bottom = Damage/dosage plots for the results in Figure 17.*

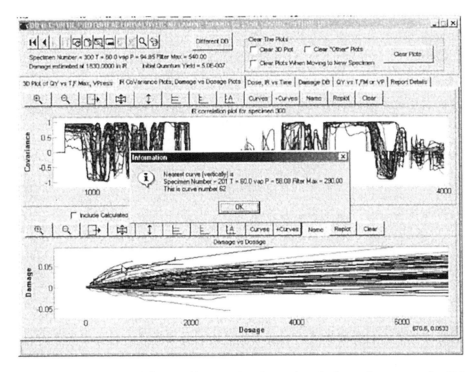

Figure 20. The zoomed damage/dosage plot shows that the least dosage was for 290 nm radiation.

Conclusions

Data documenting exposure conditions and the resulting changes in clear films have been assembled into a system of relational database tables. Programs have been provided to create and assemble the tables, to examine and edit the tables, and to use the tables to provide estimates of apparent relative quantum yields of damage production in the films. Other programs allow the evidence for degradation/damage to be examined and the quantum yields to be compared in ways that make sense to the analyst.

Processing the data for a single specimen once the data tables have been assembled takes about 20 seconds on a 300 MHz PC even when all data are used. A batch of specimens comprising one set of exposures typically consists of 88 specimens. The entire batch of specimens can therefore be processed in 30 minutes or less. It is feasible and desirable to add each set of new data to the database tables as soon as it has been measured and rerun the calculations to assess whether a suitable stopping point has been reached or whether something unexpected is happening. The new data gathered at each stage of the investigation are spectra from the lamps, filters and specimens. It may take half a day to measure the spectra at each intermediate stage. It may take another hour to add them to the spectral tables including setting up the procedure, which then proceeds automatically. Rerunning the calculations requires specifying and reading in an initialization file and clicking on a couple of buttons in the program interface. The calculations take up to about 30 minutes per set of 90 specimens. The entire process, including spectral measurements, takes significantly less than a day and, given that the exposure period is typically 3 months, is in this sense essentially real-time.

References

1. J.W. Martin, E. Byrd, E. Embree, T. Nguyen and J.A. Lechner, "A collimated, spectral ultraviolet radiation, temperature, and relative humidity controlled exposure chamber", (1998).
2. B. Dickens and E. Byrd, "Programs to estimate UV dosage and damage", in review, (1999).

Chapter 21

A First Step toward Photorealistic Rendering of Coated Surfaces and Computer-Based Standards of Appearance

Fern Y. Hunt[1], Egon Marx[1], Gary W. Meyer[2], Theodore V. Vorburger[1], Peter A. Walker[2], and Harold B. Westlund[2]

[1]National Institute of Standards and Technology, 100 Bureau Drive, Gaithersburg, MD 20899
[2]Department of Computer and Information Science, University of Oregon, Eugene, OR 97403

In an effort to understand the physical basis for coating and surface appearance we are combining the results of optical and surface topographical measurements, mathematical modeling and computer graphic rendering. We seek to explore the feasibility of producing computer graphic images to visualize the color and gloss of surfaces using measured data and models so that rendering becomes a tool to identify important parameters in the material formulation process that contribute to appearance. Here we report on a study of gloss variation in a series of samples with controlled roughness. The work makes use of the sample preparation, characterization and measurement described in [3]. Modeling based on that data was used to produce computer graphic images of the samples. We also briefly describe work on the rendering of data from some early measurements of colored metallic paint.

Goals and General Description of the NIST Appearance Project

As advances in material science and technology have enhanced the ability to manufacture coatings that are exciting and attractive in appearance, customer expectations for these products have increased and with this the challenge of characterizing and predicting appearance at the coatings formulation level. The ability to do this will require a much better understanding of the microstructural basis for coating appearance and the development of tools that can be used to firstly to identify important parameters in the formulation process that contribute to a desired appearance and secondly allow designers to visualize the surface appearance of a proposed formulation as part of a virtual formulation and manufacturing process. The ability to view a virtual end stage product will eventually pave the way to a computer graphics based standard for appearance. This is the vision guiding an ongoing project at the National Institute of Standards and Technology. The project was initiated in response to recommendations by industry [1], and the Council of Optical Radiation Measurements [2] and its purpose is to apply the technical advances in microstructural analysis, optical metrology, mathematical modeling and computer graphic rendering to the development of more accurate methods of modeling and predicting the appearance of coatings and coated objects. The specific research goals are:

- Develop advanced textural, spectral and reflectance metrologies and models for quantifying light scattering from a coating and its constituents and use the resulting measurements to generate scattering maps and validate physical models describing optical scattering from a coating and the relationship that the scattering maps have to the appearance of a coated object

- Integrate measurements and models in making a virtual representation of the appearance of a coating system that can be used as a design tool capable of accurately predicting the appearance properties of coated objects from the optical properties of its constituents.

In this paper we will discuss our progress towards these goals through our investigation of a special case - that of gloss in a black glass sample with a topcoat of clear epoxy. In the course of this discussion we will illustrate the issues involved in achieving them.

Reflectance and Topographical Measurements of Coated Epoxy Samples of Varying Roughness

Sample Preparation

The samples used in this study were made of a clear epoxy coating on a base of black glass. As detailed in [3], the samples had a controlled degree of rms roughness varying from 3 nm to 805 nm . The sample preparation procedure except for the smoothest case began with a mold made of a square steel panel with rms roughness ~1 μm. A new mold with the desired degree of roughness was created from the base mold by a spin coating procedure that applied varying amounts of a surface-modifying polymer to the base material. Roughness decreased as the thickness of the polymer coat increased. The mold was used to create the epoxy sample. See [3] for more details and discussion. Although the assumption of sample isotropy was not needed in the calculations, the optical reflectance of the sample was compared with measurements of a rotated sample . The differences were not large and this remained true for reflectances obtained from topographical measurements as well.

Optical Scattering and Surface Topographical Measurements

Optical measurements and surface topographical measurements were performed on each of the samples. To understand the nature of the optical measurement it will be helpful to review some definitions. We can assume a ray of light is incident on a sample surface that lies in the x–y plane. If the z-axis is considered to be the normal direction, the direction of the incident ray can be taken to be $(\theta_i, 0)$ where θ_i is the angle between the ray and the z-axis and 0 is the value of the azimuthal angle. If the direction of a scattered ray is defined by the angles (θ_s, ϕ_s), where θ_s is the angle between the ray and the normal direction z and ϕ_s is the azimuthal angle, then the bidirectional distribution function (BRDF) can be defined as:

$$\text{BRDF} = \frac{\text{differential radiance}}{\text{differential irradiance}}$$

Here the irradiance is the light flux (in watts) incident on surface per unit illuminated surface area. The radiance is the light flux scattered through a

440

solid differential angle per unit of illuminated surface area and per unit of projected solid angle [4]. If the solid angle is Ω_s the projected solid angle is $\Omega_s \cos(\theta_s)$. In general terms, the BRDF gives the fraction of incident light reaching a surface from a given direction that is reflected in a specific outgoing direction. Arguments in [4] show that for an isotropic surface we can make the following approximations,

$$\text{BRDF} ; \frac{dP_s\big/d\Omega_s}{P_i \cos(\theta_s)} ; \frac{P_s\big/\Omega_s}{P_i \cos(\theta_s)}$$

where P_i and P_s are incident and scattered light fluxes. The actual quantity measured by the optical instruments is called the reflectance ρ the ratio of the fluxes, where

$$\rho = \frac{P_s}{P_i} = \text{BRDF} \cdot \Omega_s \cos(\theta_s).$$

In-plane measurements (i.e. $\phi_i = \phi_s = 0$) were performed on each of the samples and a black glass replica (included for comparison), using the NIST spectral tri-function automated reference reflectometer (STARR) [3]. Incident light had varying wavelengths in the visible spectrum 500 nm, 550 nm, and 600 nm for incident angles of 20 degrees, 45 degrees, 60 degrees, and 70 degrees and scattering or viewing angles ranged from –75 degrees to 75 degrees in steps of 1 degree. These measurements formed the basis for comparison to computations of the sample reflectance that were based on topographical data i.e. measurements of the surface height that were obtained by interferometric microscopy. See [3] for details of the procedures used.

Light Scattering Calculations and Comparison with Optical Measurements

Calculations of the surface reflectance based on surface topographical measurements were done in two ways. The first uses the Kirchhoff approximation of the solution of the wave equation as developed by P.Beckmann [10]. The second approach- the ray method- is based on the assumption that incident light is specularly reflected by the local tangent plane. The results of reflectance computations using the two methods were compared to measurements carried out with STARR. The agreement, as shown in Figures 1 and 2, for the rough and smooth samples was good, demonstrating the adequacy of the approximations used in the ray method. In addition, its implementation was also computationally efficient enough to generate the large number of incident directions and scattering intensities required to determine a BRDF suitable for a rendering program. We will therefore briefly describe some of its details and follow it with a brief description of the Kirchhoff method.

The Ray Method

A surface topography map of a rectangular patch was obtained using an interferometric microscope. The z or height coordinate was determined for a grid of x- and y-coordinates to represent the sample surface. The local normal to the surface was computed either by fitting a cubic spline to the data or from a tangent plane through an interior point; the method used made little difference for the scattered intensity [3]. The approximate tangent plane was obtained by a least-squares fit of a plane through an interior point by minimizing the sum of the squares of the distances to that plane from the eight nearest neighbors. Once the local unit normal $\hat{n}$ is determined, the wave vector of the incident ray $\vec{k}_i$ can be decomposed into the sum of a component along the local normal and a perpendicular component.

$$\vec{k}_i = \vec{k}_i \cdot \hat{n}\hat{n} + (\vec{k}_i - \vec{k}_i \cdot \hat{n}\hat{n}).$$

The wave vector specularly reflected ray $\vec{k}_s$ is in the plane defined by $\vec{k}_i$ and $\hat{n}$ and it is obtained by simply changing the sign of the component of $\vec{k}_i$ along $\hat{n}$, that is,

$$\vec{k}_s = \vec{k}_i - 2\vec{k}_i \cdot \hat{n}\hat{n}.$$

Therefore the magnitudes of $\vec{k}_i$ and $\vec{k}_s$ are the same, that is, $\vec{k}_s^2 = \vec{k}_i^2$. Dividing by the common magnitude, we obtain the corresponding relationship between unit vectors,

$$\hat{k}_s = \hat{k}_i - 2\hat{k}_i \cdot \hat{n}\hat{n}.$$

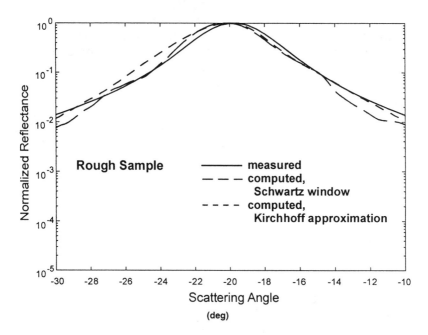

Figure 1. Comparison of measured and computed intensity distributions for the rough sample.

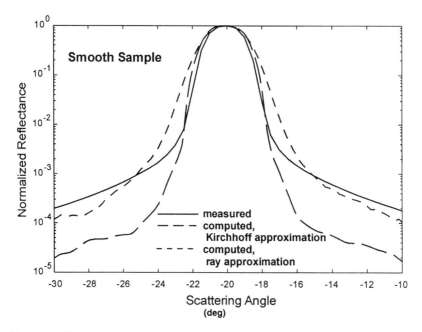

Figure 2. Comparison of measured and computed intensity distributions for the smooth sample.

For each direction of incidence, uniform illumination is simulated by computing a reflected ray at each of the interior points. We further assume that a number of detectors are located at various positions to measure the intensity of scattered light. The position of each detector is defined by a unit vector $\hat{r}_j$ and its aperture by the half-angle α. The angle β between the direction of the jth detector and a reflected ray can be determined from

$$\cos \beta = \hat{k}_s \cdot \hat{r}_j.$$

The scattered ray reaches the detector and a "hit" is recorded if $\alpha < \beta$, i.e., if $\cos\beta > \cos\alpha$.

The number of detector "hits" is proportional to the radiance (watts/sr) in that scattering direction for the given incident direction. The unnormalized BRDF can be estimated by dividing the computed intensity by $\cos\theta_s$. For the comparison with optical measurements, α was taken to be 1.4 degrees, the half angle of the aperture of the STARR. When this procedure was used for the rendering program, 0.2 degrees was used, based on an estimate of the half angle for a detector representing the human eye.

The Kirchhoff approximation

This approximation of the solution of the scalar wave equation corresponds to a field that is equal to the incident field plus the field reflected by the local tangent plane to the surface in the illuminated region and that vanishes outside this region [3]. A calculation of the phase integral uses the same surface topography map used for the ray approximation, interpolated by a two-dimensional spline function to obtain the surface at about ten points per wavelength. The assumptions used in the Kirchhoff approximation lead to jump discontinuities on the boundary of the illuminated region, which causes the solution to oscillate and vanish like a sinc function for a flat illuminated surface. To avoid the difficulties with discontinuities at the edge of the illuminated region, a windowing function is introduced in the resulting phase integral. We have found that a good windowing is a smooth function that vanishes outside a finite region based on the test function used by Schwartz in his theory of distributions [3].

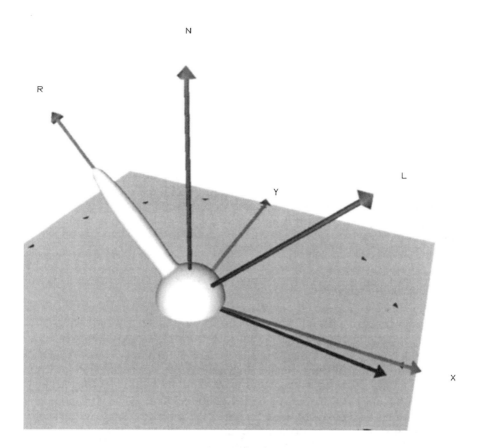

Figure 3. Phong reflection model

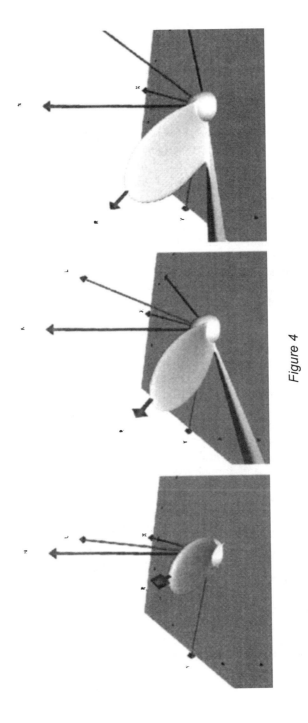

Figure 4

Computer Graphic Rendering

Rendering is the creation of a synthetic image based on information about the scattering and reflection of light. The specific requirements for rendering a scene are:

- A geometric description of objects in the scene
- The location and strength of light sources in the scene
- A description of how surfaces in the scene reflect incident light as provided by the BRDF or other angle dependent scattering function.
- The location of a detector or observer's viewpoint.

There are therefore three basic steps in the creation of a computer graphic picture. First the size, shape, and position of the objects in the scene must be defined. Next the light reflection properties of all surfaces must be specified, and the location of the illumination sources in the environment must be given. Finally, a local reflection model is used to determine the color of each surface point, each point is projected onto the picture plane, and the color of the surface location is stored into the appropriate position of the image raster. When each picture element (pixel) has been computed, the image is complete.

The model that is used to characterize light reflection from an object's surface determines the appearance of each object in the scene. There are many parameterized light reflection models that have been developed for use in computer graphic rendering (see e.g. [3]). As shown in Figure 3, most of these models have a general diffuse component and a specular lobe in the mirror direction. The entire reflection distribution is usually a linear combination of the diffuse and specular portions, and the shape of the diffuse and specular parts is controlled parametrically. These models have been extended to include such effects as anisotropic reflection (Figure 4), polarization, and Fresnel effects.

Allowing for variation with wavelength, the BRDF allows a completely general specification of surface reflection. However, in the context of image synthesis, the use of the BRDF makes it more difficult to compute the color of each point on an object's surface. To achieve a realistic approximation of the reflectance from any differential surface patch, all light impinging on a patch must be gathered. The gathered irradiance is then scaled by the magnitude of the BRDF in the viewing direction (the solid angle leading back to either the eye or the image plane). Of course it would be impossible to sample all incident directions, so a reasonable approximation is achieved by stochastic sampling. A discrete number of rays is cast, each of which is scaled by the magnitude of the BRDF in its particular direction. Then the results are integrated. Increasing the sampling of the BRDF improves the accuracy of the approximation. The technique of stochastically sampling the BRDF is known as Monte Carlo sampling. By reciprocity, one could equally well sample reflecting directions from a surface patch instead of incident directions.

We have developed an efficient method of performing this Monte Carlo integration. Instead of casting rays in a uniform distribution about the hemisphere and weighting the returned value by the reflectance, the ray distribution itself is weighted by the reflectance. This can be done in a straightforward manner when the BRDF is composed of invertible functions such as Gaussians. When the BRDF is represented discretely, either by taking measurements over the hemisphere or by sampling a non-invertible functional form, another method must be used to generate random variates for Monte Carlo integration. This can be accomplished by first subtracting the smallest hemisphere that fits within the BRDF data. This removes the diffuse or uniformly varying portion of the BRDF and leaves only the highly directional specular part. The alias selection method [7] can be employed to create random variates from these remaining specular reflectances.

This Monte Carlo integration scheme has been implemented using the public domain Radiance rendering program [9]. Radiance was designed to facilitate illumination engineering studies and tests have shown that it provides radiometrically correct global lighting simulations. Accurate Radiance surface reflection calculations however, are limited to reflection distributions that can be approximated using the Ward model built into the program [8]. To handle general BRDF data that might only be available as discrete measurements or that may have been approximated using a non-invertible function it was necessary to extend the shading capability of the program. This was accomplished by developing the iBRDF shader that utilizes the efficient Monte Carlo integration technique described in the preceding paragraph.

Computer Graphic Images From Spectral and Reflectance data

To create photorealistic pictures of coated surfaces, a synthetic image must accurately depict both the spectral and spatial distribution of the light reflected from a surface. Therefore, to test the rendering capability of the iBRDF shader, BRDF data with primarily spectral or spatial variation was employed. The spectral data was obtained from a metallic paint and the spatial data from the coated epoxy samples discussed earlier in the paper.

Figure 5. Synthetic image made using *iBRDF* extension to *RADIANCE* software and spectrogoniophotometric data from reference5.

449

Figure 6. Rendering from reflectance data generated using the Ray method and a surface topographical map of coated epoxy samples with rms roughness values 201 nm (left) and 805 nm (right).

J.G. Davidson made one of the earliest attempts to completely characterize the BRDF of metallic paints. He constructed a spectrogoni-
ophotometer and he used it to make complete in plane reflectance
measurements for a sequence of incident light directions. This was
repeated for several different metallic paint speciments. Figure 5 was
synthesized using the data that Davidson published in his PhD
dissertation [5]. To make these pictures, the spectral data in Davidson's
tables were reduced to CIE tristimulus values. Then a smoothing
interpolation function was fitted and the interpolated data was re-
sampled. The Monte Carlo integration technique described above was
then used to turn the re-sampled data into random variates and the
image in Figure 5 was created.

To make pictures of the coated epoxy samples a reflectance model
based on the Ray method, provided BRDF values that were used to
produce the images depicted in Figure 6. The model used surface topographical data from coated epoxy samples with roughness values 201
nm (smooth) and 805 nm (rough) respectively. The model offered
enough ease of computation so that sufficient number of BRDF values
for incident and outgoing directions were generated. The BRDF data was
converted into random variates by the iBRDF shader and then the
RADIANCE software was used to perform a Monte Carlo simulation and
the image in Figure 6 was created. Thus we have demonstrated a direct
link between the material properties of the coated epoxy sample and its
predicted appearance.

Conclusions and Future Work

We presented the results of a study of the relationship between
surface topography and rendered appearance for an isotropic sample of
coated epoxy. Previous work shows good agreement between in-plane
optical reflectance measurements and values of the reflectance
predicted by a geometric reflectance model. The latter was used in
iBRDF, an enhancement of the rendering program RADIANCE and an
image was created by a Monte Carlo sampling procedure. Spectral data
for metallic paints obtained by J.G. Davidson, was also used in iBRDF to
obtain renderings of colored metallic painted objects.

As we turn our attention to more complex materials such as metallic
and pearlescent coatings, the challenge will be to develop optical
reflectance models that adequately capture their appearance effects and
retain enough computational efficiency to be suitable for use in rendering

programs. At the same time, we need to gain a better understanding of the role of out of plane measurements and of the optical properties of materials that display subsurface scattering. We can only hope to make partial progress on these difficult issues but the rapid advances in optical metrology, modeling and computer graphics make it likely that the prediction and computer design of coated objects will become a reality in the near future.

References

1. McKnight, M.E.; Martin, J.W.; Galler, M.; Hunt, F.Y.; Lipman, R.R.; Vorburger, T.V.; Thompson, A.E., NIST Technical Report 5952, National Institute of Standards and Technology, Gaithersburg, MD, February 1997
2. CORM Sixth Report, *Pressing Problems and Projected National Needs in Optical Radiation Measurements,* 1995
3. McKnight, M.E.; Vorburger, T.V.;Marx, E.; Nadal M.E.; Barnes, P.Y.; Galler, M., "Measurements and Predictions of Light Scattering by Coatings", to appear Applied Optics
4. *Optical Scattering: Measurement and Analysis;* Stover, J.C.; SPIE Optical Engineering Press, 1995.
5. Davidson, J.G.,*The Color and Appearance of Metallized Paint Films,* PhD Thesis, Rennsselaer Polytechnic Institute, NY 1971
6. Phong, B., Communications of the ACM, **1975**, Volume 18, 311-375
7. Walker, A.J., ACM Transactions on Mathematical Software, **1977,** Volume 3, 253-256.
8. Ward, G.J., SIGGRAPH'92 Conference Proceedings 26, 265-272
9. *Rendering with Radiance: The Art and Science of Lighting Visualization.* Ward Larson, G.J.; Shakespeare, R.; Morgan Kaufmann Publishers Inc. 1998.
10. *The Scattering of Electromagnetic Waves From Rough Surfaces,* Beckmann, P; Spizzichino, A., Artech House Inc. 1987

Chapter 22

Computer Simulations of the Conformational Preference of 3' Substituents in 2-(2'-Hydroxyphenyl) Benzotriazole UV Absorbers

Correlation with UVA Photopermanence in Coatings

A. D. DeBellis, R. Iyengar, N. A. Kaprinidis, R. K. Rodebaugh, and J. Suhadolnik

Ciba Specialty Chemicals, Additives Division Research Department, 540 White Plains Road, Tarrytown, NY 10591-9005

Derivatives of 2-(2'-hydroxyphenyl) benzotriazole (BZT) are widely used in the coatings industry as effective UV stabilizers. Many commercial stabilizers in this class contain a substituent ortho to the phenolic hydroxyl group. This substitution serves to sterically protect the hydroxyl substituent, which is the structural feature responsible for the photostability of the molecule. A rotamer of BZTs, in which the intramolecular hydrogen-bond has been disrupted, has been implicated in its photodegradation, which occurs in basic, highly polar media. The existence of this form in ortho-unsubstituted BZTs can be detected spectroscopically by differences in both the absorption and emission spectra of the chromophore. Unfortunately, very small spectroscopic differences are seen in differently ortho-substituted BZTs. This contrasts the differences in photostability observed in polar media. As a result, the quantitative use of absorption and fluorescence spectroscopy in the search for more photostable BZTs is precluded. In this work, the superior photostability of a 3'-alpha-cumyl substituted BZT is rationalized in terms of the pronounced conformational preference of this group, first suggested from the results of molecular dynamics (MD) simulations. This result is supported by both solution NMR and solid-state x-ray crystallographic data. In addition, predictions made from the results of further MD runs on the photostability of a differently substituted compound have been experimentally confirmed *via* loss rate measurements in weathered automotive clearcoats.

Introduction

The service lifetime of an organic coating can be dramatically extended through the use of an appropriately selected stabilization system (1). In virtually all cases, especially involving outdoor applications, stabilization involves the incorporation of ultraviolet light absorbers (UVAs) into the coating formulation. The absorption of light by these chromophores provides a protective screening effect to the system being stabilized. It is well established that the screening of incident radiation is the primary stabilization mechanism by which UVAs function, although additional stabilization mechanisms, which involve energy transfer, have been suggested (2-4). Aside from suitable secondary properties (e.g. compatibility, color, etc.), and in addition to having high absorptivity at the appropriate wavelengths (those to which the substrate is sensitive), a fundamental requirement of a UVA stabilizer is to possess the ability to absorb radiative energy without substantially undergoing photodegradation itself. If this were to happen, the stabilizer would eventually be depleted to a level which is not effective in stabilizing the system, resulting in premature coating failure. Thus the longevity (as measured in terms of physical property retention) of a formulated coating system will, in general, depend upon the photostability of the UVA used in the formulation.

UVAs based on the 2-(2'-hydroxyphenyl) benzotriazole (BTZ) chromophore fulfill the above mentioned requirements in a number of ways. They can be derivatized to provide compatibility in a number of different polymeric substrates. Their high extinction (typical $\varepsilon \approx$ 16-20,000 liter/mol-cm at λ_{max}) extends to around 370 nm, just short of the visible range, which provides maximal wavelength coverage without imparting color to the substrate. In their electronic ground state at 70°C, they have been shown to be unreactive with thermally generated free radicals, both in the presence and absence of oxygen (5). This finding suggests that BZT UVAs would be unreactive in their ground state with radical species which may be generated in the photodecomposition of a coating system. As a result of an excited-state intramolecular proton transfer (ESIPT) process, the excited state of BZT UVAs possess a very efficient pathway for the dissipation of radiant energy *via* internal conversion. Interestingly, Estévez *et al* (6) have suggested the potential role of a conical intersection in the deactivation process. An important manifestation of these processes lies in the extremely short excited state lifetime, which is on the order of 1 picosecond (7). As a result, the reactivity of this potentially sensitive species is negligible by virtue of its short lifetime. Quite surprisingly, in actual practice, a small but finite amount of photoloss can be observed in spite of the above mentioned properties (8-11). Furthermore, as the quantum yield for photoloss φ_{loss} is typically on the order of 10^{-6} or 10^{-7} (12), small absolute differences in loss rates can have a measurable impact on coating service life.

Valuable information about the photoloss has been obtained by the elegant work of several authors (5,12-14). Their investigations have led to a number of suggestions regarding the important aspects of the process. To a greater or lesser extent, all workers have essentially implicated a form of the UVA in which the intramolecular H-bond has been disrupted. Figure 1 depicts the equilibrium involved, in which the

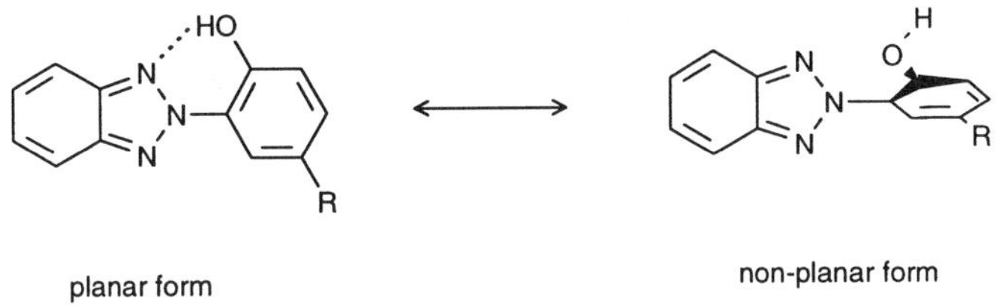

Figure 1. Rotomeric equilibrium between planar and non-planar form of BZT UVAs.

intramolecular H-bond of the planar form has been replaced by an intermolecular H-bond to a solvent molecule or the polymeric matrix. Gerlock *et al* have suggested a reaction of the excited state of this form with free radicals generated in the coating matrix. Their conclusion was based on the finding that UV light, oxygen, free radicals, and a polar matrix are all necessary to effect a loss of UVA (5). Turro *et al* report the detection of the corresponding phenoxide ion of Tinuvin® P (a BZT UVA) in the strongly H-bond accepting solvent DMSO, using laser flash photolysis experiments (13). In addition, these authors have determined the fluorescence lifetime of the excited singlet of Tinuvin® P in argon purged DMSO to be less than 20 nanoseconds at room temperature. Port *et al* have measured this lifetime more precisely and have reported it to be 170 picoseconds (7). Under Turro's conditions, even the triplet was tentatively assigned a lifetime of only 130 nanoseconds. Subject to the same conditions, the lifetime of the phenolate was reported to be approximately 80 microseconds, a factor of greater than 4000 longer than the H-bond disrupted singlet. Catalán *et al* have reported on the facile oxidation of the phenolate ion of Tinuvin® P by singlet oxygen (12). Given this information, Turro *et al* have argued that the phenolate is a more likely source of irreversible photochemistry because of the short excited singlet and triplet lifetimes of Tinuvin® P in DMSO. In any event, all authors agree that a key structure involved in the photoloss is the non-planar, H-bond disrupted form of the UVA, it being a common intermediate in all proposed mechanisms. As such, we have focused our attention on this intermediate, and on ways to reduce phenoxide formation in a effort to design more robust UVAs. The present work attempts to outline these efforts and to offer a rationalization of the photopermanence behavior of some structurally varied BZT UVAs.

Experimental

Clearcoat Preparation

Acrylic melamine clearcoats were prepared from a 6 to 4 mixture of an experimental acrylic polyol and hexamethoxymethyl melamine (Resimene® 747, Solutia), to which was added 0.25% of a flow modifier (Modaflow®, Solutia) and 0.7% dodecylbenzene sulfonic acid (Nacure® 5225, King Industries). The viscosity was then reduced with xylene for spin coating onto quartz discs to give a coating thickness near 50 microns. UVAs were incorporated at typically 1-3% followed by addition of 1% of a hindered amine light stabilizer (Tinuvin® 123, Ciba). The coatings were then cured by heating at 127° C for 30 minutes.

UV Absorbers

All UVAs were synthesized and purified by the Synthetic Research Department of the Additives Division of Ciba Specialty Chemicals, Tarrytown, NY.

Loss Rate Determination

Loss rates were determined by plotting the intensity of the UVAs long wavelength absorption near 345nm as a function of exposure time. The data were analyzed using the method of Iyengar and Schellenberg (11). The coatings were exposed in an Atlas Ci65 Xenon Arc Weatherometer run at 0.55 W/m^2 irradiance at 340nm using both inner and outer borosilicate Type S filters. Automotive Exterior Cycle SAE J1960 was used for all experiments.

UV-Vis and Fluorescence Spectra

All solution spectra were measured in a 1x1 cm quartz cell. Coatings spectra were taken either on free-standing films or spin-coated quartz discs. The film thickness was typically near 50 microns. UV-Vis spectra were determined on a Perkin Elmer Lambda-2 spectrometer with 2 nm resolution. The UVA concentrations were typically 20mg/ml in solution and near 1 % (by weight) in coatings. Fluorescence spectra were measured with a Spex FluoroMax-2 spectrometer. The concentration and/or film thickness for all fluorescence samples were such that the optical density was near 0.3 at the excitation wavelength of 330 nm.

Calculations

All software was run on an IBM RS/6000 model 35T computer. Molecular mechanics and Monte Carlo/stochastic dynamics were performed using MacroModel Version 5.0 (28). Ab initio calculations were done using Spartan Version 4.0 (29).

NMR Spectra

Proton NMR spectra were recorded at room temperature on a Varian UNITY 500 spectrometer using dilute $CDCl_3$ solutions. Chemical shifts are reported relative to TMS, where a positive shift is downfield from the standard. The identity of the aromatic methoxy signal was established in each case by observation of the $^1J_{C-H}$ coupling, via the ^{13}C satellite sidebands, or by difference NOE experiments.

Crystal Structure Determination

A Philips PW 1100 automatic diffractometer was used for data collection with Mo Kα radiation and a graphite monochromator. The structure was solved by direct methods (SHELXS-86). Parameters were refined by full-matrix least-squares techniques (SHELXL-93) with anisotropic displacement parameters for all non-hydrogen atoms. Most hydrogen atoms were located in difference electron density maps, the remaining ones calculated assuming standard geometry.

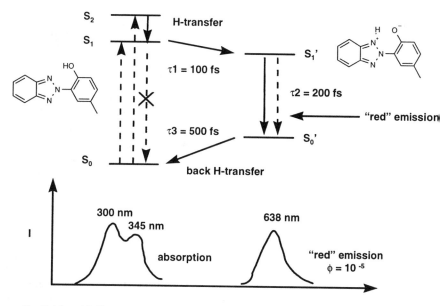

Figure 2. *Jablonski diagram and spectral characteristics of the planar form of BZT UVAs.*

Results and Discussion

Correlation of Absorption and Emission Spectra with Photostability

As mentioned previously, BZT UVAs exist in an equilibrium between a planar and a non-planar form, as illustrated in Figure 1. The position of this equilibrium depends upon the polarity and H-bonding ability of the medium in which the UVA is immersed. The two forms of the UVA exhibit very different photophysical behavior (7), which can be rationalized with reference to Figures 2 and 3. The planar form of the UVA, with the intramolecular H-bond intact, absorbs at around 300 and 345 nm in the UV-Vis spectrum. The long wavelength absorption is attributed to the co-planarity of the two pi systems. This band can be removed from the spectrum *via* O-methylation of the phenolic hydroxyl group which, in the solid state, has been observed to rotate one ring by approximately 55° with respect to the other (15). The absorption at 300 nm is attributed solely to the benzotriazole nucleus. If the H-bond remains intact, the molecule can participate in the ESIPT mechanism. Furthermore, as a consequence of the ultrafast kinetics of this process, the planar form is not generally luminescent. Only under very specific conditions of low temperature and restrictive molecular environments can a red fluorescence, with very low quantum yield, be detected at around 640 nm (16,17). The large Stokes shift relative to the absorption is indicative of emission from the proton-transferred form of the molecule. More recently, this fluorescence has been detected at room temperature in tetrachloroethylene solution and polystyrene films using sophisticated time-resolved spectral techniques (7). Obviously, efficient non-radiative pathways dominate the excited-state relaxation of this form and the molecule can therefore function as an effective UV stabilizer. In contrast, the non-planar form exhibits emissions in the red and blue regions of the visible spectrum as a result of the disrupted H-bond. The blue fluorescence, at around 430 nm, arises because radiative deactivation now becomes a competitive pathway for relaxation whereas previously, the radiative rate was too slow relative to the rate of proton transfer. This blue emission has been used as a sensitive diagnostic to detect the presence of the non-planar form. Additionally, a structured red phosphorescence (most intense peak around 540 nm) can also be detected as a result of intersystem crossing which is now competitive with fluorescent emission.

Substitution proximate to the BZT phenolic hydroxyl can have a significant effect on the spectroscopic properties of the chromophore. Catalán et al have reported the dramatic effect of a 3' tert-butyl substituent on the solvent dependence of the absorption spectrum (18). Turro et al have shown an analogous effect on the 430 nm fluorescent emission utilizing a 3'-alpha-cumyl substituent (13). Both effects arise as a result of steric "protection" of the critical intramolecular H-bond as was suggested by Heller (19). Given the above information, one might reasonably suggest the utilization of the absorption and emission spectra in the molecular design of more robust BZT UVAs, through a systematic variation of 3' substituents. Increased photostability should correlate with a decreased solvent dependence of the absorption spectrum and a decreased fluorescent emission at 430nm. Unfortunately, the dependence of the photostability of the UVA on the spectroscopic effects of 3' substitution is highly non-linear. This can be seen through inspection of Tables I and

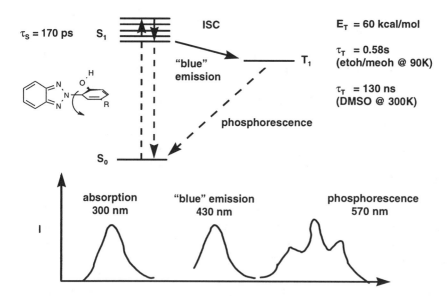

Figure 3. Jablonski diagram and spectral characteristics of the non-planar form of BZT UVAs.

Table I.   Ratio of the long to short wavelength absorption intensity for 3'-substituted BZT UVAs in various solvents.

3'-substituent	Hexane	Ethyl Acetate	DMSO	Percent Decrease
-H	1.12[a]		0.65	42%
-tert-butyl	1.16		1.03	11%
-α-cumyl	1.08	1.00	0.89	18%, 11%
α-methyl-benzyl		0.99	0.86	13%

[a] cyclohexane

II and Figure 4.   In Tables I and II it can be seen that each substituent has approximately the same effect on reducing the solvent dependence of the absorption spectrum, in spite of the fact that each compound exhibits a significantly different rate of loss. Furthermore, regardless of 3' substitution, the ratio of the intensity of the long to short wavelength absorption of BZT UVAs is virtually unaffected by incorporation into model acrylic melamine clearcoats. The results reported in Table II have been obtained on multiple photolysis experiments and were analyzed using the method of Iyengar and Schellenberg (20), which is analogous to Pickett's "infinite absorption" zero order kinetic scheme (21). The corresponding effect in emission is illustrated in Figure 4.   Fluorescence spectra of BZT UVAs obtained from a model acrylic melamine clearcoat system show that a significant emission is only obtained with 3'-unsubstituted compounds. The slightly greater emission observed for the 3'-cumyl substituted compound versus the 3'-tert-butyl is just the opposite of what would be expected based on the photopermanence data. The difference here might arise as a result of the low signal-to-noise ratio of the instrument in the very low intensity region. These facts illustrate the difficulties in using the UV-Vis absorption and emission data as quantitative tools in the search for more photopermanent BZT UVAs. As a result, computational modeling along with structural information from NMR spectroscopy and x-ray crystallography were brought to bear in an attempt to rationalize differences in photopermanence behavior for these systems.

Table II. Relative loss rates for 3'-substituted BZT UVAs in a model acrylic melamine clearcoat subjected to accelerated weathering.

3'-substituent	Relative loss rate
-H	1.00
-tert-butyl	0.37
-α-cumyl	0.24
α-methyl-benzyl	0.30

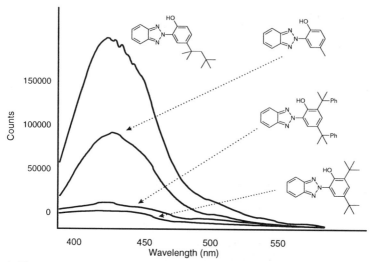

Figure 4. *Fluorescence emission spectra of BZT UVAs in a model acrylic melamine clearcoat.*

Computational Results

Consistent with our present level of understanding about these compounds, we can rationalize the superior photostability of BZT UVAs containing the 3'-α-cumyl substitution in terms of a further shifting (relative to tert-butyl) of the operative equilibrium towards the intramolecular H-bonded form in a coating matrix. The results from the UV-Vis solution spectra seem to indicate *trends* in photopermanence behavior in coatings but do not provide a truly *quantitative* discrimination between structures. In general, there is no fundamental reason to expect a quantitative correlation between solution and in-coating behavior. In fact specific interactions, which may exist in either medium, would argue against such a relationship. Furthermore, the total response to changes in solvent may occur through multiple mechanisms, which may also be substituent dependent. Ghiggino *et al* (22) have used principal component analysis to resolve the absorption spectrum of Tinuvin® P and 5-sulfonated Tinuvin® P into separate components due to the planar and non-planar forms. Their results show that each form makes significant contributions to both the long and short wavelength absorptions (i.e. the resolved spectra overlap significantly), and the relative contribution that each form makes to either absorption is a function of substitution. Using Ghiggino's data for Tinuvin® P, we calculate a 4% decrease in the long to short wavelength absorbance ratio upon halving the nonplanar concentration from 8% to 4%. A similar calculation, *keeping the nonplanar concentration constant* at 8%, yields a 7% decrease in this ratio upon 5-sulfonation. This illustrates that substituents can affect this ratio *via* alternate mechanisms (in addition to affecting the equilibrium concentration of the nonplanar form) and the magnitude of these effects can be comparable to the concentration effect. Furthermore, when considering the emission spectra, we must note that variations in

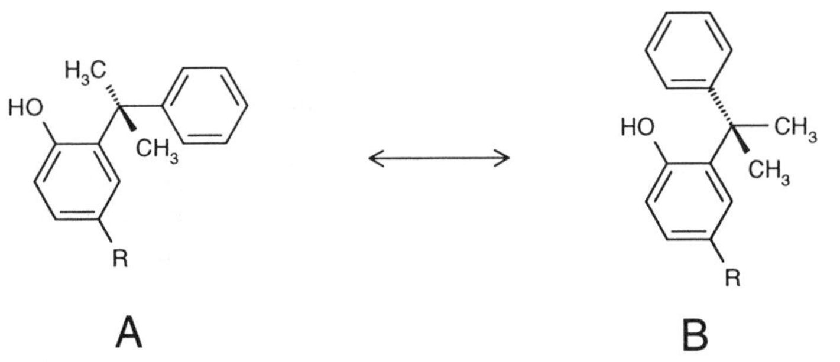

Figure 5. *Relevant conformations of the model compound 2- α-cumyl phenol.*

quantum yields among differently substituted BZTs (e.g. the top two spectra in Figure 4), as well as the low emission intensity at 430 nm (for 3'-substituted BZTs), contribute to making the fluorescence data non-quantitative. In light of the above arguments, we suggest that gross comparisons are possible and useful, but a truly quantitative discrimination is precluded.

Operating with the above assumption, we proceeded to investigate the conformation of the α-cumyl group in these systems. In the model compound, 2-α-cumyl phenol, the cumyl substituent can exist in only two distinct, stable conformations. These are illustrated in Figure 5. It is understood that each conformation has an equivalent contributing enantiomeric structure. In conformation A, the local environment of the hydroxyl group is identical to that induced with an ortho tert-butyl substituent. Conversely, conformation B places a phenyl ring in close proximity to the hydroxyl. This conformational aptitude may potentially provide a degree of steric "shielding" to the hydroxy group, provided that this conformer of the molecule is sufficiently populated. To investigate the stability of this form we performed energy optimizations starting from each conformer of the model compound using various molecular mechanics and molecular orbital methods. The results of these calculations are given in Table III. It can be seen that for all methods conformer B is most stable by at least 0.9 kcal/mol. Interestingly, the solvent-accessible surface area of the hydroxyl group, which provides a measure of accessibility, is reduced by 20% in the most stable conformation.

Table III. Relative stability of conformers of 2-α-cumyl phenol calculated by various methods.

Conformer	Amber*[a]	MM2*[a]	MM3*[a]	HF/6-31G(d)[b]
A	11.58	17.13	17.44	-652.192056
B	9.74	15.97	16.56	-652.194905
E(A)-E(B)[a]	1.8	1.2	0.9	1.79

[a] in kcal/mol [b] in hartrees, not corrected for zero-point energy or temperature

The results of Table III apply strictly only to motionless, non-vibrating, isolated molecules at zero K. Furthermore, the results are for the model compound, not for a full BZT UVA structure. Therefore, in an attempt to perform more realistic calculations, molecular dynamics simulations were initiated. In various ways, the dynamics simulations account for finite temperature, molecular vibrations, solvent, and conformational entropy, both in terms of the number of equivalent conformers, as well as the width of the conformational potential energy well. In addition, the simulations were performed on structure 1 (see Figure 6), which is more similar to the full UVA structure (1 contains the benzotriazole group) after a proper parametrization of the intramolecular potentials. The primary difference between the studied structure and a BZT UVA was in the substitution of a methoxy group for the phenolic hydroxyl. This has the effect of rendering the molecule non-planar, and was done to assist in the comparison to and utilization of NMR data, as will be outlined in the next section. We used chloroform as the simulated solvent within the GB/SA model (23), and applied the Monte Carlo/stochastic dynamics scheme (24) at 300 K using the Amber* force field. The details of the simulation work have been previously published (25), so we will only summarize the results here. Figure 6 displays a histogram of the substituent phenyl C(ipso) to methoxy oxygen distance sampled every 20 picoseconds during the 14.5 ns dynamics simulation. The intense peak near 3.0 Å is due to the enhanced population of a conformation analogous to B in the model compound (2-α-cumyl phenol). The small population of the conformation analogous to A can be seen from the peaks at 3.9 Å and above. The preference for the hydroxyl-"shielded" conformation is quite pronounced. Integration of the peaks yields a ratio in excess of 95:5 in favor of the shielded conformation.

Molecular dynamics simulations were also performed on structure 2, a slight modification of 1, in which one of the cumyl group methyls has been replaced by

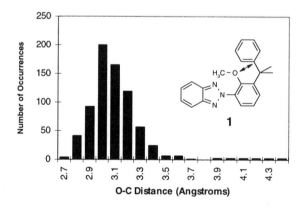

Figure 6. Histogram of substituent phenyl C(ipso)-O(methoxy) distance for 2-(2'-methoxy-3'-α-cumylphenyl) benzotriazole. The dynamics were run for 14.5 ns. The structure was sampled every 20 ps during the simulation. (Reproduced with permission from reference 16. Copyright 1997 John Wiley & Sons, Ltd.)

hydrogen (an α-methyl-benzyl substituent). This slight modification has a dramatic effect on the results of the molecular dynamics, shown in Figure 7. In this compound, there is a preference for the conformation similar to conformation A of Figure 5, which places the phenyl ring in a more distant position relative to the oxygen atom. Based upon these results, one would predict a greater rate of loss for BZT UVAs similar to **2**, relative to those similar to **1**. This behavior has been experimentally confirmed, as can be seen in Table 2.

NMR Results

Proton NMR spectra of model compounds **3**, **4**, and **5** were recorded to provide a more direct investigation of the structural predictions emerging from the molecular dynamics simulations. Figure 8 shows the room temperature spectra of all three compounds in the 2.0-4.0 ppm range. In the spectrum of **3**, a singlet is observed at 3.86 ppm, which is assigned to the protons of the methoxy substituent. The observed chemical shift is within ±0.1 ppm of the chemical shift for the methyl protons of anisole. We thus conclude that, in compound **3**, the methyl group of the methoxy substituent resides in a position distant from the benzotriazole ring *via* rotation of the O-C(aromatic) bond. Conversely, in the spectrum of **4** the methoxy signal has moved upfield to 3.08 ppm. This dramatic shift is almost out of the standard range of expectation for any methoxy resonance. The observation is interpreted by noting that the aromatic rings, separated by the C-N single bond, are not co-planar in these systems. In addition, the 3'-tert-butyl group hinders rotation about the O-C(aromatic) bond such that the methyl protons reside in the shielding region of the benzotriazole aromatic system. The spectrum of compound **5** indicates an even more dramatic upfield shift for the methoxy protons to 2.04 ppm. *This constitutes a virtually unprecedented chemical shift for the protons of a methoxy group.* A reasonable

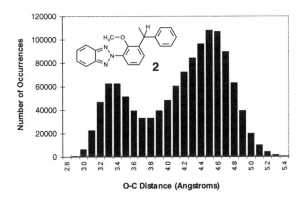

Figure 7. Histogram of substituent phenyl C(ipso)-O(methoxy) distance for 2-(2'-methoxy-3'-[α-methyl-benzyl]phenyl) benzotriazole. The dynamics were run for 12.6 ns. The structure was sampled every 10 fs during the simulation.

464

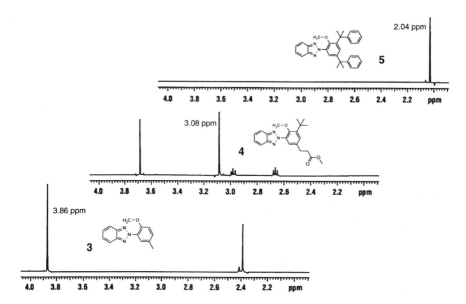

Figure 8. *Proton NMR spectra of model compounds in the 2.0-4.0 ppm range. The chemical shift of the methoxy proton signal is indicated. (Reproduced with permission from reference 16. Copyright 1997 John Wiley & Sons, Ltd.)*

explanation of this observation is that the shift is due to an overwhelming conformational preference, even in solution, of the 3'-α-cumyl substituent which places the methyl substantially in the shielding region of both aromatic systems. Interestingly, the simulated time-averaged distance between the centroid of the phenyl ring and the centroid of the methoxy protons is 3.72 Å. According to a derived shielding function based on free electron theory (26), the predicted change in chemical shift due to the presence of the phenyl ring is about 1.0 ppm. The observed difference is 1.04 ppm from the already shielded methoxy protons of **4**.

Figure 9 shows the spectrum of **6**, which has a 3' substituent identical to **2**. The methoxy proton signal has shifted downfield relative to **5**. In fact, the position of the resonance is quite similar to that observed for **4**. We thus conclude that **6** exists in a conformation such that the phenyl ring of the 3' substituent is spatially removed form the methoxy group.

Solid-State Structure Determination

As an additional check on the structural predictions from the molecular dynamics simulations, we have determined the conformation of a related compound in the solid state. Figure 10 shows a ball-and-stick representation of a single molecule in the crystal structure of the commercial BZT UVA, Tinuvin® 928, 2-(2'-hydroxy-5'-tert-octylphenyl)benzotriazole. It can be seen that the orientation of the α-cumyl substituent is analogous to that predicted for model compound **1**. Interestingly, this

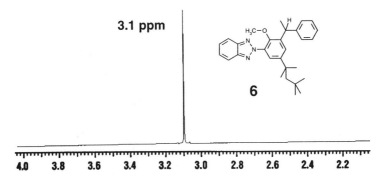

3.1 ppm

6

Figure 9. Proton NMR spectrum of 6 in the 2.0-4.0 ppm range. The chemical shift of the methoxy proton signal is indicated.

orientation has also been observed in the solid-state structure of a related compound, 2,4,6-Tri(α,α-dimethylbenzyl)phenol (27).

Conclusion

UV-Vis absorption and fluorescence emission spectroscopies can be used as general tools in the molecular design of more photopermanent BZT UVAs. Unfortunately, the correlation between spectroscopic behavior and photopermanence in a given medium is not truly quantitative. The structural results of molecular dynamics simulations provide a more robust correlation to the photostability data obtained in model acrylic melamine clearcoats.

The conformational preference of the 3'-α-cumyl substituent in appropriately substituted BZT UVAs has been directly observed in the solid state. Proton NMR results on methylated derivatives of BZT UVAs show that this preference, which was first predicted from molecular dynamics simulations, also exists in solution. It is likely that a 3'-α-cumyl substituent in hydroxylic BZT UVAs also exhibits this behavior. Given the decreased accessibility of the hydroxyl in 3'-α-cumyl substituted UVAs (relative to e.g. tert-butyl), one can rationalize the enhanced photostability observed for compounds of this type in polar media. The predicted increase in accessibility of the hydroxyl group of the hydroxylated derivative of model compound **2** (relative to model compound **1**) correlates with its measured photopermanence in a model clearcoat.

Acknowledgments

The authors thank Dr. G. Rihs for the crystal structure determination, and Ms. C. Hendricks-Guy for the synthesis of compound **6**. We also thank Dr. S. Pastor, Dr. R. Ravichandran, and Ms. N. Cliff for helpful discussions and review of the manuscript.

466

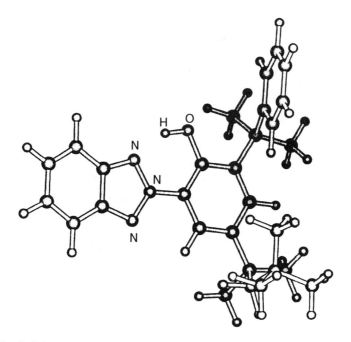

Figure 10. Solid-state structure of Tinuvin® 928.

References

1. Valet, A. *Light Stabilizers for Paints*; Curt R. Vincentz: Hannover, Germany, 1997.
2. Werner, T. *J. Phys. Chem.* **1979**, 83, 320.
3. Werner, T.; Kramer, H. E. A. *Euro. Polym. J.* **1977**, 13, 501.
4. Werner, T.; Woessner, G.; Kramer, H. E. A. in *Photodegradation and Photostabilization of Coatings*, ACS Symposium Series No. 151, Pappas S. P. and Winslow F. H. Eds.; American Chemical Society: Washington, DC, 1981.
5. Gerlock, J. L.; Tang, W.; Dearth, M. A.; Korniski, T. J. *Polym. Degrad. Stab.* **1995**, 48, 121.
6. Estévez, C. M.; Bach, R. D.; Hass, K. C.; Schneider, W. F. *J. Am. Chem. Soc.* **1997**, 119, 5445.
7. Wiechmann, M.; Port, H.; Frey, W.; Lärmer, F.; Elsässer *J. Phys. Chem.* **1991**, 95, 1918.
8. Gupta, A.; Scott, G. W.; Kliger, D. in *Photodegradation and Photostabilization of Coatings*, ACS Symposium Series No. 151, Pappas S. P. and Winslow F. H. Eds.; American Chemical Society: Washington, DC, 1981.

9. Pickett, J. E.; Moore, J. E. *Polym. Degrad. Stab.* **1993**, 42, 231.
10. Bell, B.; Bonekamp, J.; Maeeker, N.; Priddy, D. *Polym. Prepr.* **1993**, 34, 624.
11. Pickett, J. E.; Moore, J. E. *Polym. Prepr.* **1993**, 34, 153.
12. Catalán, J.; Del Valle, J. C.; Fabero, F.; Garcia, N. A. *Photochem. Photobiol.* **1995**, 61, 118.
13. McGarry, P. F.; Jockusch, S.; Fujiwara, Y.; Kaprinidis, N. A.; Turro, N. J. *J. Phys. Chem. A* **1997**, 101, 764.
14. Pickett, J. E.; Moore, J. E *Die Angew. Makromol. Chem.* **1995**, 232, 229.
15. Woessner, G.; Goeller, G.; Rieker, J.; Hoier, H.; Stezowski, J. J.; Daltrozzo, E.; Neureiter, M.; Kramer, H. E. A. *J. Phys. Chem.* **1985**, 89, 3629.
16. Goeller, G.; Rieker, J.; Maier, A,; Stezowski, J. J.; Daltrozzo, E.; Neureiter, M.; Port, H.; Wiechmann, M.; Kramer, H. E. A. *J. Phys. Chem.* **1988**, 92, 1452.
17. Flom, S. R.; Barbara, P. F. *Chem. Phys. Lett.* **1983**, 94, 488.
18. Catalán, J.; Pérez, P.; Fabero, F.; Wilshire, J. F. K.; Claramunt, R. M.; Elguero, J. *J. Amer. Chem. Soc.* **1992**, 114, 964.
19. Heller, H. J. *Eur. Polym. J.-Suppl.* **1969**, 105.
20. Iyengar, R.; Schellenberg, B. *Polym. Degrad. Stab.* **1998**, 61, 151.
21. Pickett, J. E. *Macromol. Symp.* **1997**, 115, 127.
22. Ghiggino, K. P.; Scully, A. D.; Leaver, I. H. *J. Phys. Chem.* **1986**, 90, 5089.
23. Still, W. C.; Tempczyk, A.; Hawley, R. C.; Hendrickson, T. *J. Am. Chem. Soc.* **1990**, 112, 6127.
24. Guarnieri, F.; Still, W. C. *J. Comput. Chem.* **1994**, 15, 1302.
25. DeBellis, A. D.; Rodebaugh, R. K.; Suhadolnik, J.; Hendricks-Guy, C. *J. Phys. Org. Chem.* **1997**, 10, 107.
26. Johnson, C. E. Jr.; Bovey, F. A. *J. Chem. Phys.* **1958**, 29, 1012.
27. Kurashev, M. V.; Struchkov, Y. T.; Veretyakhina, T.G.; Shklover, V.E. *Izv. Akad. Nauk SSSR, Ser. Khim.,* **1986**, 1843, CASREACT 107:39304e.
28. Mohamadi, F.; Richards, N. G. J.; Guida, W.C.; Liskamp, R.; Lipton, M.; Caufield, C.; Chang, G.; Hendrickson, T.; Still, W. C. *J. Comput. Chem.,* **1990**, 11, 440.
29. Spartan Version 4.0, Wavefunction, Inc., Irvine, CA 92715 USA, Copyright 1995

Chapter 23

The Pulsing of Free-Amine during Polymer Hydrolysis

Sam C. Saunders

S C S Inc., 218 Main Street, Pmb 184, Kirkland, WA 98033-6108

We explicate a question in physical chemistry concerning hydrolysis in acrylic melamine which is analogous to the oscillation of water levels in coupled tanks, under related conditions. But polymer hydrolysis is more complicated; it involves slower reaction rates, partial-yields and moisture infusion due to the ambient relative humidity. Convolutions of the different concentrations are used to represent successive first-order reactions in the governing coupled differential system, rather than characteristic equations and eigenvalues obtained from a computer program. We give explicit formulae from hydrolytic degradation for the concentrations of certain compounds which are of chemical interest. This analysis makes transparent when and why pulsing occurs, to the confoundment of others besides physical chemists.

Introduction

At a recent conference on the service life of polymer coatings, a preliminary report was made on quantitative prediction of concentrations in polymers by chemical degradation due to hydrolysis, e.g., humidity and acid rain, affecting car paint. The mathematical analysis showed "pulsing" could occur, under some conditions, in the generation of free-amine (and other degradation products) in such polymers as acrylic-partially-alkylated melamine coating. But the audience was sceptical. The physical chemist's questions about the mathematical demonstration were pointed: "Please explain to us the *chemical reason* there should be acceleration and deceleration of the chemical activity, with a period as long as three weeks, in polymers such as these?" This should be interpreted as: "This mathematical tom-foolery seems unrelated to the chemistry and is so esoteric that until a chemical explanation, which is physically intuitive, can be provided it is insufficiently convincing."

The experimentalist's explanations of the origin of the observed variations in the FTIR responses, see Figure 1, in the chemical hydrolysis of melamine, see Figure 2, were:

1. During the early measurement period in FTIR analysis, when percentage changes were lower than the measurement error a systematic error of over and under compensation often resulted in the early oscillation that later disappeared.

2. The mechanism that controlled humidity during exposure may have fluctuated with a common period which generated the pulse. (Unfortunately no records were found in the humidity control that exhibited such pulsing.)

Representation of Hydrolytic Degradation

To analyse Nguyen's degradation scheme, see [3], which is quoted in Figure 2 and depicted in Figure 7, we construct a mathematical model for two important scenarios by denoting $- \rightarrow, \longrightarrow, \rightleftharpoons$ respectively, as zero-order, first-order and reversible first-order reactions, namely,

$$\boxed{A} \overset{\mu}{-\rightarrow} \left(\boxed{A} + \boxed{B} \right) \overset{\eta_0}{\rightleftharpoons} \boxed{C} \overset{\eta_2, p_2}{\longrightarrow} \boxed{D_1} \overset{\eta_1, p_1}{\longrightarrow} \left(\boxed{E_1} + \boxed{F} \right),$$

$$\boxed{A} \overset{\mu}{-\rightarrow} \left(\boxed{A} + \boxed{B} \right) \overset{\eta_0}{\rightleftharpoons} \boxed{C} \overset{\eta_2, \bar{p}_2}{\longrightarrow} \boxed{D_2} \overset{\eta_3, p_3}{\longrightarrow} \left(\boxed{E_2} + \boxed{A} + \boxed{F} \right).$$

The subscripts indicate possible choices for D_j for $j = 1, 2$ and E_k for $k = 1, 2$ where $\boxed{A}$ represents water, $\boxed{C}$ is melamine methylol. Here $\boxed{B}$ is a unknown mixture of the three chemical complexes below, each of which is assumed to have the same reaction constant. This is an approximation to the average of three different reaction constants for:

$$\boxed{B_1:} - N \overset{\text{CH}_2\text{-OCH}_3}{\underset{\text{H}}{\big<}} \qquad \boxed{B_2:} - N \overset{\text{CH}_2\text{-O-R}}{\underset{\text{H}}{\big<}} \qquad \boxed{B_3:} - N \overset{\text{CH}_2\text{-OCH}_3}{\underset{\text{CH}_2\text{-O-R}'}{\big<}}$$

It is the accumulation of end-products $\boxed{E_1}$ and $\boxed{E_2}$ which are of interest; $\boxed{E_1}$ is $-NH_2$, primary amine, and $\boxed{E_2}$ is $-N-CH_2-N$ while $\boxed{D_1}$ is $\bullet$-C and $\boxed{D_2}$ is C-C.

Here η, λ, with affixes, are first-order reaction rates and p, with affixes, represents fractional yield. The problem is to obtain formulae which represent the concentration over time of E_1 and E_2, the two ultimate degradation products, involving all combinations of intermediate products. Some additional

470

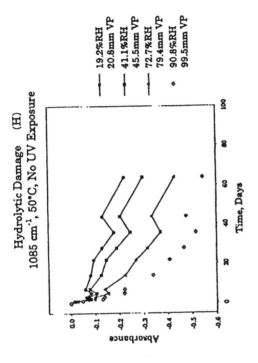

Hydrolytic Damage   (H)

1085 cm⁻¹, 50°C, No UV Exposure

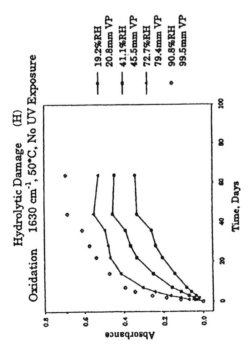

Hydrolytic Damage   (H)

Oxidation 1630 cm⁻¹, 50°C, No UV Exposure

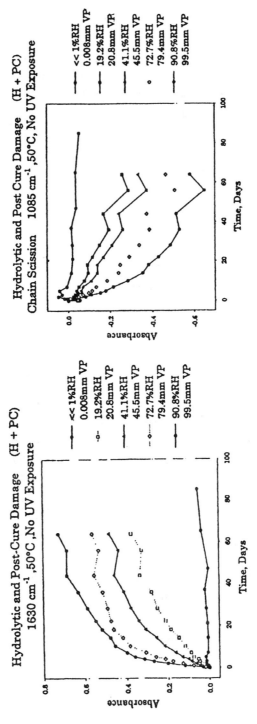

Figure 1: Selected Figures of Hydrological Changes from FTIR Spectroscopy

analysis, not given here, would be required to determine formulae for the concentrations of carbonylic acid and amide.

However the actual problem in the chemistry of hydrolysis is more complicated than that of three coupled water tanks. Hydrolysis involves a sequence of reactions which begin with $\boxed{H_2O}$ and at the end of the chemical sequence a fraction of the $\boxed{H_2O}$ with which it began is resupplied. As a by-product, the hydrolysis cycle contributes to the production of free-amine which is the initiator of the chemical degradation. Chemical behavior during hydrolysis resembles that shown in Figure 7.

A quantitative analysis of this situation cannot depend only on a linear system of differential equations because of two important complications: (i) partial-yield reactions, i.e., only a fraction of the reagent consumed becomes the chemical species of interest, (ii) the infusion of water due to relative humidity. To attack these problems we adopt a different method.

A Single Tank

The bottom pressure, due to the weight of water, is proportional to the depth of water. The volume (amount) of water is $V = \pi r^2 \cdot y$ where r is the radius of the tank and y is the depth of water. With outflow but no inflow the rate of change of the volume of water is

$$\frac{dV}{dt} = -\kappa(\pi \varepsilon^2) \cdot y \quad \text{for some } 0 < \kappa < 1.$$

Here κ is the nozzle efficiency and ε is the drain radius. Since $\pi r^2 \cdot y' = -\pi \varepsilon^2 \cdot \kappa \cdot y$ it follows that $y' = -\eta y$

Figure 3 A draining tank with inflow   for some $\eta > 0$. Therefore the water level of a draining tank, as in Figure 3, is an exponentially decreasing function of time, namely, $y(t) = y(0)e^{-\eta t}$ for all $t \geq 0$.

This type of behavior often arises elsewhere in science, e.g., in (i) Newton's law of cooling, (ii) the electrical charge draining from a capacitor, or (iii) Carbon C_{14} decaying to C_{12}. Such representation of "flow" using the appropriate parameters, namely,

WATER VOLUME,   WATER DEPTH,   TANK BASE AREA,   and   DRAIN AREA

can be reinterpreted in other situations, respectively, as

HEAT,   TEMPERATURE,   SPECIFIC HEAT,   THERMAL CONDUCTIVITY, or

CHARGE,   VOLTAGE,   CAPACITANCE,   ELECTRICAL CONDUCTIVITY

and, most importantly as we argue subsequently, *the same model applies in all first-order chemical reactions during hydrolysis in polymers,* where correspondingly we have the concepts

MATERIAL MASS, SPECIES CONCENTRATION, REACTION CONSTANT.

We assert the reason for the counter-intuitive behavior of coupled tanks is not due to any electrical or hydrologic properties but follows by mathematical implication from the common governing law, as expressed, e.g., in eqn(1) below, of a tank with constant inflow and gravitational outflow:

$$y' = \mu - \eta y; \quad \text{therefore} \quad y(t) = \frac{\mu}{\eta} + \left(y(0) - \frac{\mu}{\eta} \right) e^{-\eta t} \quad \text{for all } t \geq 0. \tag{1}$$

Two Coupled Tanks

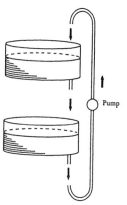

Figure 4 Two coupled tanks

The differential system for flow-rate is

$$y'_1 = -\eta_1 y_1 + \eta_2 y_2,$$

$$y'_2 = +\eta_1 y_1 - \eta_2 y_2.$$

The solution is checked to be of the form:

$$y_i(t) = y_i(0) + y'_i(0) \int_0^t e^{-(\eta_1 + \eta_2)x} \, dx, \tag{2}$$

for $i = 1, 2$. This can be easily obtained using $\mathcal{LS}$-transforms defined by:

$$G^\dagger(s) = \int_0^\infty e^{-st} \, dG(t).$$

Transforming the differential system and using

$$sG^\dagger(s) = g(0) + g^\dagger(s), \text{ where } G' = g, \text{ one obtains}$$

$$y_i^\dagger(s) = \frac{y'_i(0)}{s + \eta_1 + \eta_2} \quad \text{for } i = 1, 2.$$

In Figure 4, which see, both water levels adjust quickly, each at the same rate, to static levels with one tank increasing and the other decreasing, depending upon $y'_i(0)$.

In the special circumstances in [2] the interpretation is in these terms:

The solution is of the form: $y_i(t) = a_i \pm b e^{-(\eta_1 + \eta_2)t}$ and we see that $-(\eta_1 + \eta_2)$ is one of the two eigenvalues, with $(\eta_1 + \eta_2)$ being the sum of the drain-hole areas. The eigenvector is $(b e^{-(\eta_1 + \eta_2)t}, -b e^{-(\eta_1 + \eta_2)t})$.

The solution given in eqn(2), involving inital values and initial rates, seems to appeal to engineering intuition. One checks, since $V_i(t) = \pi r_i^2 \cdot y_i(t)$ is the water volume of the ith tank for $i = 1, 2$ that total water volume is conserved, i.e., $V_1(t) + V_2(t)$ is constant.

Three Coupled Tanks

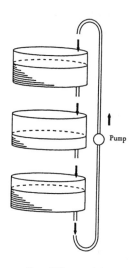

Figure 5 Three coupled tanks

Clearly the differential system, for the flow-rate in Figure 5, is,

$$
\begin{aligned}
y_1' &= -\eta_1 y_1 + \quad\quad +\eta_3 y_3, \\
y_2' &= +\eta_1 y_1 - \eta_2 y_2 \quad\quad , \\
y_3' &= \quad\quad +\eta_2 y_2 - \eta_3 y_3.
\end{aligned}
\tag{3}
$$

In matrix notation this is a linear system:

$$
y' = \mathcal{K} \cdot y
\tag{4}
$$

Following [2] we let $\lambda_1, \lambda_2, \lambda_3$ be the three roots, one being zero, of the characteristic polynomial

$$
\det(\mathcal{K} - \lambda I) = \lambda(\lambda^2 + \sigma_1 \lambda + \sigma_2) = 0,
$$

where we have written, for short,

$$
\sigma_1 = \eta_1 + \eta_2 + \eta_3, \quad \sigma_2 = \eta_1 \eta_2 + \eta_1 \eta_3 + \eta_2 \eta_3.
\tag{5}
$$

The general solution, when the eigenvalues λ_i are distinct (here negative) and ν_i are the corresponding linearly independent eigenvectors, see [2], is given by $y(t) = c_1 e^{\lambda_1 t} \nu_1 + c_2 e^{\lambda_2 t} \nu_2 + c_3 e^{\lambda_3 t} \nu_3$. Alternatively, the transform of eqn (3) is $y^\dagger(s) = (sI - \mathcal{K})^{-1} y'(0)$. The solution, when $\sigma_1^2 < 4\sigma_2$, gives a percentage change, for $k = 1, 2, 3$, of

$$
\%y_k(t) = \frac{y_k(t) - y_k(0)}{y_k(0)} \propto \upsilon_k [1 - e^{-\alpha t} \cos \beta t] + \varpi_k e^{-\alpha t} \sin \beta t,
$$

where $\sum_{k=1}^3 \upsilon_k = \sum_{k=1}^3 \varpi_k = 0$. Then syncronized pulsing, with diminishing oscillation, occurs in each tank. Such pulsing will occur unless one drain is so large relative to the others, that it rapidly empties its tank, essentially reducing the system to two coupled tanks. The solution exhibits no oscillation, whenever the barycentric coordinates of η_1, η_2, η_3 (letting $u_i = \eta_i / \sigma_1$) fall within the region $\sum u_i^2 > 2 \sum_{i<j} u_i u_j$. In Figure 6 this region is exterior to the circle within the triangle, where each vertex is a point at which some $u_i = 1$.

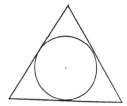

Figure 6 Region of Oscillation

of variation in service life caused by the actual infusion of moisture due to stochastically fluctuating humidity from diurnal and seasonal variation. Only by utilizing such a result can the concentration of a degradation species over time be used to predict accurately the useful life of polymers under the various environmenal conditions of exposure arising in service.

Acknowledgements

This research has been supported by the *Consortium on Prediction of Service Life in Organic Coatings* directed at NIST by Jonathan Martin. The data cited herein are from experiments conducted by NIST in that effort. Other persons at NIST who have beneficially reviewed, commented or contributed to this study include Tinh Nguyen, Brian Dickens and Eric Byrd. Some fitting of the formulae was done by Nairanjana Dasgupta to whom special thanks are extended.

References

[1] D. R. Bauer, Degradation of Organic Coatings, I. Hydrolysis of Melamine Formaldehyde/Acrylic Copolymer Films, *Journal of Applied Polymer Science*, **27**, p3651-3682, 1982.

[2] K.M. Kendig, When and Why Do Water Levels Oscillate in Three Tanks?, *Mathematics Magazine*, **72**, p22-31, 1999.

[3] T. Nguyen, J. Martin, E.Byrd, B. Dickens, Photolysis, Hydrolysis, and Moisture-Enhanced Photolysis of Acrylic/Melamine Coatings: Effects of Humidity, Temperature and Spectral UV, proprietary report NIST, *Coating Service-Life Consortium*, January 1998.

[4] S.C. Saunders, The Relationship between Chemical Kinetics and Stochastic Processes in Photolytic Degradation, in *Service Life Prediction of Organic Coatings*, p233-45, ACS Symposium Series 722, American Chemical Society: Washington DC, 1999.

[5] S.C. Saunders, On Confidence Limits for the performance of a System when few Failures are Encountered, Proceedings of the Fifteenth Conference on the Design of Experiments in Army Research Development and Testing, ARO-D Report 70-2, 797-833.

Chapter 24

Summing Up: Where We Are Now; Next Steps to Take

F. Louis Floyd

PRA Laboratories, Inc., 430 West Forest Avenue, Ypsilanti, MI 48197

The proceedings of the first conference were published as ACS Symposium Series #722, "Service Life Prediction of Organic Coatings: A Systems Approach," Bauer and Martin, editors, 1999 (ACS), Oxford University Press. (ISBN 0-8412-3597-X). Unfortunately, the end-of-conference summary and planning session chapter was omitted from that proceeding. It is included here to complete the record.

The first conference focused on setting the stage for the service life prediction debate. It defined the need for better laboratory testing, for better quantification of the environment ("natural" weather), for the use of different (reliability-type) protocols, and for construction and use of verified materials databases in the coatings industry.

Attendees. There were 70 participants from the U.S. and 9 foreign countries. The top four employers of participants were government labs, coatings companies, raw materials suppliers, and universities.

- Government labs: **USA** -- National Institute for Standards and Technology (NIST), National Renewable Energy Lab (NREL), Wright Patterson Air Force Base (WPAFB), Federal Hwy Adm, and Forest Products Labs (FPL); **foreign** -- Swedish National Testing and Research Institute, NILU (Norway), Fraunhofer Institute for Solar Energy Systems (Germany), Singapore Productivity & Standards Codes, and Japan Atomic Energy Research Institute.
- Companies: DuPont, Ford, Shell, Cook Composites & Polymers, Duron, Reichhold, Rohm & Haas, 3M, Standard Products, GE, Owens-Corning, Monsanto, Elf Atochem, Cytec, Americhem, Atlas Weathering Services, Courtaulds, Akzo-Nobel, BHP Coated Steel, SC Johnson, Dow, and Sherwin-Williams.
- Universities: Washington State, Iowa State, Univ of Belgium, Univ of Colorado, Univ of Cincinnati, Pascal Univ (France).

The last day of the conference was given over to a free-form discussion that attempted to capture a consensus of the conferees understanding of the subject matter, and where they felt we as a coatings community should go next. Following is a summary

476

of that discussion, arranged in three broad subject areas: what did you learn, what techniques did you feel were being underutilized in SLP studies, and what is needed in order to improve our chances of succeeding in our attempts to predict service life.

A. Based on your previous knowledge, the formal presentations that you've heard this week, and the informal discussions that you've had with other symposium attendees, what *broad conclusions* have you reached concerning the current status and future direction of SLP methodology?

1. Time is the currency of today. Accelerated testing must be utilized to make commercial decisions. We no longer have the luxury of waiting for outdoor exposures. This makes SLP methodologies essential, if we are to avoid major blunders.

2. Time, however, is not the proper abscissa for service life comparisons. Cumulative-damage-events is the proper basis for comparison. This circumvents the problem resulting from varying damage event frequency over time.

3. Biggest single driving force today in product development programs: risk. Business is driving hard to do more in shorter times. R&D needs to determine how to accomplish this without incurring unacceptable risk levels. Unfortunately, risk is not being given the forum for discussion it deserves in corporate venues.

4. "Failure" is a defined event, and may be either loss-of-protection or loss-of-appearance. Failure may evolve from either continuous or discontinuous processes. Failure definitions will vary based on the end-use market served.

5. There are actually two different kinds of failure: infant mortality (at beginning of life -- frequently from errors in application), and true product capability (late in life). The former is the source of most product liability claims, while the latter is the focus of product development efforts. Much more attention needs to be paid to the early failures – in particular to diagnosing their causes and eliminating them.

6. Our current condition of replication-deficit prevents us from understanding our current systems, and knowing with reliability whether new systems are better.

7. We are heavily focused on chemical changes during weathering, and overlook the role of physical stresses and changes.

8. Degradation occurring during accelerated aging tests must occur by the same mechanism(s) responsible for natural aging degradation. Any acceleration that produces chemistry that is different from natural aging is inherently flawed, and potentially misleading.

9. It is essential that the end-use environment be completely characterized. "Micro-environment" differences are real, and quite significant for product performance.

10. Useful models do not necessarily require perfect understanding.

11. We can, and should, learn from other industries already using reliability methodology.

12. We can improve data credibility with image analysis techniques.

B. What are some of the currently available techniques that you think are underutilized?

1. Cyclic experiments; measuring cyclic stresses
2. Utilizing fracture energy measurements to predict cracking behavior
3. Adequate replication / population descriptions
4. Acoustic emission for sub-visible mechanical damage measurement
5. Image analysis for quantification of results
6. Thermal analysis (DSC, DMA, TGA, etc.)
7. Electrochemical measurement of corrosion
8. Chemiluminescence for early detection of oxidative processes

C. What capabilities and/or knowledge are needed in order to improve our ability to make more reliable service life predictions?

1. Spectral UV data (dose vs. wavelength) on light sources (natural and artificial)
2. Spectral UV responses of materials.
3. Venue to discuss hardware, software, and best practices in this field at future conferences.
4. Some kind of lab / testing certification process for SLP protocols.
5. Investigation of "induction periods." This is currently viewed more as an artifact than a behavior worthy of study.
6. Internet discussion group on Reliability Theory, where we can exchange ideas and results between now and future SLP workshops and symposia
7. We need to be able to get to point where we truly trust our test data as an industry. Until that time, we'll continue wasting substantial time confirming each other. This means that we need to carefully develop the field of "meta data": complete disclosure of experimental details under which data are collected. We also need to develop "meta data" standards for use in peer review and publication. This will lead to industry standardization. [Sounds like a natural role for NIST]
8. Workshop / short courses / tutorials needed: reliability theory, statistical methods, experimental design, image processing, meta data, techniques for utilizing "old" data. Should definitely be part of agenda for future conferences.
9. R&D must educate general management regarding reliability methodology.
10. We need to vastly speed up the approval process for new standards.

Summary of the Second Conference (1999)

The second conference focused heavily on the photochemical aspects of weathering, with special emphasis on automotive clear topcoats, since that is the coatings industry's biggest single product liability event in recent memory. While there was some recognition that cyclic exposures were important, there were few papers treating this subject.

There were 65 attendees at the second conference from the U.S. and 6 foreign countries:

- **Suppliers:** DuPont, Millennium, Kerr-McGee, Rohm & Haas, DSM, Dow Chemical, Dow-Corning, Elf Atochem
- **Academics:** North Dakota State, U Missouri@ KC, Iowa State, Colorado School of Mines
- **Agencies & independent labs:** U.S. Fed Highway Admin, U.S. National Renewable Energy Lab, U.S. National Institute for Standards & Technology, Aspen Research Corp, PRA Labs (Ypsilanti, MI), USDA Forest Products Lab (Madison, WI), Japan Atomic Energy Research Institute, Swedish Institute for Wood Technology, EMPA – Switzerland, Singapore Productivity & Standards Board, KTH Center for Building Environment – Sweden, Inst of Macromolecular Chemistry – Czechoslovakia, Federal Inst for Materials Research – Germany, Fraunhofer Inst. for Solar Energy Systems – Germany
- **End-users:** Ford, Daimler-Chrysler, 3M, GE, Northrop-Grumman, PPG
- **Paint manufacturers:** Sherwin-Williams, Carboline, BASF, Visteon, Akzo-Nobel
- **Testing & equipment:** Atlas (exposure services; accelerated testing equipment)
- **Software:** Galactic of Salem, NH
- **Consultants:** several retired folks from various companies, still keeping their hand in

The last day of the conference was given over to an open discussion that had as its goal the capturing of the attendees perspectives regarding where we are now, and where we should go next on the issue of service life prediction. Following is a summary of the discussion and conclusions drawn by the attendees during that session.

A. What did you learn at this conference that was of particular significance to you?

1. the state of the art world-wide in weather monitoring is much more extensive than realized before the conference;
2. substantial variations occur in our weather over even short time intervals;
3. photo degradation behavior of materials as a function of depth into polymer films and as a function of UV wavelength;
4. UV absorber performance and permanence over time;
5. the commonalities we have with other industries;
6. the need to embrace a wider range of variables in order to adequately describe "weathering" (e.g. acid rain, fracture energy, cyclic processes, temperature, humidity, time-of-wetness, corrosion, adhesion, and of course UV dosage);
7. the surprising possibility that today's coatings materials are so much better in UV resistance than their predecessors that photochemical resistance may no longer be the most important component in "weathering" resistance; the possibility that hydrolysis may now be as important as UV resistance in determining "weathering" resistance. [presumption: the first property to

"fail" is considered to be the mode of failure. UV degradation was that first property to fail in the past, but may no longer be.]

8. the possibility that in the future, we will be spending a lot more time and resource on feeding verified data into data bases, and then mining the data bases instead of running new experiments [while this is certainly hard to accept in the coatings world today, other industries have already made this conversion -- materials science, electronics, aerospace];

9. reliability theory offers a route to shorter R&D cycle times without increasing risk – but it requires very different mind sets and techniques to be successful.

B. "Weathering" is a complex issue. What sticks-in-the-sand can we place today regarding what we think we know about the weathering process, and our ability to predict the effects of weathering on coatings? [These can be used to measure our progress during future conferences]

1. **These appear to be the principal components of weathering, in the form of [cause (manifestation)]:**
   o UV resistance (gloss loss, color change, chalking)
   o Water resistance (blistering, wrinkling)
   o Adhesion - particularly under wet conditions (blistering, peeling, flaking)
   o Hydrolysis resistance (erosion, color change, chalking, gloss loss - even under mild pH conditions)
   o Fatigue failures (primarily cracking) from cyclic stresses: wet/dry, hot/cold, freeze/thaw, light/dark (implicit hot/cold and wet/dry in diurnal cycle). Fracture energy, fatigue resistance, hysteresis (in stress-strain tests), and permeability probably all play a role in fatigue failures.
   o Corrosion resistance (blistering, rusting, delamination, peeling)
   o Surface fouling (dirt pick-up, mildew)
   o Staining bleed-through from substrate (tannin, stains)

2. **UV resistance is strongly related to chemical bond energy** (higher is more resistant). The bonds (or functional groups) susceptible to UV degradation are the same ones susceptible to free radical grafting reactions, and in turn are inversely related to published chemical bond energies in handbooks. This can be compensated for by using screening agents (UV absorbers) which prevent the UV light form reaching the susceptible polymer, or by free radical traps (HALS) which function to intercept degradation reactions at early stages, thereby preventing the chain of events which lead to physical property changes.

3. **UVA and HALS effectiveness is very sensitive to its (micro) environment.** They appear to be far more useful in clear coats than monocoats, perhaps due to the considerably thicker path length for clear

coats (mils) relative to monocoats (tenths of a mil for clear glossy surface layer). Certain systems do not benefit from the presence of HALS (e.g. p-MMA). Acid rain and acidic components in the coating can de-activate HALS, which may explain why waterborne systems tend to have less success in utilizing HALS than solvent-borne and powder systems.

4. **Hydrolysis resistance probably follows the rules set down by Turpin, et al. for water-reducible polymer systems:** hydrolysis is reduced by increases in polymer hydrophobicity, increases in steric hindrance, and reduction in anchimeric (neighboring group) effects. Hydrolytic degradation may be as significant as UV-induced degradation in determining the overall appearance durability of a coating (gloss loss, chalking, color change).

5. **Cracking of paint films are of two kinds:** drying (or curing)-related, and cyclic stress (fatigue) related. The drying/curing-related types are known as
    o   mud-cracking (related to solids and wet film thicknesses which are excessive -- characterized by cracks in surface skin which do not usually reach all the way through a film),
    o   film-formation cracking (too high T_g or too low coalescent level -- characterized by "star" appearance with cracks radiating from central flaw),
    o   checking, alligatoring, (related to shrinkage caused by continuing reaction of residual reactive functionality -- may start as star-cracking, but takes on a cellular structure appearance once cracking is well-advanced),
    The cyclic stress fatigue related types are usually known as
    o   grain-cracking (related to film permeability (initiation) and fracture energy (propagation) -- appearance is one of long linear cracks, which mimic the grain pattern of the substrate),
    o   cold-checking (typical of furniture lacquers exposed to hot/cold cycles), and
    o   blister-cracking (related to blistering/wrinkling of water-sensitive films -- cracks mimic wrinkling patterns when film dries out).

6. **Surface fouling is probably related most strongly to surface hardness and surface energy (wetting).** Surface hardness probably has a thermal (softening) component and water (softening) component. There is also a wetting/adhesion (surface energy) component, but that is far less well studied. Fouling probably starts with environmental "dirt" becoming embedded in the surface, followed by active growth of whatever species is present. In the case of marine paints, an additional component is organisms that seek and adhere to surfaces, such as barnacles. A laboratory test for dirt resistance that correlates well with exterior results utilizes independent exposure to both water and heat.

C. **What issues were not adequately treated during this conference?**
1. Attendees are not sure that all the necessary tools are yet in place to accomplish SLP. Perceived gaps: characterization of environmental causative agents; deficiencies in accelerated testing devices, particularly in light source filtration. As a result, the coatings community is not comfortable today with any leaps to prediction.
2. Not enough on wavelength-dependence of materials degradation.
3. Want more on actual prediction techniques (case studies; specific examples of how to do it)
4. Statistics (how to deal with probability distributions, apart from discrete values)
5. Cyclic processes and mechanical integrity issues were not well treated -- may be at least as important as photochemistry
6. Is there an ISO standard on SLP? Need references and linkage. Building materials -- NIST should have references.
7. Want more on effect of water on degradation processes.
8. Need better ways to cross-communicate on SLP issues. Web site? Who, how, etc? [NIST may well take the lead on this]
9. Train of variability through the whole SLP process. Critical step in accomplishing SLP. Need to quantify each step.
10. Cumulative distribution vs. differential distribution; framework for choosing which to deal with.
11. What is the role of matrix in weathering chemistry? ("matrix" here is used to describe the local micro-environment that a species sees, which may be substantially different from any overall composition)
12. Do we know enough about the light to get it right? Many think that existing accelerated testing light sources may not be accurate enough. But it is unclear whether better filtration is the answer, primarily because the experiments haven't been run as yet.
13. What is the role of cycles in getting it right?
14. Lots of other issues in "weathering": cycles, acid deposition, hydrolysis, mechanical failure, corrosion.
15. Need simple models that work. How good is good enough?

D. **What steps should we take to improve the next conference?**
1. Hold the next conference only when enough work has been completed to warrant it -- perhaps 3 years' hence (2002).
    o   Site the conference on the east coast of the U.S., to facilitate foreign travelers, who largely come from Europe.
    o   Use the "unanswered questions" from the following section to guide the content and organization of the next conference.
    o   Need to broaden foreign participation. How does work in US stack up with the rest of the world?
    o   Consider publishing extended abstracts in advance of next meeting.
2. Extend subject matter to include:
    o   plastics;

- o sealants;
- o automated real-time measurements of performance;
- o influence of lifetime costs on assessment of service life;
- o beta testing as a means of reducing risk;
- o other modes of failure, such as corrosion, cracking, blistering;
- o role of pigments other than TiO_2;
- o risk: measurement of, tolerance of, definition(s) of.

3. Include more papers on role of:
   - o water,
   - o cycles,
   - o mechanical integrity,
   - o corrosion,
   - o fouling (dirt, mildew, etc),
   - o acid rain,
   - o micro-structural characteristics.

4. Need an actively managed web site for communication between conferences [NIST will set this up during 2000]

5. Use many more case studies. They are an excellent means of communicating issues. Helps clarify suitability of competing approaches.

6. To facilitate the measurement of progress on such a complex issue, place some sticks in the sand today, representing what we think we know, and then compare future work/findings/learning to them to measure our progress? [see section B of this discussion]

7. Clearly define (and publicize) the scope of the conference in advance in order to achieve a better match between attendees' expectations and actual conference content.

8. Add sessions on:
   - o effect of quality of artificial light source on reliability of results vs. natural weathering. Press providers of accelerated weathering devices to develop and report on UV filters for their existing artificial weathering devices that "exactly" match the spectrum of sunlight.
   - o effect of cyclic stresses on coating performance: hot/cold, light/dark, wet/dry, freeze/thaw.
   - o fracture mechanics of polymeric materials, and how such measurements aid in predicting lifetimes.
   - o effect of residual (unreacted) cure functionality on survivability of a coating exposed to substantial swings in temperature during its lifetime, like clearcoats on automobiles (summer days cycle between 60° F and 180° F)

9. Form a study group to explore the implications to reliability theory for the coatings industry. Ask them to develop a list of suggested roles for various players in the food chain. Ask them to prepare a paper for the next conference, describing what they've found or concluded.

10. Invite appropriate software companies to participate via a poster session, with hands-on, live-time demonstrations of their wares. This includes

statistics, design-of-experiments, database, and artificial intelligence software.

11. *Put out a call for papers now in the key topical areas noted above, to encourage research on these topics on a time-scale useful for the next conference.*

12. Provide some sort of tutorial process to bring newcomers up to speed. Possibilities: plenary lectures that review key areas; encourage attendees to procure and read prior proceedings.

E. What important questions remain largely unanswered? What needs haven't been adequately met?

1. Chaos theory suggests (among many other things) that such things as weather may be deterministic in nature, but that it has multiple causes, those causes interact in unknown fashion, and initial conditions can never be completely specified. The result is a lack of predictability, or chaos, or randomness, depending on one's background. How much of our variability in service life (and resultant lack of predictability) is simply due to such things as our inability to
   - adequately measure initial conditions?
   - truly reproduce materials for "replicate" tests?
   - adequately measure all causative factors during test?

2. How can we detect the onset of unexpected (different) failure modes when pushing a particular mode of acceleration? What about changing modes of failure over time?

3. What role should each level of the "food chain" (e.g. raw material suppliers, manufacturers, users of paint) play in reliability testing? What should suppliers be obliged to do before approaching coatings companies? What information can/will be believed between levels in the chain? Goal: stop duplicating effort.

4. Given that *time* is not the correct X-axis for comparisons (cumulative damage events is the correct one), how can we translate SLP information back into a time-scale, which the commercial world both needs and demands?

5. If accelerated weathering device manufacturers were to develop a "perfect" filter for their light source (i.e. one that reproduces natural sunlight "exactly"), would that result in an improvement in our predictive capability? Would that be sufficient to develop a simple model that works well enough?

6. Is there a relatively simple combination of short-term lab tests today that together yield an adequate (for commercial purposes) prediction of service life? For example:
   - fracture mechanics for mechanical integrity
   - cycling of physical conditions (wet/dry, hot/cold, freeze/thaw)
   - photochemical degradation
   - permeability (water, oxygen) for substrate protection

- fouling resistance (dirt, mildew, etc; hardness, bio-resistance)
- adhesion (probably with a wetting/penetration component)

7. What is the environment really like (how variable is it) on the micro scale? To what extent does this variation contribute to question #1 above? [Even today, monitoring of environment is broad and crude, compared to typical monitoring of lab conditions.]

8. Are there "healing" processes that occur and are significant during the weathering process?

9. Is the ultimate arbiter of performance simply one of comparison to some "known" system?

10. What is the rate-limiting step for various failure modes?

11. What are some good commercially-available software packages for DOE/AOE? Criteria: user-friendly, simple, easy, minimal learning curve, readily understandable (and of course competent). [One liked ANOVA TM from ASD. A few like Stat-Ease.]

12. What are the minimum number of properties that need to be specified to guarantee the "quality" of a coating?

13. What are the minimum number of environmental conditions that need to be specified/monitored in order to account adequately for "weathering" processes?

14. What is the role of residual unreacted cure capability on the performance of coatings (particularly clearcoats)?

15. How much risk are we willing to accept, and how do we determine that level? [relates to acceptable failure rate vs. time].

16. Approach the polymeric binder community with the need for more robust adhesion-promoting technology. Current technology is strikingly deficient in robustness: too easily poisoned by other ingredients; too-easily defeated by environmental condition swings; no significant penetration (binding) of porous (unsound) substrates.

Author Index

Subject Index